Human Centric Intelligent Transportation Systems

Fundamentals, Applications, Analysis, and Challenges

Editor:
Amit Kumar Tyagi
National Forensic Sciences University
Gandhinagar, Gujarat, India

CRC Press
Taylor & Francis Group
Boca Raton London New York

CRC Press is an imprint of the
Taylor & Francis Group, an **informa** business
A SCIENCE PUBLISHERS BOOK

First edition published 2026
by CRC Press
2385 NW Executive Center Drive, Suite 320, Boca Raton FL 33431

and by CRC Press
4 Park Square, Milton Park, Abingdon, Oxon, OX14 4RN

CRC Press is an imprint of Taylor & Francis Group, LLC

Library of Congress Cataloging-in-Publication Data (applied for)

ISBN: 978-1-041-09241-4 (hbk)
ISBN: 978-1-041-11406-2 (pbk)
ISBN: 978-1-003-65982-2 (ebk)

DOI: 10.1201/9781003659822

Typeset in Times New Roman
by Prime Publishing Services

Preface

Today Human Centric Intelligent Transportation Systems (H-ITS) utilize advanced technologies and human-centered design to improve urban mobility, safety, and efficiency for all users. These systems integrate Artificial intelligence (AI), Internet of Things (IoT), Cloud Computing, 6G, and other technologies for providing a smart infrastructure with focusing on practical implementations like real-time traffic management, autonomous vehicles, and public transit optimization, etc.

This book introduces the core idea of focusing on the human element in intelligent transportation systems (ITS). It would emphasize that ITS integrates advanced sensors, real-time data analytics, human behavior modeling, and AI technologies to improve traffic efficiency, safety, comfort, and environmental impact. The human-centric approach ensures that transportation systems not only rely on technical innovations but also consider human factors such as behavior, comfort, trust, and safety in shared driving, navigation, and interaction with intelligent vehicles.

This book also highlights the importance of combining technical and human factors knowledge to address challenges in complex urban environments, provide seamless human-machine interaction, and enhance the acceptance and effectiveness of ITS. It might introduce key themes such as planning and control of intelligent vehicles, traffic management, real-time decision-making, fault tolerance, and the integration of AI for predictive and adaptive transport solutions with a focus on human safety and experience.

This book is aligned with themes found in literature discussing human-centric ITS fundamentals, applications, analysis, and challenges, emphasizing human behavior, AI integration, and system reliability within intelligent transportation frameworks. This book is dedicated to focused identified issues real-time traffic management, autonomous vehicles, and public transit optimization with addressing future challenges and applications for more sustainable, accessible, and user-focused transportation solutions to the research community.

In summary, this book sets the stage for exploring ITS from a multidisciplinary human-centric perspective, aiming to bridge the gap between technical capabilities and human needs, with practical applications and challenges in modern urban mobility and automated transportation systems.

Happy learning.

Contents

Section 4: Open Issues and Future Research Opportunities

List of Contributors

A. S. Renugadevi
Department of Artificial Intelligence and Data Science, Kongu Engineering College, Erode, Tamil Nadu, India.

A. Vasuki
Department of Electronics and Communication Engineering, SRM Institute of Science and Technology, Vadapalani, Chennai, India.

Ambeshwar Kumar
Department of Computer Science and Engineering, GITAM University, Visakhapatnam, India.

Amit Kumar Tyagi
National Forensic Sciences University, Gandhinagar, Gujarat, India.

Aswathy Ravikumar
AI/ML Senior Consultant, Sustainable Living Lab, Chennai, India.

Bandi Rambabu
CVR College of Engineering, Hyderabad, India.

Banu Priya Prathaban
Department of Networking and Communications, School of Computing, SRM Institute of Science and Technology, SRM Nagar, Kattankulathur, Chennai, India.

Ben Sujin Bennet
Department of Engineering, University of Technology and Applied Sciences, Nizwa, Sultanate of Oman.

Brindha S.
Department of Computer Science, Vellore Institute of Technology, Chennai, Tamil Nadu, India.

C. Madan Kumar
Department of Computer Science and Engineering, Chaitanya Bharathi Institute of Technology, Hyderabad, India. or
Department of Computer Science and Engineering, SR University, Warangal, India.

Chandrasekar Venkatachalam
Department of CSE (AIML), Faculty of Engineering, Jain (Deemed-to-be University), Bangalore, India.

E. Padma Latha
Department of Computer Science and Engineering, Chaitanya Bharathi Institute of Technology, Hyderabad, India.

Gaurav Singh
School of Computer Science and Engineering, Vellore Institute of Technology, Chennai, Tamil Nadu, India

Gayathri S.
Department of Information Technology, A.V.C College of Engineering, Mannampandal, Mayiladuthurai, Tamil Nadu, India.

Georgy Prasad Philip
School of Computer Science and Engineering, Vellore Institute of Technology, Chennai, Tamil Nadu, India

Girisha G. S.
Department of Computer Science and Engineering, Dayananda Sagar University, Bangalore, India.

Gousia Thahniyath
Department of Computer Science and Engineering, Dayananda Sagar University, Bangalore, India.

I. Gusti Bagus Yosia Wiryakusuma
Assistant Professor, Department of International Business Management, Universitas Ciputra.

Jabanjalin Hilda
School of Computer Science and Engineering, Vellore Institute of Technology, Vellore, India

Jani Anbarasi L.
School of Computer Science and Engineering, Vellore Institute of Technology, Chennai, Tamil Nadu, India

Jarin T.
Department of Computer Science & Engineering, Jyothi Engineering College, APJ Abdul Kalam Technological University, Kerala, India.

Jyotsna E.
Department of Computer Science & Engineering, Sahrdaya College of Engineering and Technology, APJ Abdul Kalam Technological University, Kerala, India.

K. Janani
Department of Artificial Intelligence and Data Science, Kongu Engineering College, Erode, Tamil Nadu, India.

Kapil Dev Tyagi
Department of ECE, Jaypee Institute of Information Technology, Noida, India.

Lyernisha
Department of Software Engineering, University of the West of England, Bristol, UK

M. Rajesh
Department of Computer Science and Engineering, Aarupadai Veedu Institute of Technology, Vinayaka Mission's Research Foundation (DU), Tamil Nadu, India

M. Venkata Krishna Reddy
Department of Computer Science and Engineering, Chaitanya Bharathi Institute of Technology, Hyderabad, India.

Manoj R.
Department of Electronics and Communication Engineering, Bannari Amman Institute of Technology, Erode, Tamil Nadu, India.

Modigari Narendra
School of Computer Science and Engineering, Vellore Institute of Technology, Chennai, Tamil Nadu, India

Mohan Singh
Department of ECE, G.L. Bajaj Institute of Technology and Management, Greater Noida, India.

Neeraja R.
Department of Computer Science and Engineering, Sri Sivasubramaniya Nadar College of Engineering, Chennai, Tamil Nadu, India

Perarasi T.
Department of Electronics and Communication Engineering, Bannari Amman Institute of Technology, Erode, Tamil Nadu, India.

Priyamvada Saxena
Department. of Electronics & Communication, Amity University, Gwalior, Madhya Pradesh, India.

R. Felista Sugirtha Lizy
Assistant Professor, Department of Computer Science, A.P.C. Mahalaxmi College for Women, Thoothukudi, Tamil Nadu, India.

R. Kanimozhi
Department of Information Technology, A.V.C College of Engineering, Mannampandal, Mayiladuthurai, Tamil Nadu, India.

R. Ravinder Reddy
Department of Computer Science and Engineering, Chaitanya Bharathi Institute of Technology, Hyderabad, India.

R. Ravinder Reddy
Department of Computer Science and Engineering, Chaitanya Bharathi Institute of Technology, Hyderabad, India

R. Saminathan
Department of Computer Science and Engineering, Annamalai University, Chidambaram, Tamil Nadu, India.

Raghvendra Sharma
Department. of Electronics & Communication, Amity University, Gwalior, Madhya Pradesh, India.

Rajesh T. M.
Department of Computer Science and Engineering, Dayananda Sagar University, Bangalore, India.

Ravi Uyyala
Department of Computer Science and Engineering, Chaitanya Bharathi Institute of Technology, Hyderabad, India

Reedhu Saji
Department of Computer Science & Engineering, Jyothi Engineering College, APJ Abdul Kalam Technological University, Kerala, India.

S. China Ramu
Department of Computer Science and Engineering, Chaitanya Bharathi Institute of Technology, Hyderabad, India.

S. Durga Devi
Department of Computer Science and Engineering, Chaitanya Bharathi Institute of Technology, Hyderabad, India.

Sai Jeevan Puchakayala
AI/ML Consultant, Sustainable Living Lab, Chennai, India.

Satyanarayana Nimmala
CVR College of Engineering, Hyderabad, India

Saveetha D.
Department of Networking and Communications, School of Computing, SRM Institute of Science and Technology, SRM Nagar, Kattankulathur, Chennai, India.

Sayyad Samee
Department of Computer Science, Al Sadara School, Al Ain, Abu Dhabi, United Arab Emirates

Seema Singh
Department of Education, Lloyd Institute of Management & Technology, Greater Noida, India.

Shaila S. G.
Department of Computer Science and Engineering, Dayananda Sagar University, Bangalore, India.

Shanmugavalli Venkatachalam
Manipal Institute of Technology Bengaluru, Manipal Academy of Higher Education, Manipal, India

Shuai Li
Faculty of Information Technology and Electrical Engineering, University of Oulu, Finland.

Siddharth M Nair
School of Computer Science and Engineering, Vellore Institute of Technology, Chennai Campus, Chennai, Tamil Nadu, India.

Sivakumar
Department of Computer Science, Vellore Institute of Technology, Chennai, Tamil Nadu, India.

Subash
Department of Computing Technologies, School of Computing, SRM Institute of Science and Technology, SRM Nagar, Kattankulathur, Chennai, India.

Subashini. N. J.
Department of Networking and Communications, School of Computing, SRM Institute of Science and Technology, SRM Nagar, Kattankulathur, Chennai, India.

Sumit Ranjan
School of Computer Science and Engineering, Vellore Institute of Technology, Chennai, Tamil Nadu, India

Surem Samuel
Department of Electronics and Communication Engineering, Kingston Engineering College, Vellore, India

Thejas V. M.
Department of Computer Science, Vellore Institute of Technology, Chennai, Tamil Nadu, India.

Thomas George K.
Department of Computer Science & Engineering, Jyothi Engineering College, APJ Abdul Kalam Technological University, Kerala, India.

Uma Maheswari J.
Department of Computer Science, Vellore Institute of Technology, Chennai, Tamil Nadu, India.

V. Padmavathi
Department of Information Technology, A.V.C College of Engineering, Mannampandal, Mayiladuthurai, Tamil Nadu, India.

Vaibhav Bhushan Tyagi
Department of ECE, ISBAT University, Kampala, Uganda.

Vengatesan Krishnasamy
Department of Computer Science and Engineering, School of Engineering, Dayananda Sagar University, Bangalore 562112 Karnataka, India.

Vijayakumar Ponnusamy
Department of Electronics and Communication Engineering, SRM Institute of Science and Technology, Kattankulathur, Chengalpattu, India.

Vineeth V. K.
Department of Mechanical & Industrial Engineering, Texas, USA.

Vinila Jinny
School of Computer Science and Engineering, Vellore Institute of Technology, Vellore, India

Y C A Padmanabha Reddy
Department of Computer Science and Engineering, Chaitanya Bharathi Institute of Technology, Hyderabad, India

Chapter 1

Digital Twins in the Modern Era of IoT and AI

Sai Jeevan Puchakayala[1], *Aswathy Ravikumar*[2] and *Vineeth V. K.*[3*]

Digital twins are virtual representations of physical entities, which emerged as transformative solutions at the intersection of the internet of things (IoT) and artificial intelligence (AI). By continuously mirroring real-world processes, digital twins leverage sensor data and advanced analytics to enable predictive insights, proactive maintenance, and intelligent decision-making. This chapter explores the foundational principles of digital twin technology, offering a deep dive into its architecture and lifecycle from conceptualization and model development to deployment and iterative refinement.

Building on these fundamentals, the study then discusses the three main types of digital twins—product, process, and system, each serving distinct functions and delivering tailored value. The chapter also examines how real-time monitoring capabilities harness live data streams for immediate insights, while simulation features support scenario-based planning and risk mitigation. Emphasis is placed on the synergy between AI-driven analytics and IoT-enabled infrastructures, demonstrating how the fusion of these technologies fuels next-generation applications. Specific examples showcase the potential of digital twins to optimize operational efficiency, enhance scalability, and reduce downtime in complex industrial and urban environments. Through this lens, the chapter underscores digital twins as key enablers for more resilient, agile systems in a rapidly evolving landscape shaped by big data, AI, and decentralized IoT ecosystems.

1. Introduction

Digital twin technology is a big step forward in how physical assets, processes, and systems are managed and optimized. Because it makes complete virtual copies of

[1] AI/ML Consultant, Sustainable Living Lab, Chennai, India. https://orcid.org/0009-0005-7776-0985

[2] AI/ML Senior Consultant, Sustainable Living Lab, Chennai, India. https://orcid.org/0000-0003-0897-6991

[3] Department of Mechanical & Industrial Engineering, Texas, USA. https://orcid.org/0000-0003-1378-0619

* Corresponding author: vineeth.vk1019@gmail.com

DOI: 10.1201/9781003659822-1

things in the real world, Digital twin helps organizations watch, study, and make their work better in an effective way. Real-time data integrates with advanced analytics in this technology, improving decision-making across many fields. As more and more industries use this new method, grasping the basic ideas, importance, and functions of digital twin becomes quite vital to using its complete capabilities. This book will examine digital twin technology thoroughly, focusing on its massive effect on today's industry, its capabilities, and the challenges encountered during its implementation.

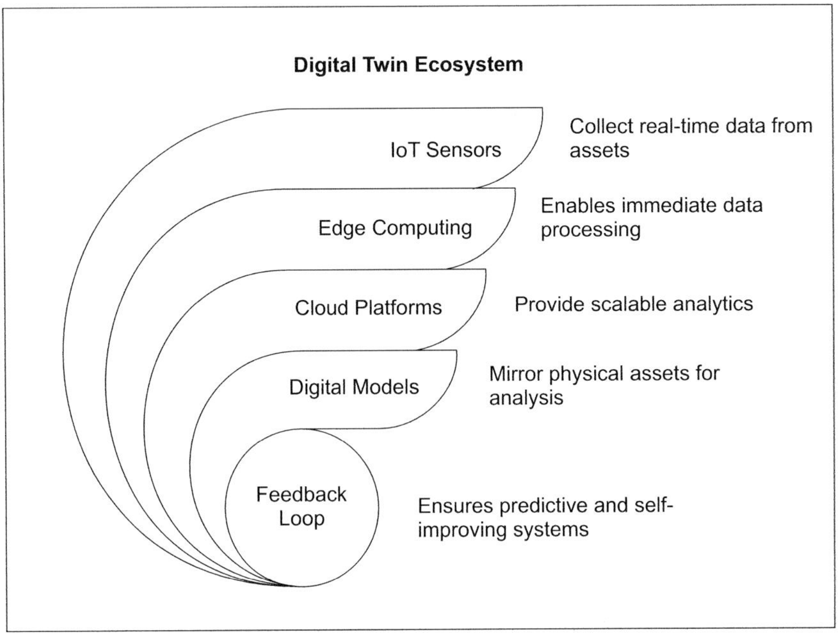

Fig. 1.1 A high-level ecosystem view of digital twin technology, illustrating how real-world physical assets are monitored and optimized through IoT data collection, cloud-edge computing, and AI-driven analytics.

Digital twin (DT) technology is a new way to handle physical assets, processes, and systems. It does this through virtual replicas. A key feature of DT is the link between current data and complex analytics. This connection sharpens observation and also improves choices. It greatly helps data control besides examination in industrial work (Schleich et al., 2017). For example, experimentable digital twins (EDTs) do thorough simulations and produce virtual versions of real systems (Liu et al., 2022). Because of this feature, DTs connect with the industrial internet of things (IIoT) to raise production effectiveness besides fresh ideas in industry (Conde et al., 2022). DT technology supports continuous fine-tuning in fields, such as video transmission. It adjusts to moment-by-moment forecasts of network state (Park, et al. 2021). Figure 1.1 illustrates the high-level ecosystem of DT systems, showcasing the interaction between IoT devices, cloud/edge platforms, and AI-driven analytics in supporting real-time feedback and operational optimization.

DTs remain vital for strengthening decision-making during different phases, mainly in asset lifecycle management. In asset management, DTs tackle combined demands for decision assistance. Ongoing efforts to develop DT technology try to unify its meaning and traits across fields. These plans to clear up questions from the stream of related writings and terms (Peitz & Bieker, 2023). DT technology changes, so it betters current work and makes new uses possible in various fields, such as the 6G system control and the complex internet of everything (IoE) (Almasan et al., 2022). The connection of DT tech with mobile edge generation (MEG) systems is aimed at making things work better as also lower delays, which makes it useful in more places (Soldan et al., 2021). For example, people look at using GenAI methods with city digital twins so that they can fix problems in the planning and running of cities (Schrotter & Hürzeler, 2020). Moreover, automating how simulation models grow via natural language processing, besides GPT-3 Codex, gives DT tech good upgrades (Chen et al., 2024). Scalable DTs link varied models and also live data, and this shows how DT tech grows (Park et al., 2019). DTs connect real things to their online copies. This shows how useful they can become (Avrilionis & Hardjono, 2021). The ability of digital engineering to boost data safety besides how systems live need more use (Bonar & Hastings, 2024). The ability of digital engineering to boost data safety, besides how systems live, need more use. Human digital twins (HDTs) are still in the early stages of development, with ongoing efforts to define what an HDT actually represents and how HDTs should be constructed. By joining predictions with live sensor data, like how heat spreads, process fault sensing goes up (Gaikwad et al., 2020). Because the Microservice System Twin Graph-based Anomaly Detection (MSTGAD) tool uses metrics, logs, and traces with a graph method, it finds anomalies with good performance (Huang et al., 2023). All these progressions add to how much DT can change many industrial and tech areas.

DT technology has several facets, which highlight its growing value in modern industrial environments. Since outfits aim for boosted output and invention, they often add DTs to their ways of doing things and this addition stands out as a key method. Tech helps with current checks besides future guesses, and it also nurtures ongoing progress and change. Continued study besides growth in this area seems set to reveal more powers, thus, solidifying the role of DT technology when pushing ahead with future industry progress.

2. The Importance of DT in Industry

DT technology is increasingly recognized as a powerful tool across various industries. DTs help improve how operations run, anticipate when equipment needs repair, and enables people to make smart choices. Within Industry 4.0, DTs join physical and virtual worlds. This integration results in a more advanced systems engineering approach by leveraging EDTs (Schluse et al., 2018) for iterative testing and validation. Figure 1.2 visualizes key domains where DTs are being widely adopted, from predictive maintenance in manufacturing to infrastructure optimization in smart cities. This is important, for example, in power grid management. DTs support better choices when things are not certain. This support is important because we rely more on renewable energy sources (Mylonas et al., 2024).

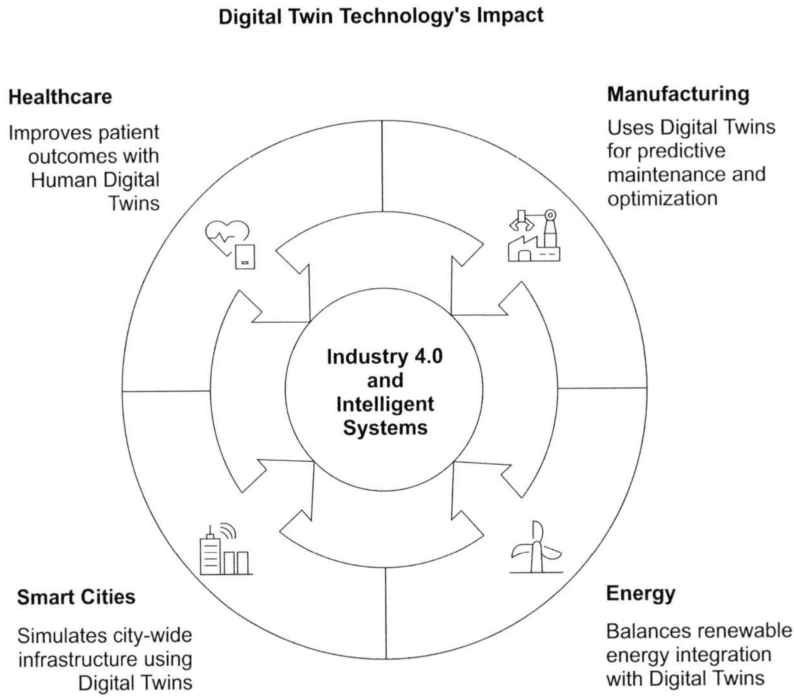

Fig. 1.2 Key domains of DT adoption across modern industries, highlighting the convergence of AI and IoT in areas such as smart cities, power grids, healthcare, and manufacturing.

Fault engine uses predictive analytics in sensitive industries. It cuts down on outages and makes operations dependable, which maintains strong security (Majumder et al., 2016). DTs simplify hard modeling procedures. This results in better output and lower project charges in different sectors (Jackson et al., 2024). About this technology's value, it boosts the electric vehicle charging setup through improved prediction. This happens because it adds current data (Soldan et al., 2021).

Within smart cities, DTs, IoT, and AI work together. This shows how they can significantly transform cities because they help with asset management and make operations better. For example, their cooperation solves information technology (IT) and operational technology (OT) convergence problems, providing answers, which deal with their requirements that do not always match (Schäfer et al., 2024). Because DTs are embedded within system-of-systems (SoS), robust engineering practices, particularly around interoperability and lifecycle management, are essential. Such practices control digital models in many areas (Michael et al., 2022).

Clarity is vital in human-in-the-loop DT systems since prior approaches had issues with open decisions. This harmed faith and precision (Zhang et al., 2024a). Because DTs are complex, a person must understand how they decide things. Especially if the decisions are automatic, explainability becomes a needed attribute to gain confidence (Zhang et al., 2024b). DT tech strengthens the collaboration between human workers and machine systems. This leads to more efficient and reliable manufacturing processes.

Additionally, GenAI's usage in city DTs highlights better data and produces scenarios, which becomes crucial for good urban control (Xu et al., 2024a). Because DTs can manage uncertainty in simulation data, they are valuable for forecasting applications across various domains (Gonçalves & Porto, 2015). DTs work well in live use. When backed by spread methods, their use grows (Xu et al., 2024b).

In manufacturing, DTs strongly improve real-time choices and optimization. They fix problems found when applying machine learning and data science (ML/DS) (Lee & Chien, 2022). The PD-CPD stands for Partially Dynamic – Change Point Detection method precisely spots changing process features in discrete event simulation models. It is vital for maintaining correctness in intricate systems. DTs support cheap, easy solutions. They merge data from MTConnect and production workers. They boost manufacturing (Urbina Coronado et al., 2018). But wireless control technologies are crucial. They are a key part of smart manufacturing, wherein they tackle flexibility and efficiency along with adaptability in production (Baumann et al., 2021).

Even though it can change many things, the construction industry uses DT technology less than other fields. This limited use means chances exist to study it more and use it to make work run smoother and to plan for repairs better (Madubuike et al., 2022). The importance of DT technology to modern industry comes from its power to join different technologies, simplify actions, and make better choices. It advances new ideas and smooth processes across several sectors.

3. Background

DT technology is a scalable structure. It brings different models together with current data. This structure builds a virtual copy of actual items and betters how systems are simulated and handled. This combination is important because it makes operations work better and react quicker. But the unclear description of DTs causes issues in creating items and making things. Inconsistencies can stop the full advantages from appearing. The idea of human digital twins (HDTs) shows why descriptions must be exact across fields. As shown in Figure 1.3, HDTs introduce additional complexity due to their reliance on biometric data, cognitive modeling, and ethical handling, unlike traditional asset-oriented DTs. For example, variations between HDTs and typical DTs cause special problems (Baby, 2014; Lauer-Schmaltz et al., 2024). Because of this fixing these problems are key for helpful uses in healthcare and tailored medicine.

"Virtual-physical representation" combines digital and physical systems. This combination matters greatly for the control of changing elements within the IoE and modern production facilities. This method gets past limits of older event engines. This improvement helps the expected requirements of logical tools inside the virtual-physical space (ViPS). It displays the need for a digital copy. This copy reacts to physical variations quickly. Because of this reaction, function becomes better. In urban design and local involvement, DTs showcase their power. For example, they let participants test choices before they happen, enabling better choices and continued growth.

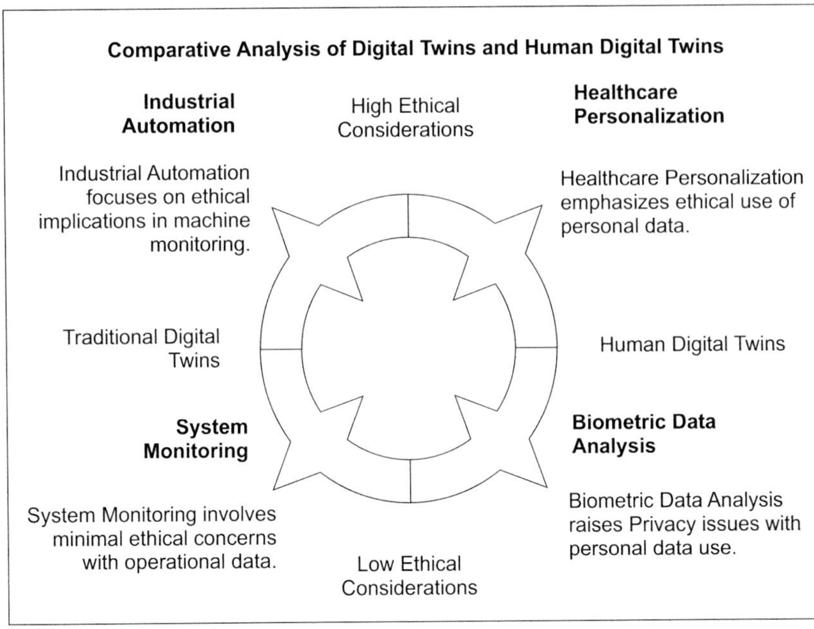

Fig. 1.3 Comparison between traditional DTs and HDTs, emphasizing domain differences, data complexity, and application contexts.

DT architecture uses a structure that contains data analytics and AI. This structure makes sure that security and efficiency exist during all parts of a product's existence. Though it is significant, problems with specifying parts and needs may stop proper use (Bonar & Hastings, 2024). Good architecture shows the way for DT progress and placement—it unites technologies with actions. Its part in improving data reach and analytics matters a lot, especially in hard industrial uses. But the joining of IT and OT makes IT fixes complex in OT places because differences exist. Definitions cover model-based engineering, digital threads, and combined product lives. They help digital twin tech grow. For example, the check of digital twin uses in fields like building shows different topics besides how far uses extend (Madubuike et al., 2022). There is a demand for a deep grasp of the DT idea and its future effects.

4. Virtual-Physical Representation

DT systems need virtual-physical representations to join digital and physical entities. This joining helps with advanced process monitoring, process control, and process optimization. Because DT makes a virtual model that reflects the physical part immediately, it improves decisions and predictions. In smart manufacturing, this representation helps quality monitoring and predictions using topological analytics. With this process, settings and quality features get better (Lee et al., 2024). Figure 1.4 depicts the real-time feedback loop within DT systems, enabling data-driven adjustments to physical operations based on analytical insights. Manufacturers watch these links in real-time and, therefore, quickly spot changes, cutting waste as well as improving product quality.

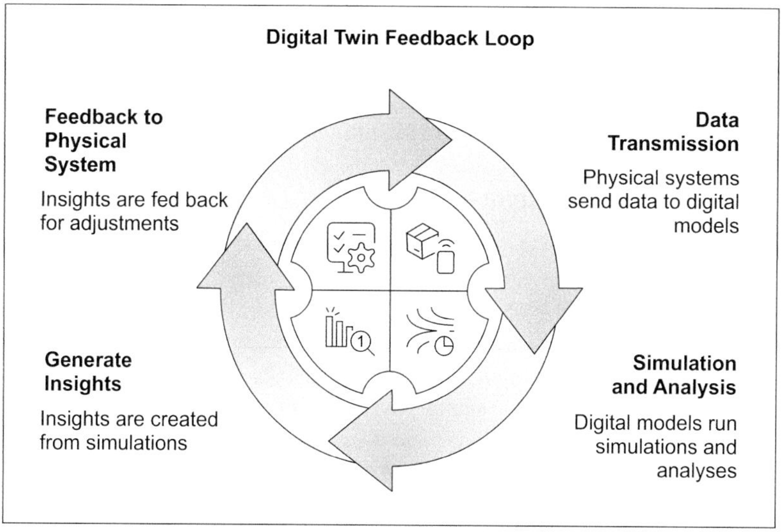

Fig. 1.4 Real-time feedback loops in a DT system, showing the synchronization between physical entities and their virtual representations through continuous data and control flows.

DT systems need a steady flow of data between the real and the virtual. This confirms that the digital model shows the current state of the physical thing precisely. Because of this dynamic depiction, there is good management of involved systems. DT systems simulate situations and assess results, without disturbing physical operations. Advanced data analytics and machine learning improve how DTs work. They give up-to-the-minute understandings of system dynamics. Maintenance happens in advance and strategies are adjusted. It goes with Industry 4.0 goals. It uses big data intelligence and modern simulations to make manufacturing run smoother and change easier. As various fields utilize these techniques, the output is better and new ideas have clarity. This underlines how much DT systems change long-standing manufacturing ways.

Outside production, virtual-physical copies offer a crucial part in city design, medical care, and other areas. These areas need current data combination and study. DT systems work like a bridge. This bridge connects digital and physical spaces. And it takes a continually matched virtual picture of real items. With this combination, operational output rises. It permits current data study and choice making. It cuts costs and boosts system work because complex math and practice runs support future care, refine actions, and enable changes according to trade wants. DTs remain important in smart production and Industry 4.0 plans. Their flexible trait illustrates their potential for broad use through fields, promoting novel ideas and output. As the area grows, an investigation of the new ways and tools involved becomes vital because it makes virtual-physical copy profits big in varied settings.

5. Digital Twin Architecture and Lifecycle Management

Knowing the DT architecture is important to realize how these systems give advantages in different areas. The architecture provides a base for DTs—it affects their ability to grow, change, and work well. This section will look at the important traits and setups of DT architecture. It focuses on how parts connect and how they change based on actual information and changes over time, which has effects on asset handling and choices (Alen et al., 2021; Ravikumar 2021; Singh et al., 2021).

5.1 Digital Twin Architecture

The DT architecture is a complex framework. It helps combine digital and physical objects. This leads to improved tracking and management along with enhancement across factories. Its modular nature permits different uses. It guarantees that systems work well together through set Application Programming Interface. These APIs are built for factory demands without sacrificing how they interact. Often architecture has many levels. These levels contain physical items, computer models, and links for data movement. For this reason, system adaptability, besides growth, improves. Real-time data merging and easy data sharing are essential. They keep digital twin systems exact and trustworthy.

Cloud and edge computing are advanced technologies that improve how fast data is processed and how systems respond. This is very important for smart city applications, as IoT, AI, and cloud computing together help manage data well. The architecture must permit two-way data exchange so it can constantly synchronize and adapt. This is vital in complicated applications, like 6G wireless systems. Frameworks that group DTs by operational context can assist in finding key parts and technologies. This allows for effective implementation across different areas.

AI models now help with choices and improvements inside DT designs. They decrease delays and quicken responses if they move to local hardware. Sizable language models improve how well one can clarify choices. They make certain that users get system results and have faith in them (Harini & Ravikumar, 2020; Ravikumar et al., 2013; Ravikumar & Sriraman, 2023). The FIWARE ecosystem gives a basis to handle data and details. It handles weak spots of present layouts to make certain strong, sizable, and safe actions in industries. Because control systems and wireless communication are tied together, this raises flexibility besides effectiveness in smart production. Active event handling shows what DT designs can do to oversee shifting locations (Wang et al., 2023).

5.2 Core Components and Structure

DT systems hold core pieces that make thorough virtual models of actual items possible. These pieces are actual items, DTs and an internet of DTs (IoDT) engine, which coordinates and connects (Wang et al., 2023). In fields like robotic assembly systems, fault detection based on AI and digital process twins make manufacturing better (Leutert et al., 2024). For traffic control, items like pan-tilt cameras and predictive models guarantee safety (Li et al., 2024).

DT systems involve definition, technical features, application cases, and architectural models. This ensures that components work together. The classification of DT systems into types, such as digital twin prototype (DTP) and digital twin instance (DTI), shows how they act in the architecture. Context data models and a context broker permit smooth device connections because of which operation remains efficient in intricate systems. Network DTs use machine learning applications as analytical methods to advance capabilities.

But core components also produce an adaptable structure. These tackle industrial requests by enhancing product lifecycle management, real-time data exchange, and developed manufacturing under Industry 4.0. With high-fidelity simulations, operational data, and predictive analytics, innovation happens. Maintenance is improved and production reaches better efficiency. This gives adapted solutions across various sectors.

5.3 Lifecycle Management and Feedback Loops

The control of a DT's existence includes the start, the active period, and the finish. Figure 1.5 presents a layered architecture of DT systems, from the physical layer to the application-level decision intelligence, illustrating modular design principles critical for scalable deployment. This makes sure the system changes when its context does. To maintain correctness, data returns consistently. This is why DT systems enable fast response, for instance, in self-steered vehicle networks (Liu et al., 2024). When feedback gets combined, the system's action gets better. For example, the models find flaws with better precision (Gaikwad et al., 2020).

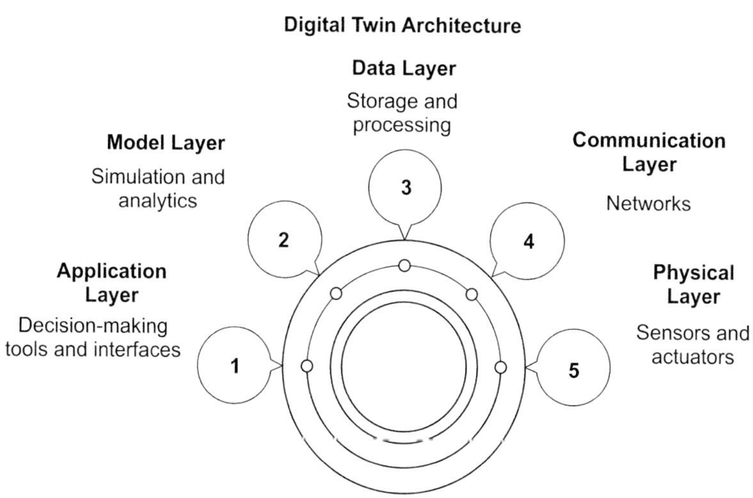

Fig. 1.5 Lifecycle stages of a DT system, from initial creation and calibration to real-time operation, optimization, and decommissioning, all supported by continuous feedback loops.

DT container innovations help with data handling and correctness (Avrilionis & Hardjono, 2021). Complex data study and quality checks, for example, distinctive simulation in robotics, make product lifespan control better. But HDTs need response paths that consider difficult human actions—this confirms value and success. One difficulty involves dependence on unchanging documents. It makes current revisions and anticipating potential problems harder. Effective systems that manage events are very important because of expandability and quick reactions. The supervision of existence cycles as well as input cycles builds a structure. This supports change and enhancement through current data and models along with analysis to improve work. It further assists fresh ideas across many areas, pushing Industry 4.0 changes forward (Durão et al., 2018).

5.4 Real-Time Data Integration and Processing

DTs require real-time data integration and processing to make certain that physical entities have correct and current displays. With this capability, industries like manufacturing, automotive, and healthcare improve decision-making, predictive maintenance, and operational efficiency. DTs reflect physical assets and give information about potential behaviors. They support strategies based on data.

DT systems use frameworks, like ServIoTicy, for real-time data work in varied user setups and they permit custom data changes (Villalba & Carrera, 2019). This data work boosts analytical predictions and simulation features. Because of these enhanced capabilities, proactive maintenance and better strategies becomes available (Farus et al., 2024). AI and machine learning integration boost analytical power. This provides precise simulations and prediction for asset lifecycle control and the decision process in many fields (Sarker et al., 2023). This fusion backs better decision processes. It also raises performance and output. For Industry 4.0's progress and intelligent production, real-time data processing in DT systems is necessary. This pushes advancement and output in business areas.

6. Types of DTs and Their Applications

IoT connectivity and AI analytical power have created complex tech, such as DTs. DT is a computer model of a real object, system, or method—it lets businesses watch, study, and improve that real entity instantly. Because IoT sends data using linked tools and sensors and because AI makes future predictions, DTs became more important for better function, less wait time, and top productiveness. By linking real and computer systems, DT tech has attracted interest across many fields.

A DT is a live, current picture of a real object. It permits observation, simulation, and refinement. DTs come in three kinds, depending on their depth and reach: product, process, and system. Each kind fulfills specific roles, which range from focused problem detection to complete management of the full picture. Figure 1.6 provides a comparative breakdown of product, process, and system twins, outlining their use cases, scale, and operational focus within industrial applications. This section examines these DT groups and their uses given the IoT and AI. It will look at product twins, used for trouble identification besides future upkeep; process twins, used for easier workflows; and system twins, used to supervise the entire setup. These uses show how DTs help industries in improving their powers and stay ahead in a world driven by data.

6.1 Product Twins

Product twins are similar to digital versions of specific tangible items. We use them to check and better how those items work, what they do, and how they are handled over their lifespan. These replicas show current conditions and copy what the item does as time passes, helped by IoT tools and sensors attached to the real products. Product twins matter a lot in areas where each item's performance is very important, such as factories, space, vehicles, and medicine. As IoT and AI become common, product twins grow better. This occurs since they depend on current data from IoT sensors. AI tools work with and study a lot of this data for a better understanding. Through the use of product twins, businesses improve their items, notice issues early, foresee when maintenance is needed, and offer nicer experiences to users.

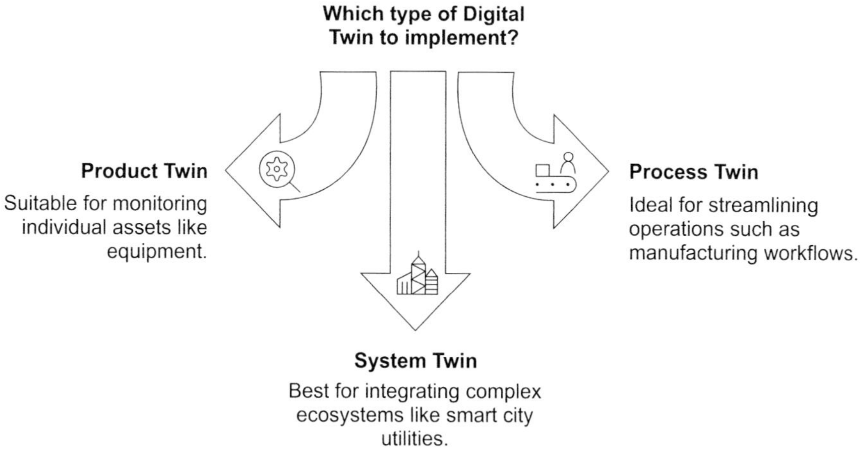

Fig. 1.6 Comparison of product, process, and system DTs, outlining differences in focus, scope, typical applications, and data complexity.

6.1.1 Fault Diagnosis

DTs have fault diagnosis as a primary use. This process constantly watches a product's state. It achieves this by comparing live sensor numbers to set limits or past trends. Through the fusion of sensors and IoT tech, fault diagnosis uses live data capture and analysis. Because of this it can spot oddities in how equipment works. When a difference shows, the system sends out warnings about the existence of a problem or unusual state. This active method supports the identification of likely failures before they happen. Downtime effects shrink considerably.

Aviation provides a good example. Turbines and engines have sensors. These sensors observe attributes like temperature, pressure, vibration, and fuel efficiency. The DT regularly checks this information. When it finds something abnormal, such as higher vibration or too much heat, the system informs technicians, who can then examine it before the equipment breaks (Tao et al., 2018). With this quick discovery, hazards decrease and safety improves. Less time is lost and repair bills get smaller.

6.1.2 Predictive Maintenance

DTs offer a very useful function in predictive maintenance. The strategy uses digital representations to foresee possible problems and then plan upkeep ahead of actual malfunctions. It depends on an examination of past information, present-day information, and machine learning to forecast when a part might stop working. Through this forward-looking method, repair is set at the right moment to halt breakdown.

In the car business, the vehicle's digital model examines data. It checks things such as engine heat, tire pressure, and how gas use changes. It then guesses when parts, for example, brake pads or the exhaust, will need work. Predictive maintenance uses past issues and live data. This information teaches AI models to predict how long parts will last. Through the use of AI and machine learning, the digital model anticipates how these parts wear down. This action supports cheaper maintenance and also increases how long the vehicle operates (Boschert & Rosen, 2016).

6.1.3 Unit-Level Monitoring

Product twins support constant checks on each product separately and this ensures effective work. This monitoring watches the state of separate parts or smaller systems to promise the best function and an agreement with planned details. It matters especially in areas like healthcare, as equipment such as magnetic resonance imaging (MRI) machines, pumps for infusions, and respirators must work very well. A product twin for a medical tool can watch various elements like heat, usage habits, and calibration details. As it uses AI, it can foresee likely faults or required adjustments. The tool stays working and precise (Negri et al., 2017). Unit-level monitoring not only betters work performance but also lowers the risk of equipment breaking down, which helps with better patient care and work speed in healthcare centers. Wind turbine blades have consistent stress and wear checks. Through this assessment, makers can adjust the materials they use for the blades and make needed design changes (Schleich et al., 2017). For setup digital twins' pair computational fluid dynamics (CFD) simulations with actual sensor data to develop better design approval.

6.2 Process Twins

Product twins target single items. Process twins stand for complete operations or setups, like assembly lines. Process twins help in the improvement of operations, enabling the spotting of shortcomings and the guarantee of good use of resources. Like product twins, IoT devices constantly take readings from several points in an operation, and AI tools examine this data. Then these tools give insights and ways to improve. Process twins matter in fields such as manufacturing, energy control, and the making of chemicals. In these areas an understanding of how an entire operation works helps raise productivity and cut costs. Process twins work to advance factory operations to make sure there is efficient work in difficult factory locations. Modeling energetic operations, process twins make it possible to settle on changes to the process in real time.

6.2.1 Workflow Optimization

Workflow optimization requires precise adjustments to operational procedures, made during execution, to better efficiency, waste reduction, and improved resource usage. Optimization employs current information and simulations to simplify industrial actions, lower waste, and increase productivity. Within manufacturing, process twins model production lines, forecast possible problems, and give knowledge about resource distribution. For example, in semiconductor creation, process twins improve lithography and wafer etching steps, and this results in reduced flaws and increased outputs (Tao et al., 2018). In the automobile industry, process twins help in procedure enhancement of component flow on the assembly line. When process twins simulate different options, they pinpoint the top scheduling for machines, labor allocation, and material handling. AI studies this data, then suggests alterations – these enhance manufacturing durations, lower expenses, and boost output (Razumowsky, 2024). Thus, process twins support ongoing advancement within manufacturing procedures.

6.2.2 Manufacturing Flow-Shop Scheduling

Flow-shop scheduling poses a big problem for manufacturing because it needs a system to sequence jobs in a way that reduces cycle time and boosts throughput. This method establishes the best sequence in which work runs on tools inside a creation system. The core goals involve a decline in creation time, lower expenses, and reduced resource use. Process twins model and boost scheduling picks—this helps manufacturers in the hunt for the finest answers. AI in process twins uses different issues, like machine uptime, process times, and breaks. For instance, process twins in a chip creation plant copy the silicon wafer flow across tools and calculate yield rates and gear breaks. AI advises the most effective schedule, which cuts idle time and improves throughput. Similarly, in automotive construction rows, DTs adjust job sequences to balance effort through many spots (Qi et al., 2021). Deep reinforcement learning and genetic algorithms modify scheduling rules to deal with real-time issues. With this efficiency rises and running costs fall. Figure 1.7 illustrates how process twins are applied to flow-shop scheduling in manufacturing, using AI to sequence jobs, reduce idle time, and optimize throughput.

6.3 System Twins

System twins form intricate ecosystems. These contain several linked products, procedures, and outside elements. Product twins and process twins center on single components. System twins, however, give a complete picture of how system segments interact and affect one another. These systems span from supply chains to complete urban infrastructures. As IoT and AI grow, system twins gain importance in handling large procedures. Many participants and parts must work together effectively here. AI serves a key purpose by examining large amounts of linked data. Because of this system twins suggest enhanced plans, foreseeing problems along with expanding the overall system output.

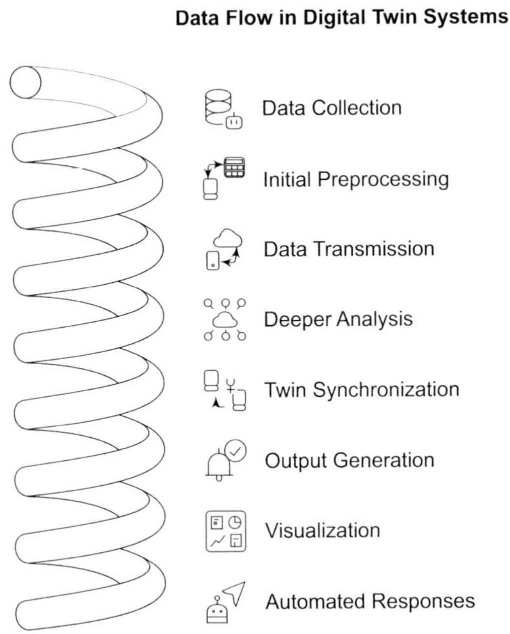

Fig. 1.8 Data flow in DT systems showing the stages from data collection to automated responses.

issues is important to increase the total possibility of DT systems when it comes to progress and efficiency in business fields.

7.2 AI and Machine Learning

AI and machine learning elevate DT systems through data handling and projections along with automated choices. They generate systems that adapt to live data and shifts, boosting efficiency and raising dependability (Weyns & Iftikhar, 2021). AI methods, for example, automatic differentiation inside DTs, help steady improvement in robotics. Machine learning approaches handle doubt and buttress projections, boosting DTs' abilities in IoT uses. Federated learning and multi-agent reinforcement learning provide privacy and effective work in group settings—therefore, safe function occurs where data is confidential.

Distributed execution frameworks help real-time machine learning application performance. As a result, DTs manage the demands of real-time data processing (Nishihara et al., 2017). Domain-specific modeling languages make machine learning model creation and use more efficiently. DT lifecycle management speeds up (Bhattacharjee et al., 2019). AI and machine learning propel advancement inside DT systems. This enhances how well operations work and how effective they are throughout sectors like manufacturing, automotive, and smart cities (Li et al., 2024). With their combination, DTs can match real-world events, improve performance, and uphold eco-friendly answers in complicated, interconnected situations.

7.3 Edge Computing and Distributed Processing

Edge computing enhances DT systems by enabling real-time data processing closer to the source, reducing latency, and improving responsiveness. This localized processing is crucial for applications requiring immediate analysis and decision-making, such as smart manufacturing and IoT environments. Edge computing supports the development of distributed systems that adapt dynamically to changing conditions, leveraging local processing power for complex simulations and analyses (Weyns & Iftikhar, 2021). Emerging trends include federated learning, which enables collaborative model training across distributed nodes, preserving privacy and enhancing scalability. The combination of edge computing and distributed processing empowers DT systems, ensuring they meet demands for real-time data processing and analysis in complex systems.

7.4 Simulation and Co-Simulation Platforms

Simulation and co-simulation platforms matter for DT systems because they have tools to model, analyze, and optimize complicated processes. These platforms make detailed virtual spaces that copy actual spaces. They improve predictive analytics and informed decision-making in areas such as education, training, and industrial applications (Zhang et al., 2024c). For example, RiotBench shows what simulation platforms do in IoT environments—it gives benchmarks to check the performance and scalability of DT systems. Co-simulation platforms connect different models at the same time, so a person can understand interactions and dependencies which help decision-making (Jiang et al., 2023). This connection of simulation and co-simulation platforms pushes DT technology forward, leading to better predictive analytics and operational efficiency.

8. Challenges and Future Directions

DT system deployment has diverse problems. These include scalability, data quality, security, and future innovations. Because industries use these models more, they must handle the intrinsic issues to promote effective implementation. Figure 1.9 presents the layered arrangement of problems and future paths in DT systems. It breaks down the main topics such as scalability, data quality, security, and innovation. Each topic contains specific problems and possible fixes, which gives a complete picture of the present state and future possibilities of DT technology. Scalability and computing requirements create important hurdles. People must get past them to use DTs completely. The following sections investigate these vital areas, with a focus on scalability, data quality, security, and the upcoming paths of DT systems.

8.1 Scalability and Computational Demands

Significant obstacles to the increased use of DT systems are scalability and computational needs across numerous fields. A main problem is model integration, so it retains reliable virtual depictions and also manages computing burdens. The need to exploit information from many data types builds upon this complexity,

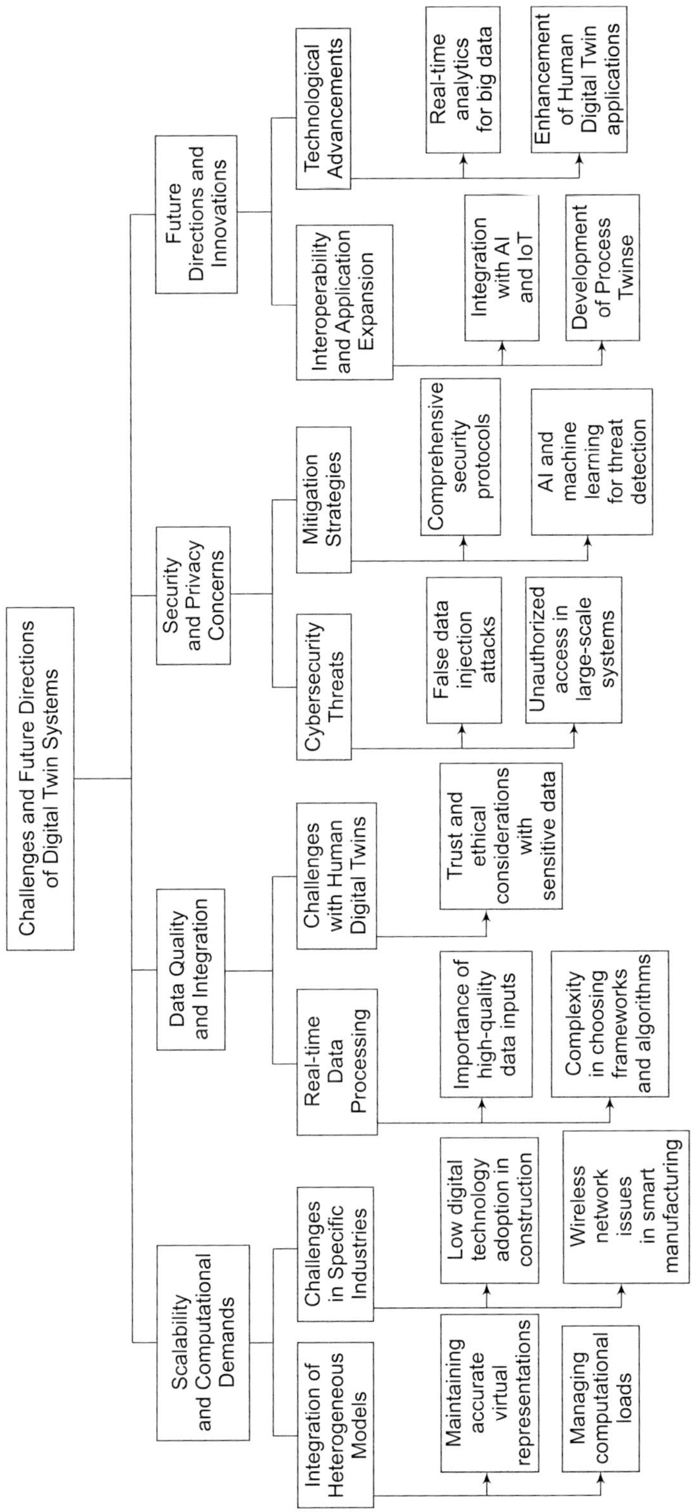

Fig. 1.9 The hierarchical structure of challenges and future directions in DT systems.

and this process lessens false positives and undetected irregularities. A shared understanding of DT definitions impedes these efforts, for it generates inconsistencies in applications. In the construction sector, infrequent use of digital tools intensifies these scalability concerns. Wireless networks in smart manufacturing encounter difficulties in preserving closed-loop stability. This is particularly true in multi-hop setups where communication and control traits must have balance. The difficulty of handling event types and guaranteeing exact alerts lifts computing needs. Because of these issues, strong structures and modern computing methods become a requirement. DTs can stimulate digital improvements in the industrial world and also improve the operating output.

8.2 Data Quality and Integration

Poor data quality and difficulties with integration seriously hurt the use and value of DT systems. Real-time data streams must be correct and easy to get, because good data supports the dependability of digital representations. Predictive models greatly depend on this correct data. Real-time processing makes the job trickier, so firms must pick the right frameworks and algorithms for different data streams. When different data sources connect, you need to make formats and rules consistent. But HDTs deal with more issues. Data quality trust and ethics grow, specifically with private data. To solve these issues, firms need clear standards and ways to share data smoothly and to keep it good. This makes DTs dependable and useful. Progress encourages new ideas, adjustment in manufacturing, and easier upkeep. This allows companies to use information to help their work run faster and to beat their competitors.

8.3 Security and Privacy Concerns

DT systems present serious security and privacy problems. This is especially true when data flows between the real and the digital versions all the time. Cyberattacks such as putting in false data endanger how complete and correct these models are. As a result, predictive maintenance is at risk (Mode et al., 2019). Because DTs connect to devices on the internet, data privacy and access without permission become worries, especially in big, automated systems. For this reason, complete security protocols and methods that protect privacy become necessary to lessen such worries (Naveiro et al., 2018). This also makes use of AI and machine learning to find threats as they happen and make smarter decisions. Such action boosts security in tough settings, like semiconductor supply chains (Shaikh et al., 2022). With robust security and privacy, one can protect private data and keep the trust of those involved. This is vital particularly in spread-out internet of DT setups.

9. Conclusion

DT tech has become a crucial part of current industry. It transforms how choices are made using data, improving operational efficiency and strengthening system resilience. DT tech makes digital versions of real systems. These twins permit ongoing oversight and review. They include predictive analysis that projects future conditions and lowers dangers. This ability is essential for things like picking suppliers. For example,

(CLIP) Contrastive Language–Image Pretraining boosts production excellence and productivity because it prevents mistakes. Although there is advancement, issues like having standards, keeping data safe, and implementation costs exist. These problems need fixes before DT tech can truly affect every field.

DTs present changes to city spaces. Planners and leaders utilize them to mimic situations, refine resources, and connect with people. Network DTs displays how these systems can help network control via models and current awareness, but existing versions sometimes do not have service abilities. Cycle detection improvements display time reductions, which shows DTs' influence in areas. For this reason, blending DTs into frameworks requires attention to growth and change.

Social sciences and engineering come together in DT projects, and this union has improved the human–machine interface design. It has encouraged deep thought about the social and technical effects of automation. This mix of fields gives better insight into what users need and how society feels the impact, making DT systems easier to use and perform better. Greater clarity in DT systems then boosts how well they work in the real world and how much users trust them—a key aspect for general approval. Though DTs support decisions based on data, there is a lot of room to expand. Limits in standardization and combining different services now hold DTs back. For DT technology to effect change across sectors, these roadblocks must disappear.

References

Agho, M.O., N.L. Eyo-Udo, E.C. Onukwulu, A.K. Sule and C. Azubuike. (2024). Digital twin technology for real-time monitoring of energy supply chains, *International Journal of Research and Innovation in Applied Science*, *9*(12), 564–592. https://doi.org/10.51584/IJRIAS.2024.912049.

Alen, S., U. Advaith, J.K. Johnson, K.M. Joies, R. Sunil, A. Ravikumar and J. John. (2021). Accident alert system with false alarm switch, in A. P. Pandian, X. Fernando and S.M.S. Islam (Eds.), *Computer Networks, Big Data and IoT*, Singapore: Springer, pp. 319–328.

Almasan, P., M. Ferriol-Galmés, J. Paillisse, J. Suárez-Varela, D. Perino, D. López, A.A.P. Perales, P. Harvey, L. Ciavaglia, L. Wong, V. Ram, S. Xiao, X. Shi, X. Cheng, A. Cabellos-Aparicio and P. Barlet-Ros. (2022). Digital twin network: opportunities and challenges. https://doi.org/10.48550/arXiv.2201.01144.

Almeida, A., S. Brás, S. Sargento and F.C. Pinto. (2023). Time series big data: a survey on data stream frameworks, analysis and algorithms, *Journal of Big Data*, *10*, 83. https://doi.org/10.1186/s40537-023-00760-1.

Avrilionis, D. and T. Hardjono. (2021). Towards blockchain-enabled open architectures for scalable digital asset platforms. https://doi.org/10.48550/arXiv.2110.12553.

Baby, K. and Ravikumar, A. (2014). Big data: an ultimate solution in health care. *Int J Comput Appl,* 975, 8887.

Baumann, D., F. Mager, U. Wetzker, L. Thiele, M. Zimmerling and S. Trimpe. (2021). Wireless control for smart manufacturing: recent approaches and open challenges, *Proceedings of the IEEE, 109*, 441–467. https://doi.org/10.1109/JPROC.2020.3032633.

Bhattacharjee, A., Y. Barve, S. Khare, S. Bao, A. Gokhale and T. Damiano. (2019). Stratum: a serverless framework for the lifecycle management of machine learning-based data analytics tasks, in, pp. 59–61. https://www.usenix.org/conference/opml19/presentation/bhattacharjee (accessed April 5, 2025).

Bonar, J. and J. Hastings. (2024). Transforming Information Systems Management: A Reference Model for Digital Engineering Integration, in *2024 Cyber Awareness and Research Symposium (CARS)*, pp. 1–9. https://doi.org/10.1109/CARS61786.2024.10778791.

Boschert, S. and R. Rosen. (2016). Digital twin—the simulation aspect, in *Mechatronic Futures*, Berlin & Heidelberg: Springer Vieweg, pp. 59–74. https://link.springer.com/chapter/10.1007/978-3-319-32156-1_5.

Chen, J., C. Yi, S.D. Okegbile, J. Cai and X. Shen. (2024). Networking architecture and key supporting technologies for human digital twin in personalized healthcare: a comprehensive survey, *IEEE Communications Surveys & Tutorials*, *26*, 706–746. https://doi.org/10.1109/COMST.2023.3308717.

Conde, J., A. Munoz-Arcentales, Á. Alonso, S. López-Pernas and J. Salvachúa. (2022). Modeling digital twin data and architecture: a building guide with FIWARE as enabling technology, *IEEE Internet Computing*, *26*, 7–14. https://doi.org/10.1109/MIC.2021.3056923.

Durão, L.F.C.S., S. Haag, R. Anderl, K. Schützer and E. Zancul. (2018). Digital win requirements in the context of Industry 4.0, in P. Chiabert, A. Bouras, & F. Noël, J. Ríos (Eds.), Cham: Springer International Publishing, pp. 204–214. https://doi.org/10.1007/978-3-030-01614-2_19.

Farus, Z., B. Searcy, T. Nassisid and K. Muhammad. (2024). Dynamic optimization of video streaming quality using network digital twin technology. https://doi.org/10.48550/arXiv.2407.00513.

Gaikwad, A., R. Yavari, M. Montazeri, K. Cole, L. Bian and P. Rao. (2020). Toward the digital twin of additive manufacturing: Integrating thermal simulations, sensing, and analytics to detect process faults, *IISE Transactions*, *52*, 1204–1217. https://doi.org/10.1080/24725854.2019.1701753.

Gonçalves, B. and F. Porto. (2015). Managing large-scale scientific hypotheses as uncertain and probabilistic data with support for predictive analytics, *Computing in Science & Engineering*, *17*, 35–43. https://doi.org/10.1109/MCSE.2015.102.

Harini, S. and A. Ravikumar. (2020). Effect of parallel workload on dynamic voltage frequency scaling for dark silicon ameliorating, in *2020 International Conference on Smart Electronics and Communication (ICOSEC)*, pp. 1012–1017. https://doi.org/10.1109/ICOSEC49089.2020.9215262.

Huang, J., Y. Yang, H. Yu, J. Li and X. Zheng. (2023). Twin graph-based anomaly detection via attentive multi-modal learning for microservice system. https://doi.org/10.48550/arXiv.2310.04701.

Jackson, I., M.J. Saenz and D. Ivanov. (2024). From natural language to simulations: applying AI to automate simulation modelling of logistics systems, *International Journal of Production Research*, *62*, 1434–1457. https://doi.org/10.1080/00207543.2023.2276811.

Jiang, Y., R. Ding, T. Qiao, Y. Zhu, S. Han and D. Zhang. (2023). A unified and fast interpretable model for predictive analytics. https://doi.org/10.48550/arXiv.2111.08255.

Lauer-Schmaltz, M.W., P. Cash, J.P. Hansen and A. Maier. (2024). Towards the human digital twin: definition and design—a survey. https://doi.org/10.48550/arXiv.2402.07922.

Lee, C. -Y. and C. -F. Chien. (2022). Pitfalls and protocols of data science in manufacturing practice, *Journal of Intelligent Manufacturing*, *33*, 1189–1207. https://doi.org/10.1007/s10845-020-01711-w.

Lee, J., D. -Y. Ji and Y. -M. Hsu. (2024). Novel topological machine learning methodology for stream-of-quality modeling in smart manufacturing. https://doi.org/10.48550/arXiv.2404.14728.

Leng, J., D. Wang, W. Shen, X. Li, Q. Liu and X. Chen. (2021). Digital twins-based smart manufacturing system design in Industry 4.0: a review, *Journal of Manufacturing Systems*, *60*, 119–137. https://www.sciencedirect.com/science/article/abs/pii/S0278612521001151.

Leutert, F., D. Bohlig, F. Kempf, K. Schilling, M. Mühlbauer, B. Ayan, T. Hulin, F. Stulp, A. Albu-Schäffer, V. Kutscher, C. Plesker, T. Dasbach, S. Damm, R. Anderl and B. Schleich. (2024). AI-enabled cyber-physical in-orbit factory—AI approaches based on digital twin technology for robotic small satellite production, *Acta Astronautica*, *217*, 1–17. https://doi.org/10.1016/j.actaastro.2024.01.019.

Li, T., Z. Bian, H. Lei, F. Zuo, Y. -T. Yang, Q. Zhu, Z. Li, Z. Chen and K. Ozbay. (2024). Digital twin-based driver risk-aware intelligent mobility analytics for urban transportation management. https://doi.org/10.48550/arXiv.2407.15025.

Li, X., H. Xu, J. Tupayachi, O. Omitaomu and X. Wang. (2024). Empowering cognitive digital twins with generative foundation models: developing a low-carbon integrated freight transportation system. https://doi.org/10.48550/arXiv.2410.18089.

Liu, F., M. Li, J. Lu, E. Su and M.C. Yip. (2022). Parameter identification and motion control for articulated rigid body robots using differentiable position-based dynamics. https://doi.org/10.48550/arXiv.2201.05753.

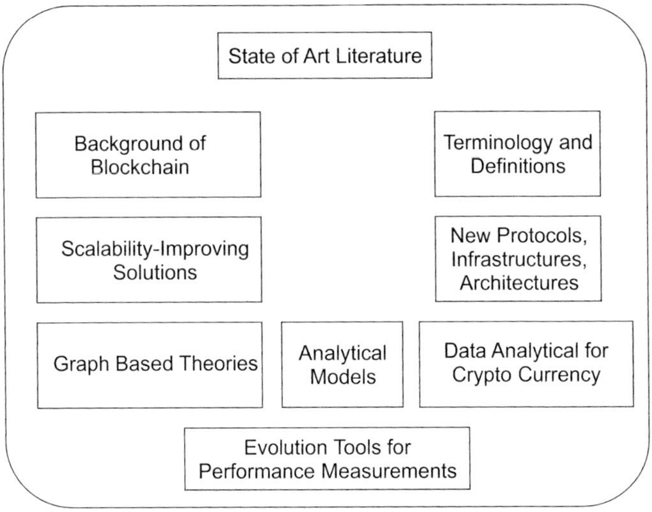

Fig. 2.1 Blockchain architecture: a comprehensive overview.

Figure 2.1 presents a comprehensive view of the blockchain concept. A hash chain allows a single set of credentials (such as an ID and password) to generate a huge number of hashes. Extra data can be successfully hashed using a hash function to record the data's existence chronologically for non-repudiation. One way to avoid potentially dangerous tabletop collisions is to utilize cording. When two hash table keys are hacked to the same index, a collision occurs (Mancy et al., 2022). Because each hatch table slot can only hold one element, collisions are an issue. Figure 2.2 shows the tree hierarchy.

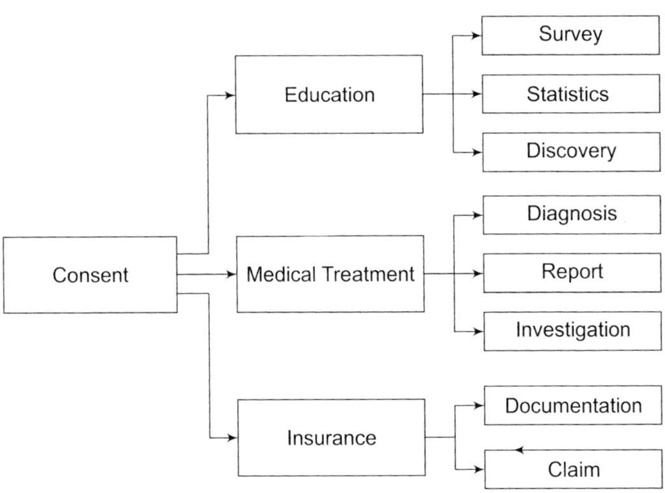

Fig. 2.2 Constant tree hierarchy.

1.3 Transaction Process in Blockchain

According to Andrew et al. (2023) secure data transfer improves diagnosis and treatment efficiency by facilitating clinical communication and the gathering of expert opinions and confirmations. There are a number of interoperability issues that arise regularly in this field. For example, there may be several logistical challenges associated with ensuring the safe and dependable transmission of patient records between healthcare facilities and academic institutions (Benchoufi & Ravaud, 2017). Significant, reliable, and positive cooperation between the entities involved is required for such exchanges of clinical data. Possible stumbling barriers include data sharing agreements, sensitive clinical data, ethical regulations, laws and procedures, and algorithms that have a hard time matching patients. The procedure for creating blocks in a blockchain is shown in Figure 2.3.

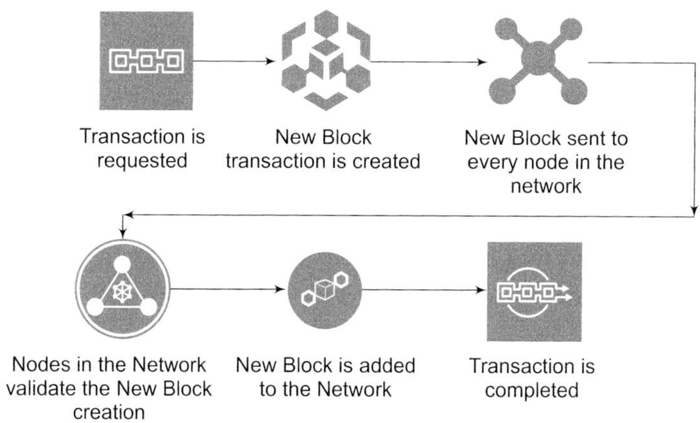

Fig. 2.3 Blockchain transaction process

1.4 Healthcare System

The healthcare business regularly generates, accesses, and disseminates vast volumes of data, which results in an abundance of clinical data. The storage and dissemination of this enormous data set is both essential and incredibly difficult because of security and privacy concerns and the delicate constitution of the data (Mamo et al., 2020).

Hospitals, pharmacies, and insurance companies are just a few of the many medical practitioners that have access to a wealth of information (Androulakı et al., 2018). Thanks to the proliferation of wearable sensors made possible by the fast growth of information and communications technology (ICT), a great number of physical things can now be linked. The use of electrical devices and software allows for the processing, collection, and generation of data on an enormous scale. Despite an abundance of medical facilities, the incidence of deadly diseases including pneumonia, influenza, cancer, and heart diseases has increased significantly (Laroiya et al., 2020). Sensing devices that can collect and analyze patient data have enabled constant monitoring and vital sign tracking, made feasible by the technology revolution. In order to provide healthcare, hospitals and other relevant departments get this data.

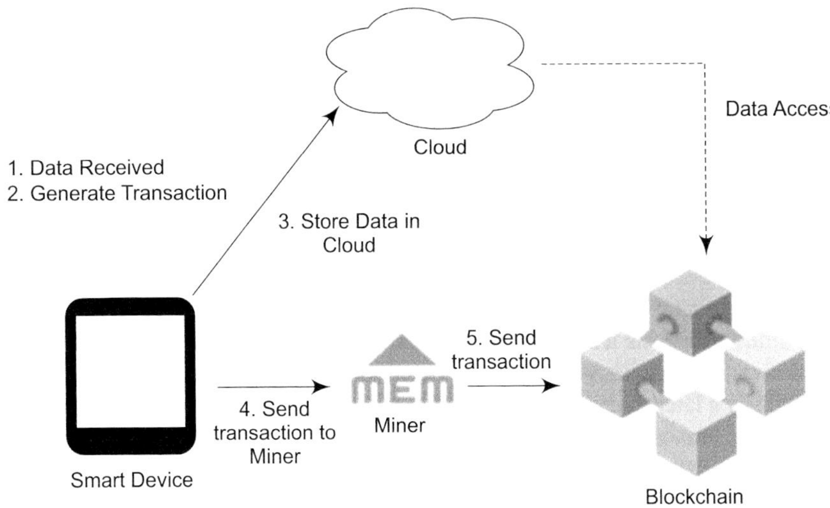

Fig. 2.4 Blockchain process with decentralized storage.

has changed the game in every business, and the healthcare industry is no exception. Both new approaches and old problems in blockchain are brought to light in this assessment.

2.1 Introduction to Blockchain Technology

According to Rantos et al. (2019), anybody with access to the internet can view a public blockchain, which is a distributed ledger that can be used to manage the access to specific data. The idea behind this method is to use blockchain technology to construct a personal health record (PHR) by keeping sensitive medical records publicly. Improved access to patients' clinical data was a direct result of their suggestions. Patients are now able to participate actively in how their data is handled, view it whenever they choose to, and distribute it to any associated medical expert. Sunny et al. (2022) proposed a method for the secure administration and exchange of patients' private medical records that used cloud computing and blockchain technology. Possible applications of the proposed method include safeguarding the transmission and storage of sensitive patient medical information. By providing patients complete access to their own health records, the suggested system cuts out the intermediary, which is a revolutionary change. By (Tong et al., 2024), a paradigm based on the blockchain that connects various devices will greatly simplify the management of EHR transactions. The design proposed utilizing the MQTT protocol to link the bio-sensors to the IoT platform. Furthermore, the InterPlanetary File System (IPFS) was incorporated into the architecture to reduce stored transaction deduplication by identifying changes to blocks or states induced by particular appended transactions.

According to Agbo et al. (2019), a provider of corporate blockchain solutions, one of the very first extensive medical operators to become a member of the Gem Health network was the Philips Blockchain Lab, a research and development division of the healthcare behemoth Philips. When it comes to healthcare application and shared infrastructure development, there is an Ethereum-powered network called Gem Health.

The work of Watanabe et al. (2019) proposed a cryptocurrency system design for medical data assessment that substitutes AI healthcare systems with parallel execution. As a clinical decision-support tool, the proposed approach simultaneously runs computational trials to assess a patient's overall health, diagnosis, treatment plan, and any relevant therapeutic processes. We have tested the proposed method on both actual and virtual healthcare networks to see how well it diagnoses patients and how well it treats their conditions. During an earlier clinical study that had shown evidence of change outcomes, Irving and Holden tested the approach. It was proven that blockchain technology can be used as a cheap, independently verifiable auditing tool and that scientific research may be trusted.

2.2 Blockchain Technology in Healthcare

An interactive setting will not be complete without an IoT–controlled healthcare system (Hajian et al., 2023). In order to give patients easy-to-understand feedback and remedies, the proposed design relies on generative data input from smart wearables and biosensors. Patients can benefit from faster and more precise advice (Liang et al., 2021), and by combining the effectiveness of sensors with human reaction recommendations, a smart hospital system might be realized. The Radio Frequency Identification (RFID), Wireless Sensor Networks (WSN), and smart wearables-enabled system is the brainchild of the plan, which aims to achieve several objectives: intelligent sensing of a patient's surroundings, ward assignment in accordance with physician placement needs, tracking of a patient's movements, and, finally, report analysis based on computed data. Jayabalasamy et al., 2024, built on blockchain technology, BloCHIE is a revolutionary platform for the sharing of medical records. Contained in many sources, the suggested platform evaluates the requirements for healthcare data interchange, specifically pertaining to EHRs and individual healthcare data. It also manages a diverse array of other data kinds. To make sure the platform was secure and private enough, authors included off-chain and on-chain verification procedures. With blockchain technology's transformational potential, the healthcare sector as a whole—from payers and providers to patients—stands to gain.

According to Saranyah et al. (2025), there are a number of existing blockchain designs that can be used to talk about the basics of blockchains and their potential applications in healthcare data storage, validation, and maintenance. Additionally, it has been determined that the consortium blockchain offers the best answer since it is almost tailor-made for storing healthcare data.

In their innovative and systematic proposal, Riggs et al. (2022) used blockchain technology to safeguard sensitive patient data, deal with critical data security issues, and incorporate blockchain software into the entire infrastructure of healthcare organizations. Privacy, confidentiality, and exchange of medical records can be substantially enhanced with the utilization of blockchain technology for healthcare providers and hospitals.

2.3 Security and Scalability in Healthcare Using Blockchain

This chapter primarily bases its investigation of blockchain technology's possible healthcare applications on the work of Nakamoto et al. (2024), who summarized many current trends in this area of research. There have been infinite non-financial uses of the underlying technology since Bitcoin's creation (Ivan et al., 2016). The data recorded on public blockchains can be seen by anybody with an internet connection because they are decentralized database systems. This approach of encrypting health data was also demonstrated by Ivan (2016). Through the public storage of encrypted healthcare data, this method generates a blockchain-based PHR. By following their recommendations, patients were able to have better access to their medical records. Patients now have the freedom to view and control their data, and they can also help keep their data up-to-date and accessible to any affiliated healthcare professional. In a separate study, Xu et al. (2019) proposed a system that combined blockchain technology with cloud storage for the management and exchange of patients' personal medical data. The proposed method may find value, for example, in guaranteeing the safekeep and transmission of individual medical records. The suggested approach is fresh because it gives patients complete control over their own health records, cutting out the intermediary.

Dey and Shekhawat (2021) proposed an IoT paradigm based on blockchain technology to simplify the management of EHR transactions. In order to connect the bio-sensors to the IoT platform, the design recommended utilizing the MQTT protocol. The IPFS was also a part of the design—it can reduce stored transaction deduplication by identifying state entries or block modifications induced by particular appended transactions.

According to a statement made by Gem, a developer of business blockchain solutions, (Prisco et al., 2016). Using Ethereum as its backbone, Gem Health is creating a platform for healthcare app and infrastructure development.

2.4 Smart Consent Management in Healthcare

By combining the traditional database storage system with the mobile agent paradigm, Kosba et al. (2016) propose a method to consolidate health records stored in multiple databases into a single central repository. A framework for the administration and analysis of large amounts of data in healthcare is presented by Lin et al. (2021), with a focus on data pre-processing and data transformation. Due to the high costs and extensive technical support needed, most organizations choose to centralize their health data storage in the cloud. They confront a plethora of security threats, including corruption, data theft, authentication, and privacy concerns.

Blockchain technology's possible uses in healthcare, according to Watanabe et al. (2025), include data interchange, administration of data access, and medical history keeping (Watanabe et al., 2025). They highlight the importance of smart contracts and voice concerns about scalability when it comes to developing applications. We discover a simple solution to this problem by studying the scalability of smart contracts for sharing health data. The issue of genomics data sharing based on consent is addressed by Zheng et al. (2018). In an effort to sidestep overly wordy agreement conditions, they suggest a one-page web-based consent form for genetic data sharing. By analyzing survey responses from 5,162 individuals, they prove that the simplified permission form results in improved data accessibility.

In particular, Attaran et al. (2022) proposed that consent-based double-blind anonymous data exchange be built on HLF. Their method details what a permissioned Know Your Customer (KYC) application on the blockchain should include. In a permission-less setting, where anyone may join the platform and begin contributing, this tactic fails, while our approach KYC works effectively.

Both ProvChain (Liang et al., 2019; Magagna et.al., 2020) and DataProv Both ProvChain and DataProv (Liang et al., 2019; Magagna et.al., 2020) are blockchain-based are blockchain-based approaches to verifying the veracity of data's provenance. However, these approaches are not mainly aimed at modeling dynamic consent. Neisse et al. (2020) offer a blockchain-based method for data accountability and provenance tracking. Their solution enables data tracking and reuse with the consent of data suppliers. Accompanying the solution is a smart contract that specifies its usage rules and provides details about the data's provenance.

3. Materials and Methods

This research addresses the advent of blockchain methodology and its decentralized, trustworthy, and open nature system for transactions and data management. As blockchain continues to evolve, researchers explore its potential applications, challenges, and innovations across sectors like healthcare and supply chain. This chapter outlines the phased methodologies employed to achieve energy-efficient and scalable WSN solutions.

3.1 Problem Identification

- The process lacks transparency and trust, and present methods for patient consent and medical data exchange are inefficient, costly, and error-prone, despite quality assurance measures.
- The existing system uses traditional lengthy blocks and public blocks.
- Existing methods are "MedRec" and Consent Management System (CMS).
- System response time and data transaction speed can be affected by a lack of scalability.

3.2 Contribution

The research contribution in the context of blockchain in healthcare highlights the value and advancements that studies provide to the field. A well-defined contribution ensures that your work is impactful and relevant, and bridges existing gaps. The major goals are:

- To provide a paradigm shift in healthcare communication by making use of blockchain technology to make it both scalable and safe.
- To use consortium blockchain technology to create a secure network for healthcare organizations.

3.3 Motivation

Regular collection, access, and sharing of huge volumes of data has resulted in clinical data being abundant in the healthcare sector. Storing and sharing this massive amount of data is important but very difficult because of privacy and security concerns and the sensitive nature of the data. Data privacy and security is an ongoing issue in the healthcare industry due to the fact that numerous stakeholders depend on it to achieve their individual goals. Society relies on the healthcare profession for patient care, ethical practice, and unwavering trust. But these admirable goals have been tarnished by recent allegations of wrongdoing by several healthcare organizations. There has been a call for greater honesty, integrity, and ethics in healthcare after patients spent too much time in intensive care units (ICUs) for reasons unrelated to their illness.

Healthcare organizations have the potential to greatly reduce or even eliminate reconciliation, rework, and repetition by utilizing blockchain technology. In a predictive healthcare context, we can use AI to determine the ideal degree of service intensity. Adapting one's communication style for different contexts is crucial while interacting with patients, professionals, and careers.

Because they provide doctors with complete patient records, EHRs are essential to efficient healthcare delivery. Yet, there are limitations to existing centralized EHR systems, such as issues with data privacy, security, and interoperability. Because of its distributed ledger and immutability, blockchain technology has become an attractive alternative. Blockchain technology gives people control over their own health data and allows for transparent information sharing amongst different healthcare providers. Potentially game-changing, blockchain-based EHR solutions still need time to mature before they can see widespread adoption in the healthcare industry.

This study introduces a patient health records management architecture that utilizes blockchain technology. The objective is to enhance the management and exchange of patient health information while ensuring that patients maintain control. There is hope that blockchain technology can overcome the drawbacks of existing centralized health records systems. As a result, healthcare providers can exchange patient data in an open and safe manner. Data privacy and security, interoperability, auditability, decentralization, and automation are some of the benefits that patients get while they have ownership over their health data. When used with EHR systems, blockchain technology offers promise for better patient data management and exchange without compromising patients' right to privacy.

3.4 Smart Contracts

Our electronic health record system is built on top of smart contracts built in Solidity. These contracts automate crucial operations like data validation, access control, and compliance enforcement. In order to meet stringent data privacy standards, these contracts are meticulously written to incorporate procedures for handling patient consent and making sure that access to medical records is dependent upon specific patient assent. It starts with the user or patient making an account and adding their medical records. Afterwards, a distinct key allows access to these records. The user and the hospital both enter this one-of-a-kind key into the system, which grants them access to the particular record or data. Patients and hospitals primarily use the system to create accounts, upload data, and access or retrieve that data. Blockchain technology manages and underpins the entire system.

3.5 Smart Wearable Devices

In the medical field, "wearable gadgets" refer to electrical smart devices containing miniature controllers that may be worn as attachments or embedded into clothing. They are small, unobtrusive, and packed with advanced features including wireless data transmission, built-in alerting systems, and real-time feedback. Through these devices, medical personnel can receive a number of vital signs, such as the rate of respiration, blood sugar levels, and pressure. The IoT includes wearable gadgets. The software, electronics, sensors, actuators, and connectivity contained in these devices allow them to connect to one another and exchange data. Ensuring secure data sharing is crucial for this infrastructure to effectively manage patient data in collaboration with other organizations. Health data is extremely sensitive, and its disclosure can raise the risk of exposure.

3.6 Body Sensors and Environmental Sensors

The implanted sensors monitor a number of vital indicators, including blood pressure and heart rate (Bs = Bs1, Bs2, Bsn). We carefully deploy environmental sensors (Es = Es1, Es2, Esn) throughout the surroundings to monitor any changes in those variables while keeping the patient's health in mind. These sensors track a variety of environmental variables for patients, including humidity and air quality.

Among the many forms of big data, medical records stand out because of the far-reaching effects they have on people's well-being Prisco, G. 2016. With the rapid growth of the IoT, there has been a marked increase in efforts to construct the internet of medical things (IoMT), especially in light of the 2019 coronavirus disease outbreak.

4. Blockchain-Enabled Secure Access Control for IoMT to Improve Patient Trust in EHRs

Patient care, ethical practice, and unwavering trust underpin the healthcare system. Recent allegations of wrongdoing on the part of healthcare groups have put a shade over these noble goals. Patients' lengthy ICU hospitalizations for non-medical reasons have prompted calls for healthcare openness, integrity, and ethics.

The rapid transmission of patients' EHRs to doctors allows for more effective diagnosis and treatment in the IoMT. The gap in understanding between doctors and their patients might be further diminished in this manner. Simultaneously, clinical treatment and medical research can benefit greatly from data gathered from the entire IoMT. Today, the medical industry is not developing healthily due to the total isolation of data between medical institutions (Mis), which causes patients to waste time and money with repetitive inquiry and examination during consultations and increases the likelihood of misdiagnosis. A major threat to patient safety is also posed by unlawful medical practices. It can be problematic in the case of medical errors because patients rarely consider whether MIs and doctors have the proper certifications and licenses. There is an immediate need to find a way to connect the data islets managed by various

used in the suggested method is somewhat limited, consisting solely of patient vital signs. Only emergency patients are eligible for the indicated cause. Our findings prove that it can successfully store data on the Ethereum blockchain. Table 2.1 shows the statistics of gas consumption related to number of patients, single transactions, and separate transactions.

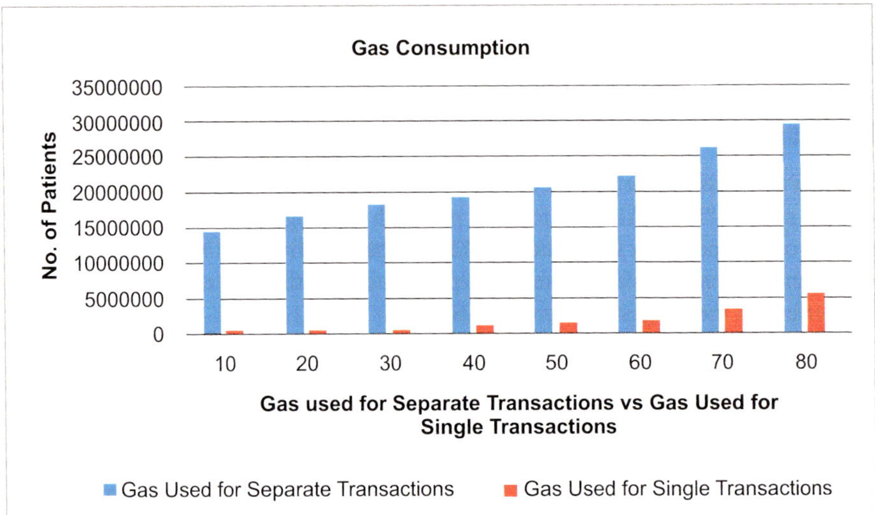

Fig. 2.9 Gas consumption comparison

Table 2.1 Statistics of gas consumption.

No. of Patients	*Gas Used for Separate Transactions*	*Gas Used for Single Transactions*
10	14,654,140	433,892
20	16,634,250	548,373
30	18,443,312	833,963
40	19,453,243	1,248,211
50	20,678,211	1,586,343
60	22,353,198	1,937,246
70	26,248,927	3,506,956
80	27,663,280	3,768,150

6.2 Access Time

Access time is the amount of time needed to access the patient's records. Access time comparison for the proposed system TSAM, the existing system TSACS, and another method ET is depicted in Figure 2.10 and the statistics are displayed in Table 2.2.

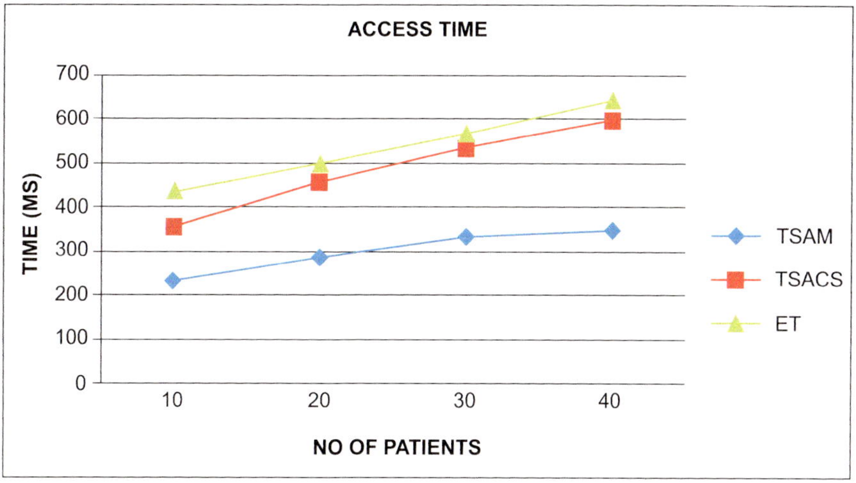

Fig. 2.10 Access time comparison

Table 2.2 Access time of proposed vs compared methods.

No. of Patients	*TSAM (ms)*	*TSACS (ms)*	*ET (ms)*
10	235	356	435
20	287	458	499
30	334	534	568
40	348	598	645

6.3 Processing Overhead

Figure 2.11 displays the results of the simulations that were used to evaluate the systems' performance in relation to processing overhead. The validation of new blocks in blockchain may require a lot of resource measures, including indirect memory, processing time, bandwidth, and excess memory. This is known as processing overhead in the technical world. This value is found to be lower in the proposed model compared to several existing methods. The proposed method TSAM stands head and shoulders above the current solutions when compared to processing speed and data security. Statistics are provided in in Table 2.3.

Table 2.3 Processing overhead of proposed and compared methods.

Processing Overhead	*Proposed System TSAM*	*Existing Method TSACS*	*Other Method ET*
No. of Blocks (10)	56	68	98

CHAPTER 3

Optimizing Vehicle-to-Cloud and Grid Interactions for Smart Energy Management of EVs

Reedhu Saji[1*], *Jarin T.*[2], *Thomas George K.*[3], *Jyotsna E.*[4] and *Shuai Li*[5]

Electric vehicles (EVs) are revolutionizing transportation by running on energy instead of fossil fuels. The full synergy of EVs with the energy ecosystem would entail optimizing the interactions of EVs with the cloud and power grid. This chapter investigates possible synergies and approaches for the smart energy management of EVs. By connecting EVs to the cloud, usage, charging behavior, and energy consumption can be logged and processed in real time to optimize the grid. EVs could be useful for the grid as distributed energy storage in case the grid comprises power plants, transmission, and distribution networks. Introduction of bidirectional electricity flow through vehicle-to-grid (V2G) technology increases grid stability and resilience. Further cloud integration provides intelligence and scalability to transportation systems. The efficiency of grid-to-cloud (G2C) interconnection is considered through virtualization, distributed energy resource integration, and advanced data analytics. Similarly, methods for vehicle-to-cloud (V2C) connectivity, including built-in and aftermarket telematics systems are also considered. V2G, G2C, and V2C connections are examined with respect to parameters like data transmission rate, latency, security protocols, cloud storage capacity, data standardization, real-time monitoring, and regulatory compliance. The results point to the fact that G2C excels in many aspects, suggesting its prospects in effective energy management. In sum, this chapter emphasizes the importance of utilizing cloud and grid connections to

[1-3] Department of Computer Science & Engineering, Jyothi Engineering College, APJ Abdul Kalam Technological University, Kerala, India.

[4] Department of Computer Science & Engineering, Sahrdaya College of Engineering and Technology, APJ Abdul Kalam Technological University, Kerala, India.

[5] Faculty of Information Technology and Electrical Engineering, University of Oulu, Finland.
https://orcid.org/0009-0007-4224-27791[1]; https://orcid.org/0000-0002-1974-0688[2]; https://orcid.org/0009-0007-8610-2997[3]; https://orcid.org/0009-0007-2004-4966[4]; https://orcid.org/0009-0004-4499-0023[5]

* Corresponding author: jarint@ieee.org

DOI: 10.1201/9781003659822-3

enhance energy efficiency, environmental quality, and the integration of EVs within intelligent transportation networks.

1. Introduction

Fueled by technological advances, the revolution in modern transportation is turning vehicles into complex entities connected to a complex network of infrastructure. This chapter discusses vehicle-to-power grid and cloud computing interactions, while exploiting the synergies of intelligent transport systems (Rashad & Ali, 2023). Increasingly, as the world demands efficient and sustainable mobility solutions, there is a need for understanding the dynamic interactions between vehicles, the network, and the cloud so that they are optimized in resource utilization, energy efficiency, and smooth integration of technological evolution (Jaworski et al., 2011).

More EVs and power grids could translate into energy management and resource sharing. With this unprecedented bidirectional flow of energy, a new paradigm in energy management and resource sharing will come into force that will demand a complete examination of all intricacies that bind together vehicles and power grids (Maimaris & Papageorgiou, 2016). The launch of cloud technology then witnessed the onset of data-driven decision-making and real-time relationships. For the intelligent traffic system, the cloud is a centralized network, where enormous data, generated by vehicles and infrastructure, is processed. This chapter presents an integrative perspective on the vehicle, power grids, and the cloud for fostering energy optimization and environmental sustainability while putting forth potential benefits and challenges (Younes et al., 2021).

1.1 Evolution of Vehicle to Grid

V2G is a lighthouse technology in terms of renewable energy and mobility. It came into existence as a concept, with the realization that the traditional approach of charging EV batteries by drawing energy from the grid could be reversed (Ntombela et al., 2023a). The realization that EVs can stand as mobile energy storage units in the parking lot, thereby adding to grid stability and resilience, was the next step that led to the advancement of V2G technology. The V2G path first needed to resolve the standardization of EV-to-grid communication mechanisms to allow seamless integration of (Ntombela et al., 2023b). The advancement of smart grids led to further development of V2G systems to embrace more sophisticated communication methods, thereby allowing real-time data exchange for the optimal directing of energy flows between vehicles and the grid. Combining renewable energy sources in this manner also makes good environmental sense, even in the face of the intermittent nature of renewables (Viswanath et al., 2023).

The enhancement in energy management systems in EVs was a landmark in the birth and evolution of V2G. These systems allow grid operators to absorb or provide energy from/to the EVs with respect to demand fluctuations and grid imbalances so that the grid never suffers from a loss of stability or reliability. Policy initiatives and regulatory frameworks were employed in propelling the evolution of V2G, as governments and utilities perceived its potential benefits in enriching grid resilience and expediting a transition toward sustainable energy (Chunhua et al. 2013). They provided for incentives and subsidies while enforcing restrictions to see to the development of V2G infrastructure so that the technology could swiftly enter conventional energy systems. Introduction of vehicle-to-everything (V2X) extended V2G beyond the usual grid-based setup, facilitating an interface for EVs with other devices (Turton & Moura, 2008).

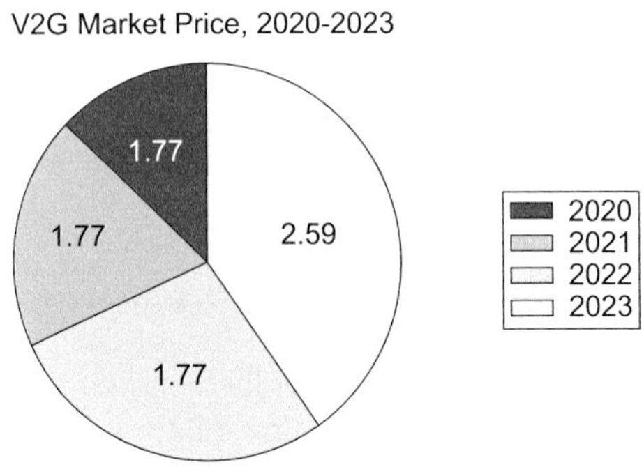

Fig. 3.1 Vehicle-to-grid market price ($), 2020–2023.

Since the large availability of V2G technologies, new partnerships have formed between vehicles, energy, and technology-based companies. Figure 3.1. represents the V2G market price ($) for the years 2020–2023. Standardization and interoperability became important in order to have smooth V2G systems entering the market from different platforms and brands. This, in a way, helped usher in a very strong V2G ecosystem. Research and development in the future, in turn, would improve the efficiency and abilities of V2G, whereas capabilities may be enhanced with battery technologies, artificial intelligence (AI), and decentralized energy systems. V2G is key to an energy scene that is sustainable and linked.

2. Literature Survey

This study handles the V2G operation from within a microgrid under a decentralized fashion by applying a non-cooperative game approach. The framework is modeled with the intent to take care of complexities caused by aspects of charging/discharging preferences, flexibility constraints, and random variables. The applied case study comprises two scenarios: one containing 5 EVs and the other containing a fleet of 100 EVs. In the analysis, the optimization framework is judged by its effects on balancing energy demand, improving system stability, and supporting smart grid control (Zerka et al., 2024).

Hybrid electric vehicles (HEVs) are simply energy management strategies (EMSs) put in place to optimize the power split between the engine and electric motors. This is achieved through EMS methods using rule-based methods, optimization algorithms, or more complex strategies, including reinforcement learning and deep learning. These strategies would benefit from machine learning, cloud computing, computer vision, and swarm technologies to predict better and improve decisions made in real-time. The main challenges integrating these methods include maintaining energy efficiency, battery life, and adapting to the adversarial driving scenarios (Mittal & Shah, 2024).

This analysis describes the AI-based approaches in EV systems, concentrating on battery management and control-related technologies. Using Machine Learning (ML), Reinforcement Learning (RL), State of Health (SOH) State of Charge (SOC) of a battery and temperature control can be predicted. These AI-based control schemes handle energy flow, regenerative braking, and power allocation; whereas, deep reinforcement learning equips adaptation intelligence for energy efficiency. Integration with the internet of things (IoT) and big data supports predictive maintenance and fleet management (Cavus et al., 2025).

Statistical analysis covers 78 pertinent publications retrieved from the Scopus database for the period of 2014 through 2023. The analyzed parameters include present-day research trends, keywords distribution focus, journals published, research categories, author collaborations, and per-country contributions. Besides this, an in-depth analysis is provided on AI applications in battery management systems for EVs by analyzing targets, contributions, advantages, and disadvantages of such approaches. Then, the challenges within advancement of Battery Management System (BMS) technologies are stated, followed by recommendations on future routes for the development of EVs (Lipu et al., 2023).

An extension of the Electric Vehicle Charging Management System (EVCMS) is proposed to address management and security issues using real-time charging parameters and an Open Charge Point Protocol (OCPP). In designing the framework, charging prices, optimization, and security are kept in consideration. With respect to key issues such as data management and security protocol, a proposal of the system model for EVCMS is developed. The experimental implementation followed client–server architecture and charge-box-based simulation. The framework is evaluated in terms of various EV charging scenarios and respective metrics, with the comparative evaluation clearly showing the efficiency of the framework in improving the EV charging ecosystem (Hamdare et al., 2024).

The research discusses the potential utilities of 6G wireless communication for smart grid systems, holding on to massive connectivity, secure communication, and time-critical operation. This includes the utilization of emerging 6G technologies to improve vital key performance indicators (KPIs) such as data rate, reliability, and latency. A modern survey of smart grid applications is also presented, highlighting areas in which improvements have been made with respect to operational efficiency and resource management. Challenges and future research directions are briefly discussed to unearth 6G potentialities in the paradigm shift for smart grid systems (Amin et al., 2023).

3. Proposed Methodology

The proposed methodology starts with an introduction to the V2G concept—a set of novel arrangements that transform electric vehicles (EVs) into energy storage systems on wheels. V2G facilitates a bidirectional flow of electricity: an EV would typically take power from the grid and use it to charge its batteries, and now it can

provides stability to the system, and facilitates demand-side control methods, which is in the best interest of grid operators.

- *Advanced Data Analytics and Predictive Maintenance*: The cloud offers efficient data storage and management for huge volumes of data coming from grid constituents such as sensors, meters, and substation devices. By applying modern data analytics, such as machine learning and AI, predictive maintenance algorithms can identify early warnings of both infrastructural breakdown and performance deterioration. This allows for efficient countermeasures, averting downtime and maximizing asset utilization. Such data-driven decisions make grid reliability possible.

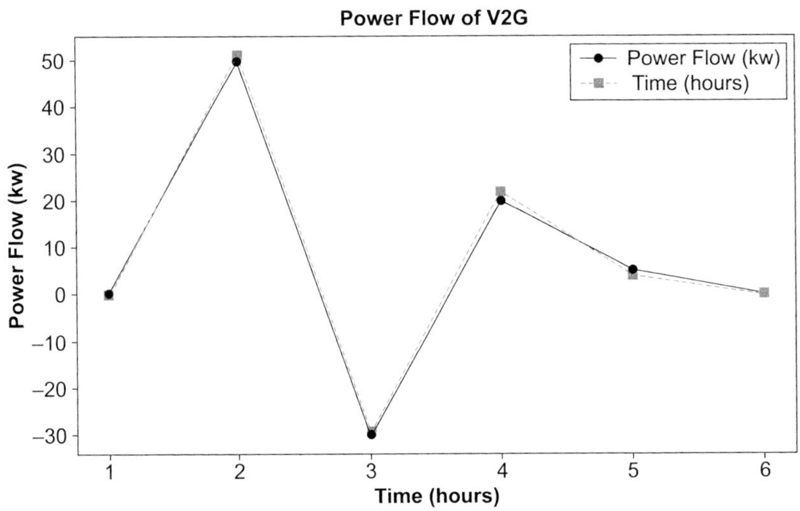

Fig. 3.4 Latency and response time graph of G2C.

Figure 3.4 shows the power flow of a V2G system for 6 hours. Starting at the 1st hour, the power flow is about 30 kW, being charged from the grid. By hour 2, maximum power flow of 50 kW is observed and is maintained till 3rd hour, where it reduces to around −30 kW, i.e., discharging. Moving to the 4th hour, the power flow appreciates again, reaching up to 40 kW, and then fluctuates. At hour 5 and 6, it is close to stabilization, near 20 kW. Therefore, the graph shows the charging–discharging cycles and how they change with time over these 6 hours.

3.5 Vehicle-to-Cloud (V2C)

In V2C communication, there is actually a bidirectional flow of data between the vehicle and the cloud services. Sensor-deployment vehicles gather multiple data points: location, speed, diagnostics, and environmental conditions. Since the data is huge and needs to be stored, processed, and analyzed, it gets sent to cloud servers. On the other hand, the cloud sends its data to the vehicle, which may include software updates, traffic alerts, and route optimization. The V2C applicability expands range from fleet management, predictive maintenance, and real-time monitoring to the construction of self-driving technology. By interweaving the operations with cloud computing infrastructures, V2C enhances the vehicle's operation in terms of safety, comfort, and economy, in addition to the driving experience, hence laying down the foundation for smart transportation.

Moreover, V2C connectivity permits the realization of advanced crowd-sourced data platforms in which fleets of connected vehicles act as a network of moving sensors around the city, constantly collecting data about traffic density, road surface condition, local weather variations, or changes in environmental pollution levels. This information is, in real-time, securely sent to municipal control centers where it may be analyzed and processed to control traffic lights in a dynamic fashion, predict congestion, coordinate emergency response, enforce environmental control, or plan better infrastructures. Innovative analytics on this data will be able to predict bottlenecks before they actually happen. Case studies of Singapore and Helsinki have shown that such systems lead to a 15–20 percent reduction in average commute times and a 10 percent cut in CO_2 emissions from idling vehicles, thus uplifting urban mobility and sustainability.

3.6 Techniques in V2C

- *Built-in Telematics System*: Many modern automobiles are supplied with built-in telematics devices that enable communication between the vehicle and the cloud. These devices commonly transmit vehicle data to the cloud via cellular networks and Global Positioning System (GPS) technologies. The data could include information about the vehicle's location, diagnosis, and operation. The integrated telematics system, which maintains a continual connection with the cloud, enables real-time data streaming and communication.
- *Mobile Data Connection*: Another method for connecting an automobile to the cloud is to use the mobile data connection of a smartphone or some other mobile device. The paired device's mobile data plan is used in this approach to connect the vehicle to the internet via Wi-Fi or Bluetooth tethering. When connected, the car is capable of sending data to and receiving it from the cloud, enabling remote vehicle management, navigation, and infotainment services based in the cloud.
- *Aftermarket Telematics Devices*: If a car is not installed with an inbuilt telematics system, it may resort to an aftermarket telematics device that connects it to the cloud. More often than not, these devices are interfaced with other vehicle systems or merely plugged into the OBD-II (on-board diagnostics) port of the car. Aftermarket telematics devices, in the same way as built-in systems, transmit data over a cellular network to the cloud. They give the owners using this telematics solution the possibility of using a number of cloud-based services and applications in support of their carriers, such as tracking, diagnostics, and remote controls.

The data transmission speed of a communication from V2C for the time period of 15 minutes is depicted in Figure 3.5. The transmission speed is anywhere between 0 Mbps and 30 Mbps. At the very start, the data transmission rate is about 5 Mbps. The rate soon goes up to around 25 Mbps at minute 5. While data is received between 6 and 10 minutes of time, the transmission rate changes from 10 Mbps to a maximum

Turton, Hal and Filipe Moura. (2008). Vehicle-to-grid systems for sustainable development: an integrated energy analysis, *Technological Forecasting and Social Change*, *75*. 1091-1108. 10.1016/j.techfore.2007.11.013.

Viswanath, Belagavi, Dheeraj Khatod and Narayana Padhy. (2023). Vehicle-to-grid (V2G) technology for optimal grid demand profile and sustainable energy management: a state-of-the-art survey, 1–6. 10.1109/SeFeT57834.2023.10245279.

Younes, Khair, Abdeslem Dennai and Youssef Elmir. (2021). A survey on cloud-based intelligent transportation system. 10.1007/978-3-030-63846-7_53.

Zerka, A., M. Ouassaid and M. Maaroufi. (2024). Strategic participation of electric vehicles in vehicle-to-grid within a microgrid system: a decentralized optimization approach, *Results in Engineering*, *24*, 103144.

Chapter 4

Integration of 6G Technology in Transportation Systems

Mohan Singh[1*], *Seema Singh*[2], *Kapil Dev Tyagi*[3] and *Vaibhav Bhushan Tyagi*[4]

The integration of 6G technology into transport systems will deliver a quantum leap improvement in transport safety, efficiency, and sustainability for different transport modes. In this chapter, we discuss the transformative role that 6G can play in revolutionizing transportation, and we discuss potential applications, including autonomous driving, intelligent traffic management (ITM) systems, enhanced safety, and connectivity among passengers and for logistics. 6G is promising transportation systems with ultra-low latency, terahertz communication, and intelligent connectivity. This will foster transport modes innovation: road, rail, air, and maritime. High-speed vehicle-to-everything (V2X) communication, real-time (RT) traffic prediction, and seamless connection in sparsely populated areas will be realized by key enablers, namely, terahertz (THz) communication, artificial intelligence (AI)–based networks, edge/fog computing, integrated sensing/communication, and non-terrestrial networks. They include driverless cars and dynamic traffic light management to improve air traffic control and smart ports. 6G integration aims to increase safety and reliability, efficient operations, environmental sustainability, and traveller customisation. Its obstacles range from technical restrictions and infrastructure expenses to the pervasive challenge of cybersecurity and regulation issues. In the future, there are new applications on the horizon (holographic communication and digital twins) with a roadmap till 2030 and also beyond commercialization. There is a need for collaborative endeavors among government, academia, and industry to ensure the ethical, equitable, and sustainable deployment of 6G in transportation.

[1] Department of ECE, G.L. Bajaj Institute of Technology and Management, Greater Noida, India. https://orcid.org/0000-0002-2973-3676

[2] Department of Education, Lloyd Institute of Management & Technology, Greater Noida, India. https://orcid.org/0009-0009-4776-2241

[3] Department of ECE, Jaypee Institute of Information Technology, Noida, India. https://orcid.org/0000-0002-4596-2321

[4] Department of ECE, ISBAT University, Kampala, Uganda. https://orcid.org/0000-0001-8153-3607

* Corresponding author: mohansingh571@gmail.com

DOI: 10.1201/9781003659822-4

1. Overview

The upcoming sixth-generation wireless communication (6G) is expected to revolutionize the way transportation systems operate and create entirely new opportunities for innovation and improvement. Since the next decade will witness an exponential growth in mobile traffic and a new type of disruptive services are being imagined, researchers have already begun to think of 6G. 6G networks are expected to provide significantly better performance compared to their 5G predecessors in aspects such as latency, bandwidth, and reliability, and can support advanced and data-hungry applications for transportation. The sixth generation networks are expected to be deployed soon following the creation of the smart infrastructure, efficient technologies, and the advent of various wireless applications (Rasti et al., 2021). These will enable new applications and services to meet the continually evolving requirements of the transportation sector. The incorporation of 6G will also help enable more connected, autonomous, and sustainable transportation systems. The real-time acquisition and processing of data from multiple data sources is critical for good transport management (El & Klein, 2016). Vehicle telematics has great potential to bring about intelligent transportation systems, which will achieve safer journeys, controlling traffic to minimize congestion of traffic and infotainment services. The chapter will address the possible uses, the main enabler technologies, the challenges, and the main research trends for 6G in transportation. The incorporation of 6G into transportation could easily be a game-changer, fundamentally changing how humans and goods are transported. At a high level, this evolution toward safer, efficient, and sustainable mobility systems can facilitate the motives society.

- **Objective:** This chapter provides a thorough investigation of the connection between 6G technology and transportation, with an emphasis on the role it can play in transforming different aspects of mobility and logistics. The scope of the study covers all modes of transportation (such as ground-road networks, rail transport, aviation, and shipping) and discusses in detail how 6G can improve safety and security, operational efficiency, and sustainability as applied to each of these modes of transportation.
- **Scope:** The chapter provides an overview of autonomous driving, intelligent traffic management, enhanced safety systems, and connected travel of passengers and logistics, explains the characteristics of 6G, methods to integrate 6G into transportation systems, examples of 6G in all types of transportation, and related technological, ethical and legal issues. In addition, the focus will be on the study of the advanced enabling technology that underlie 6G, including terahertz communication, AI-assisted network intelligence, edge/fog computing, integrated sensing and communication, and non-terrestrial network, and their functions to support high-speed vehicle-to-everything (V2X) communication, real-time traffic prediction, and uninterrupted coverage in remote areas.
- **Significance:** The significance of 6G for transport is huge in terms of enabling safer, greener, and more efficient mobility in all kinds of transport modes. In a more complex and integrated transportation system, advanced communications such as 6G are inevitable. The chapter summarizes 6G technology and discusses its significance for future transportation and existing/

future mobility problems. It highlights challenges and opportunities, providing perspectives for researchers, policymakers, and industry to promote innovation and commercialization initiatives. Ultra-low latency, terahertz communication, and intelligent connectivity are expected to become key enablers for 6G to fulfill its transformative capabilities in the future mobility ecosystems.

2. 6G and Transportation Systems

Wireless 6G technology is a giant step ahead of its predecessors, with a pledge to provide unprecedented levels of connectivity, speed, and reliability. Though in early development, 6G aims to overcome the downsides of existing 5G networks while ushering in a new wave of applications in all industry sectors (Chataut et al., 2024). This will provide data rates much higher than 5G with speeds up to terabits per second to support bandwidth-hungry applications such as augmented reality, virtual reality, and HD video streaming. Latency, or lag time, in data transfer should also decrease to the low levels of less than one millisecond, making real-time communication and control possible for use cases, including autonomous driving or remote surgery. The higher the reliability of 6G networks, the more consistent and uninterrupted the connections, which is necessary for applications demanding strong availability, for instance, emergency services and industrial automation, as shown in Figure 4.1.

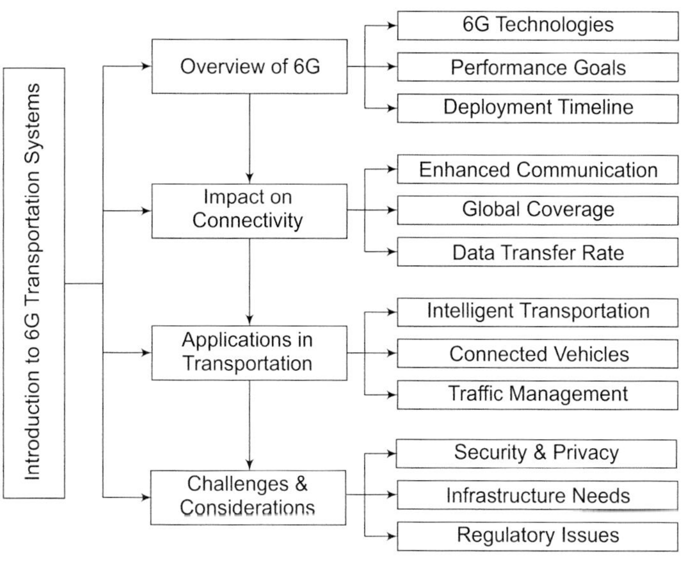

Fig. 4.1 Introduction to 6G and transportation system.

The transport industry, i.e., road, rail, air, and shipping, will greatly leverage the potential of 6G for meeting the challenging connectivity, automation, and safety needs (Fodor et al., 2021). The participation of millions of sensors in smart cities, smart homes, and smart factories, combined with AI on fog/cloud systems, will bring to life new applications where networks are the main actors of these smart systems (Wu et al., 2024). 6G can be used by intelligent transportation systems to provide real-time communication between vehicles, infrastructure, and control centers so that traffic flow is improved, congestion is reduced, and safety is increased. Table 4.1 shows the essential part 6G will have to play in transforming transport networks, introducing new levels of speed, connectivity, and intelligence to transport solutions.

Table 4.1 The impact of 6G on transportation.

Aspect	*6G Technology Overview*	*Impact on Transportation Systems*
Definition & Evolution	New wireless communication that follows after 5G: 6G promises high-speed connectivity and low delay, AI-powered networks, holographic communication, and quantum technology integration	Provides fully autonomous and intelligent transport networks, enabling smooth V2Xcommunication, an innovation for logistics, aviation, and smart mobility solutions
Speed & Latency	Speeds beyond 1 terabit per second (Tbps) are anticipated, provides subscription to a topic at least once to receive a response in less than 1 millisecond, and enhances real-time data processing	Instant response of the car in autonomous mode, enhances real-time traffic management, reduces congestion, saves lag time in traffic monitoring systems
Key Technologies	AI-driven network management, terahertz (THz) frequencies as they support high-speed data transmission, quantum communication for secure networks, integration of edge computing for distributed processing	AI-based smart monitoring and traffic analytics, secure communication for automotive cybersecurity, predictive maintenance enabled through edge processing
Autonomous Vehicles & Smart Infrastructure	Supports AI-powered transportation decision-making, enables vehicles to communicate with intelligent infrastructure, ensures high-speed and reliable spacecraft connectivity for autonomous systems	Improves automotive safety with the help of predictive analysis, makes "smart highways" and dynamic route optimization possible, and minimizes travel-related incidents through immediate identification of hazards
Sustainability & Energy Efficiency	Uses less energy than 5G, AI-optimized power management, and environment friendly communication technologies	Eco-friendly smart mobility solutions, energy efficient technology implementation in electric vehicles (EVs), supports smart cities that are sustainable and reduce carbon output
Challenges & Implementation	Requires massive infrastructure investment, security and privacy issues of the data, and the necessity of universal networking deployment	Difficulty to adopt in current transportation systems, high public transit network implementation costs, and well-defined cybersecurity for connected mobility parties should be implemented

One of the key applications of 6G in transportation lies in the development of autonomous driving technology, in which the ultra-low latency and reliability of 6G play a key role in allowing the vehicle to make decisions in real-time processing of data from sensors, cameras, and other vehicles (Javed et al., 2018). 6G can also enable next-generation driver-assistance systems, in which drivers receive real-time updates on road conditions, traffic, and potential hazards, improving safety and preventing accidents. Moreover, 6G will target massive machine-type communications in addition to connecting humans, computing resources, vehicles, devices, wearables, sensors, and robotic agents (Giordani et al., 2020). These characteristics are envisioned to support universal connectivity, high data rate, and low latency to enable a vast range of new and improved applications and to fulfill the vision of self-controlled and intelligent networks able to provide everything as a service and new experiences (Xiao et al., 2020). In the aviation domain, with 6G, the real-time health status of aircraft can be supported to achieve predictive maintenance tasks and to improve communications between pilots, air traffic control (ATC), and the ground (Kabashkin, 2024).

2.1 Evolution of Wireless Technology (1G to 6G)

Wireless communication technology, which has improved with time by means of various generations, provides remarkable speed, capacity, and latency enhancements, and has changed the way we communicate and interact with the world. From 1G to 5G, the evolution has prepared the ground for 6G to revolutionize transportation systems, as shown in Figure 4.2. 1G, which ushered in the era of wireless communication, was the first generation of wireless technology and served as an analog cellular communication that supported phone calls but did little to support data use. Digital 1G service brought the first widely used cellular phones, where analog service was no longer available. 2G digital communications systems enabled much clearer phone calls, improved battery life, and introduced a new technology called Short Message Service (SMS), which made using a phone more than just calls. When 3G was introduced, with its wireless technology, people were able to access the internet from their cell phones, leading to widespread use of smartphones and other mobile apps, and people could use their phones to send emails, view the web, and live stream video content. 4G brought mobile broadband at high speeds, enabling video streaming, online gaming, and cloud-based services to be used everywhere. 5G, already launched throughout the world, provides further improved data rates, lower latency, and higher capacity to enable applications such as augmented reality, virtual reality, and the internet of things (IoT) (Lynch et al., 2024).

A new generation of mobile communication technology emerges about every ten years, with each crop of technology improving features and capabilities. Work is ongoing on the deployment of 5G networks in many countries, representing a major milestone towards the full deployment of interconnected and intelligent enabling systems. The theory is that as 5G further proliferates and evolves across the world, it will allow for a diverse set of applications and services that 6G can build upon as its premise, strengthening some features and enabling others as new possibilities.

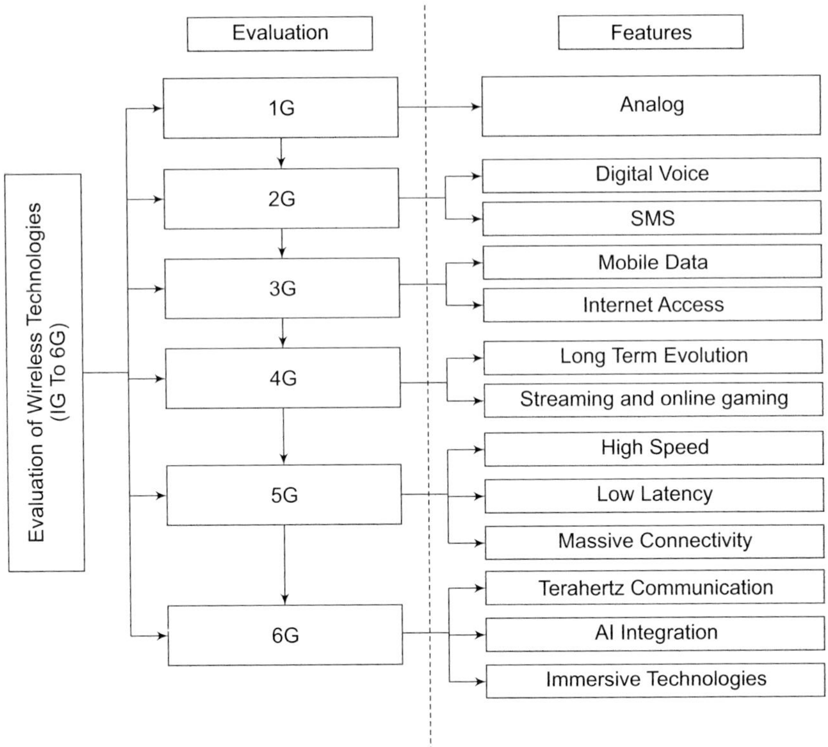

Fig. 4.2 Evaluation of wireless technologies.

Key characteristics of each generation:

- 1G: Analog voice calls.
- 2G: Digital voice calls, texting.
- 3G: Enabled for internet use, multimedia downloads, and online streaming.
- 4G: Broadband mobile technology, cloud services.
- 5G: Faster speeds, lag-free, connected.
- 6G: Terabit throughputs, sub-millisecond latency, and more reliable connections

How did each step move us toward mobility and communication in transportation?

- 1G and 2G provided basic mobile communication for logistics and fleet management.
- 3G-enabled tracking and communication for transportation.
- 4G connected advanced infotainment and connected car services.
- These include self-driving cars, real-time traffic management, and improved safety systems.
- 6G provides ultra-reliable low-latency communication (URLLC) for autonomous vehicles and advanced transportation systems such as self-driving cars and smart transportation.

2.2 What is 6G?

6G is the further developed form of some processes in wireless technology when compared with its predecessor, 5G, considering terabit per second data rates, latency in the sub-millisecond range, and very high reliability, making it possible to realize new transportation systems. The 6G network is also expected to operate at much higher frequencies, using the terahertz range of waves to deliver unprecedented data-transfer speeds capable of facilitating applications such as holographic communication, digital twins, and ultra-immersive virtual reality. These advances will enable transportation systems to be more efficient, safe, and sustainable, and provide for vehicles, infrastructure, and users to be seamlessly connected. 6G is expected to integrate communication, computing, control, and caching in a unified technology that harnesses the power of AI and resolves the challenges in today's communication system (Chowdhury et al., 2020). In addition, 6G is expected to connect more devices than ever before, and its IoT connection would cover almost everything, which allows the traffic system to communicate with all sectors smoothly. 6G would also have boosted security capabilities, protecting transportation systems from cyberattacks and securing sensitive data.

- *Projected features*: Terahertz (THz) band, AI-native networks, holographic communications, ultra-low latency (<1 ms), high reliability, and enhanced density of connections
- *6G objectives*: 1 Tbps data rate, less than 1 ms latency, worldwide "seamless" coverage, and ubiquitous connection, including non-terrestrial operation

2.3 Transportation System Overview

The transportation system is a complicated network of interconnected elements that together allow people, goods, or materials to move from one place to another with a variety of modes of transportation, such as road, rail, air, or water, each with its infrastructure and systems of operations. For instance, the road system comprises vehicles, roads, traffic control, and support systems, and the rail system comprises trains, tracks, signalling, and stations. Air transport comprises aeroplanes, airports, air traffic control systems, and navigation aids, and water transport consists of vessels, harbors, rivers, and marine traffic management systems. Optimal deployment of these multimodal transport options requires sophisticated coordination and management to be both effective and safe, and for the service to be sustainable. Modern technologies such as IoT-based connectivity can improve the efficiency of transportation systems (Islam et al., 2021). Intelligent transportation systems use emerging technologies to enhance the safety, efficiency, and sustainability of transportation systems. Smooth and efficient traffic management requires smart, managed traffic, comprising coordinated infrastructure, vehicles, and intelligent control systems that depend on reliable, high-speed communication networks to enable the exchange of real-time information, provide coordination, and control capabilities. Transportation systems as a whole can change dramatically with the evolution toward 6G, including ubiquitous deployment, real-time data handling, and more sophisticated automation.

- Modes of transport: Road, rail, air, sea, and multimodal networks.
- Role of communication technologies in intelligent transportation systems (ITS) and vehicle-to-vehicle (V2V)/vehicle-to-infrastructure (V2I) application areas: These technologies support traffic management. The challenges and opportunities in the field of transportation are congestion, safety, environmental impact, and efficiency, which technology can help to tackle.

3. Key Enablers of 6G for Transportation

Table 4.2 shows how 6G will transform transportation networks, providing higher connectivity, intelligence, security, and sustainability for the future of mobility.

Figure 4.3 depicts a structured flowchart to visualize the four pillars of enabling 6G in transportation: URLLC for real-time application and latency minimization, and always-on connectivity; edge computing for local processing and ensuring bandwidth economy and data privacy; network slicing for dynamic resource provisioning and customized quality-of-service via virtualized networks; and AI for predictive behaviors and decisions, and situational awareness. These technologies collectively build a comprehensive foundation for intelligent, responsive, and secure transportation systems in the era of 6G.

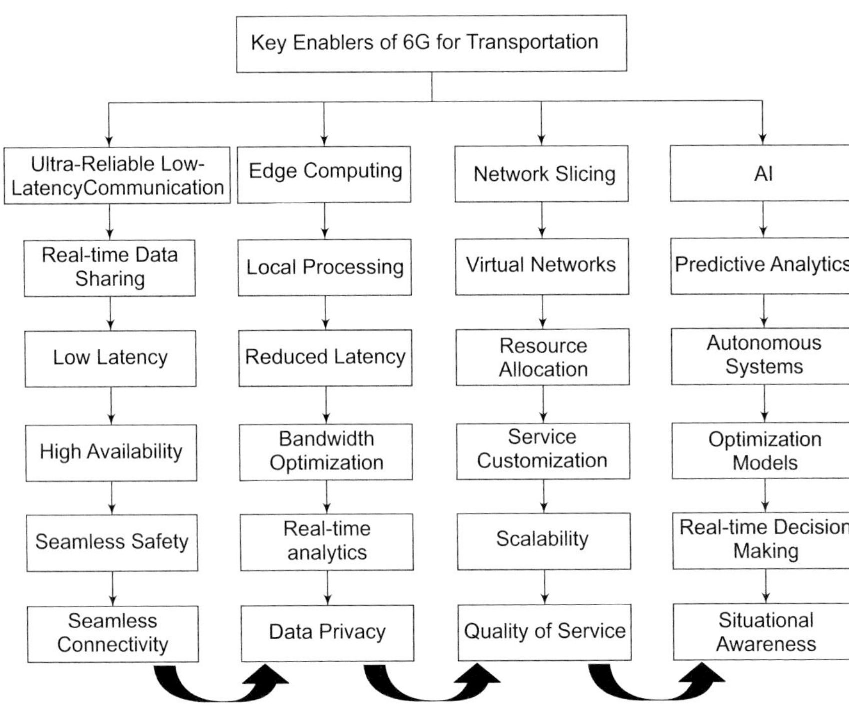

Fig. 4.3 6G enablers for transportation.

Table 4.2 Key Enablers of 6G for Transportation

Enabler Category	*Specific Enabler*	*Description*	*Impact on Transportation*	*Key Technologies/Concepts*
I. Enhanced Connectivity & Communication	**1. Terahertz (THz) Communication**	Operation at ultra-high frequency (0.1 THz to 10 THz) for very high data rates up to 1 Tbps with ultra-low latency	Supports V2X, real-time autonomous driving, high-definition (HD) road monitoring	THz transceivers, advanced antenna technologies, ultra-wideband signal processing
	2. Ultra-Reliable Low-Latency Communication (URLLC)	Offers close to perfect reliability (99.99999%) and ultra-low latency (under 1 ms)	The key to self-driving vehicle decision-making, teleoperation, and emergency intervention in intelligent transportation systems	Grant-free access, multi-connectivity, redundant communication paths, short packet transmission
	3. Massive Machine Type Communication (mMTC)	Able to serve billions of IoT devices per square kilometer, well beyond what 5G can handle	Allows for the widespread deployment of smart sensors in roads, traffic lights, and real-time city mobility analysis	Low-power connectivity, simplified network protocols, scalable data architecture
	4. Integrated Sensing and Communication (ISAC)	In 6G networks, it is anticipated that the communication will be a collaborator in the sensing; as an active prober, it will sense the communication environment through radio waves; the communication and sensing will be integrated into the mobile wireless connectivity	Enables real-time mapping, obstacle detection, pedestrian tracking, and traffic flow analysis with very high accuracy, while enhancing situational awareness for autonomous vehicles (AVs) and intelligent traffic control	Joint waveform design, AI/machine learning (ML) for signal processing, multi-antenna arrays, millimeter-wave and THz sensing
	5. Reconfigurable Intelligent Surfaces (RIS)	Passive, or semi-passive, surfaces embedded with metamaterials with responsive reflect, refract, or absorption ability of radio waves to enhance their propagation and coverage	Solves line-of-sight constraints in high-density urban areas, increases the quality of signal for V2X, minimizes energy consumption, and enhances coverage in tunnels or underground parking	Metamaterial design, dynamic beamforming, programmable surfaces

Contd.

Table 4.2 *Contd.*

Enabler Category	*Specific Enabler*	*Description*	*Impact on Transportation*	*Key Technologies/Concepts*
	6. Non-Terrestrial Networks (NTN)	Integration of satellite/high-altitude platform station (HAPS)/ unmanned aerial vehicle (UAV) communication into the 6G ecosystem	Enables comprehensive global access for vehicles in rural areas, over oceans, and during maritime and aviation transport, ensuring global navigation, emergency and fleet communications beyond the reach of terrestrial cellular networks	Satellite constellations (LEO, MEO, and GEO), HAPS deployment, and UAV relays for it were proposed to overcome this system limitation
II. Advanced Intelligence & Computing	**7. Artificial Intelligence (AI) Native Network**	Jam-packed with AI-powered algorithms covering the entire 6G infrastructure to enhance transportation ecosystems	Facilitates intelligent management, vehicle autonomy learning, and optimal route planning in real time	ML algorithms, federated learning in edge AI-tech B4, AI-powered network slicing
	8. Edge Computing/Mobile Edge Computing (MEC)	Allows large transportation data sets to be processed locally, at the edge of the network, instead of in the centralized cloud	Enables low latency for autonomous vehicle navigation, enhances traffic observation, and enhances privacy of paths and travel	Distributed computing architectures, high-performance edge processors, AI-enabled anomaly detection
	9. Digital Twins	Virtual representations on the fly of physical transportation assets, infrastructure, and whole-city mobility networks	For creating highly detailed traffic scenes, to simulate traffic in your scenes easily.	Real-time data integration, AI/ ML-driven modeling, high-fidelity simulation engines
III. Security, Trust & Sustainability	**10. Blockchain and Distributed Ledger Technology (DLT)**	Acts as a decentralized, immutable ledger of data for guaranteed and transparent transactions from the network	Ensures V2X communications for smart contract-based toll payment and cybersecurity for new autonomous mobility	Cryptographic hashing, consensus mechanisms, secure transactions with smart contracts
	11. Advanced Cybersecurity and Privacy Mechanisms	Includes post-quantum cryptography, AI-based threat detection, and privacy-preserving technologies	Safeguards transportation-related systems from cyber threats, preserves connected vehicle data securely, and protects infrastructure integrity	Post-quantum cryptography (PQC), AI-powered security monitoring, zero-knowledge proofs for privacy
	12. Energy Efficiency and Sustainability	6G infrastructure designs, environmentally friendly network protocols, hardware architectures, and power-efficient AI models	Decreases the carbon footprint of trips, promotes green transportation, and improves energy-efficient communication	Energy-harvesting technologies, dynamic power management, green AI optimization

3.1 Terahertz Communication

Terahertz communication, which works in the frequencies from 0.1 (100 GHz) to 10 THz, is considered one of the enablers for 6G technology, and it has the potential for a much higher data rate and bandwidth than current wireless communication systems. As terahertz waves have short wavelengths, they allow the design of small-sized antennas and high communication densification, and the ability of terahertz signals to penetrate many kinds of materials results in better signal coverage in compromised areas. The use of terahertz-based implementations in vehicular networks can facilitate advanced applications (e.g., high-definition video streaming, real-time sensor data processing, and holographic communications) to be realized; however, the main impediments that need attention are the atmospheric absorption, poor range communication, and sophisticated hardware device requirements (Nguyen et al., 2022). Dealing with these challenges demands innovative techniques, e.g., beamforming, adaptive modulation, and error correction.

- High-speed V2I and V2V links
- Challenges: Propagation loss, atmospheric absorption, and hardware complexity
- Solutions: Beamforming, advanced code, adaptive modulation

3.2 AI-Driven Network Intelligence

AI-based network intelligence is key to optimizing the performance and efficiency of 6G-connected transport systems by processing machine learning algorithms and analyzing massive amounts of data on vehicles, infrastructure, and users. It can improve the reliability and security of the transportation networks, including predicting traffic patterns, resource allocation optimization, anomaly detection, proactive network management, self-healing, and traveler-oriented services. Through advanced AI algorithms, traffic flow performance can be enhanced, decongestion achieved, and transport efficiency improved. AI technology allows predictive maintenance on the transport infrastructure for preemptive actions regarding the point and causation of the failure in specific use cases, based on sensor data information performing on the spot or the edge analysis and implementation. AI combined with a communications network improves the efficiency, scalability, and intelligence of transportation systems, which brings better real-time information decision-making and resource allocation. In addition to achieving more efficient network autonomous driving, AI will also help optimize the network online, so that the system performance and user experience can be improved (Wu et al., 2021).

3.3 Edge and Fog Computing

Edge and fog computing are decentralized computing models that bring computation and storage capabilities close to the network edge, offering low latency and enhancing the performance of real-time transportation applications. By taking action at the source, edge and fog computing can achieve faster response times, ease network congestion, and enhance the reliability of critical services such as autonomous driving and collision avoidance. Edge servers can be a potential solution for RT apps, since these servers that are used behind the Content Delivery Network (CDNs) are distributed at the edge of the internet and are equipped with storage and computation resources, which can reduce the response time. Fog computing scales out the edge computing model for hosting applications in billions of connected devices at the edge of the network, providing lower latency, context awareness, and support for real-time applications (Mach & Becvar, 2017). In combination, edge and fog computing support a dynamic and scalable framework to meet the diverse computing requirements of contemporary transportation systems. These technologies reduce latency and aid in real-time decision-making, which is important for autonomous driving, traffic management, and passenger safety, by shifting data processing closer to the source. Furthermore, the decentralized nature of fog computing is suitable for dealing with the sporadic connectivity issues found in transportation.

3.4 Integrated Sensing and Communication (ISAC)

Integrated sensing and communication is an essential technology in 6G, which co-integrates communication and sensing functions into a single system, for traffic systems to gather and process their environment information more effectively. By exploiting communication signals to perform sensing, intelligent transportation systems can have a constant overview of the traffic status for objects (e.g., obstacles) and track, at any time, the location and dynamics of vehicles/pedestrians. This makes it possible to have healthier, faster, and more current information for decision-making, increasing safety, efficiency, and automation. This system can achieve environment perception, imaging, and localization as well as communication at the same time. For example, using reflections of communication signals, transportation systems can form a rich view of their outdoor environment, spotting hazards, and optimizing traffic flow. The integration of sensing and communication allows for instant environmental awareness: by making it possible for communication signals to be used likewise for the purpose of communication and sensing, it provides an overall understanding of the environment.

3.5 Non-Terrestrial Networks (NTN)

Non-terrestrial networks (NTNs), like satellite or aerial communication systems, may provide services with both ubiquity and full transmit range coverage for transportation systems, especially in remote and underserved regions. These networks offer a dependable and robust communication platform that complements their terrestrial counterparts, delivering pervasive connectivity for vehicles, passengers, and infrastructure even in challenging environments with limited traditional cellular coverage. Satellite networks can cover regions that lack terrestrial networks, and skyborne networks can provide temporary coverage during emergencies or public events. These kinds of networks are very appealing and have a lot of potential when used in applications such as emergency communication, fleet management, or self-driving cars, which require uninterrupted communication. Further, messages from ground-based wireless communicating devices, such as cellular phones and mobile broadband cards, can be received by airborne vehicles (Deaton et al., 2011). The combination of terrestrial and non-terrestrial (hereafter n-terrestrial) networks allows wireless communication and data exchange in a transparent way between all the places on Earth.

4. Application Domains of 6G in Transportation

Introduction of 6G-enabled transportation systems brings quite a few application domains, which may bring significant changes in the current transportation of people and goods. These applications are pertinent in different transportation contexts such as autonomous driving, traffic control, public transportation, and logistics. These can harness 6G technology to improve efficiency, safety, and sustainability. The integration of communication technology, sophisticated sensors, and pervasive communication networks can allow for fluid movement without the need to drive, deliver, or even have a vehicle, as shown in Figure 4.4.

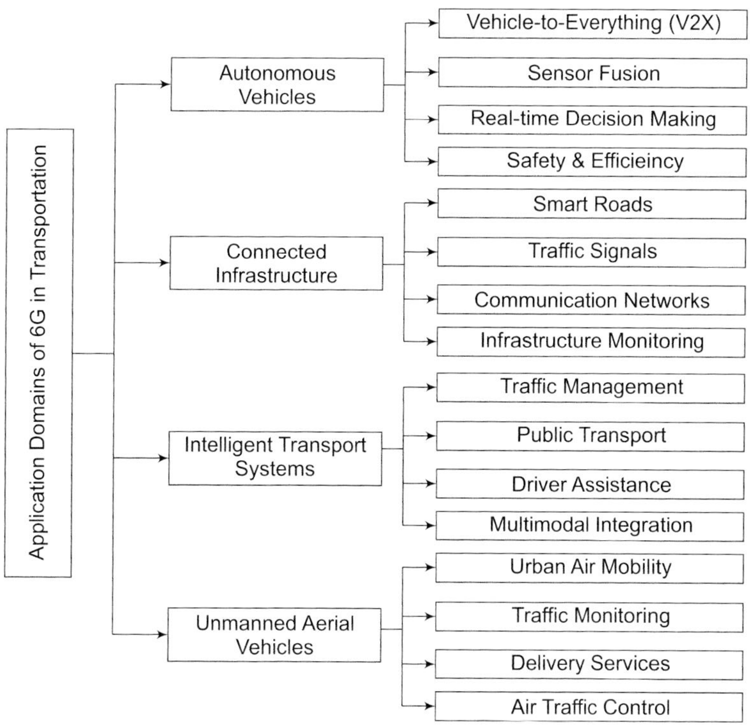

Fig. 4.4 Application domains of 6G.

The next-generation wireless network (6G) will revolutionize transportation by enabling ultra-high bandwidth, super-low latency, ubiquitous connectivity, and sophisticated intelligence. It is more than just better broadband as listed in Table 4.3. It is about immersive, self-sustaining mobility.

The capabilities of 6G will not only improve today's transport systems but also enable entirely new visions of mobility and support a future that is more secure, efficient, and sustainable.

4.1 Smart Road Transportation

The 6G road transport networks are going to transform road transportation by empowering a myriad of applications that make it safer, efficient, and convenient for both drivers and passengers. Traffic control systems are another prime example, whereby 6G's high throughput and low latency could be used to optimize traffic light control, reroute vehicles, and provide drivers with real-time information about local traffic, accidents, and road closures. It may be possible to make use of the stored traffic flow data by traffic management organizations and use it to predict traffic patterns. The advanced driver assistance systems (ADAS) we have today are only a glimpse of what is possible with 6G—these interconnected systems are able to, for example, access real-time data from other vehicles, infrastructure/roadside sensors, and cloud-based services to deliver collision avoidance, lane keeping, and adaptive cruise control with even greater precision and responsiveness. Massive vehicular communications based on 6G should provide vehicles with information to make them aware of other vehicle locations, speed, heading, and possible obstacles through which preventive actions may be taken to avoid accidents. In addition, 6G provides the possibility of HD mapping and localization services for self-driving cars (SDCs), and can hence facilitate the SDC's navigation in complicated road scenes with remarkable precision and safety, as they can acquire more information available within an even shorter period, and operate more efficiently and safely when in motion (Malec et al., 2021).

- **Autonomous driving and V2X communications:** 6G will contribute to a new era of autonomous driving with advanced URLLC and mMTC to enhance V2X communications, sensor fusion, and instantaneous data processing.
- **Real-time adaptive traffic light control rendered by 6G-based AI:** 6G empowers instant adaptive traffic light control through edge computing and AI, so as to minimize congestion and adapt dynamically to traffic reply. Traffic signal controllers can also communicate with each other using 6G in order to synchronize signals to assist with controlling the flow of traffic across networks of intersections.
- **Real-time accident detection and emergency response:** 6G technology can aid with real-time accident detection and automated emergency response, which can result in immediate dispatching of emergency response services with possible shorter response times, leading to saving lives. 6G-equipped drones can be used to investigate accident sites, communicate information to first responders, and generate situational awareness to assist in the process of searching for survivors.

4.2 Railways

The employment of the 6G technology in railway systems is particularly promising in shaping railway services, safety, and passenger comfort. Moreover, train location, speed, and track conditions of 6G-supported train control systems can be real-time monitored for accurate train control and collision avoidance so that a safe operation can be secured, and collision accidents can be further minimized. The high-speed 6G data transmission supports systems for passenger information such as real-time train schedules, platform information, and congestion forecasts, resulting in increased convenience and reduced anxiety among passengers, and also makes it possible for passengers to use high-bandwidth entertainment and communication services as they go. Moreover, 6G-connected sensors can monitor, to a certain extent, railroad infrastructure, for instance, tracks, bridges, and tunnels, to early identify issues and perform preventive maintenance, reducing interruptions to the service as well as improving safety (Jiang et al., 2021).

- **URLLC for train control system:** URLLC for train control systems of 6G will make the railways safer due to real-time monitoring of trains, exact railway control, and avoidance of collision through reliable communication.
- **Passenger experience improvement:** 6G-enabled passenger information systems can provide timely information on train schedules and platform information for passengers, increasing convenience. With 6G, passengers can enjoy streaming, online gaming, and other data-dependent activities without interruption. The vehicle becomes a good place to use for work and leisure time through seamless connectivity.

4.3 Aviation

6G will revolutionize the aviation industry by providing improved safety, efficiency, and passenger experience. If data can be exchanged between aircraft and ground control centers in real-time, responses to the traffic can be managed better, making flight paths more effective, utilizing less fuel, and taking more advantage of green and sustainable practices. Advanced communication features possible with 6G can deliver high-speed internet access for passengers and in-flight entertainment, as well as connect people's devices to surf the web, enhancing passenger experiences and boosting productivity on flights. Moreover, 6G-enabled sensors can remotely monitor and diagnose aircraft systems, which means that predictive maintenance can be achieved, downtime minimized, and aircraft security and reliability enhanced (Sallahaj, 2024). 6G-based applications can provide better surveillance around airports and flight paths for improved security and safer air travel.

4.4 Maritime Transport

6G paves the way for breakthroughs that will revolutionize maritime transportation, making it both safer and greener. 6G communication systems can, through live monitoring of vessel position, weather, and condition of cargo, provide improved routing planning and less fuel and emissions through optimized voyage, hence, making maritime operations more sustainable. Ships can be remotely controlled or take on autonomous navigation, with 6G, and the result will be unmanned or remotely operated ships, decreasing the cost of crew, improving safety, and gaining the ability to go to areas that are dangerous and/or far away. Improved communication services can give sailors better access to the internet, voice, and video services and news from home, boosting morale and keeping them more productive during long sea crossings. With 6G-enabled sensors, it is possible to monitor the state of a vessel's structural elements and equipment, which makes it possible to detect potential issues in advance and carry out proactive maintenance, so that downtime can be reduced and safety can be increased.

a logical structure for higher-level requirements that is technology independent, meaning that it can be implemented in current and future technologies (Rahman et al., 2021). Standards should be drawn up to cover every dimension of 6G-based transportation, including data format, communication protocol, and security demands. These protocols are intended to have opportunistic communications to provide robust and stable end-to-end network connectivity back to neighboring vehicles, Roadside Units (RSUs), and remote ITS servers. Open source technologies and software-based networking could enable systems to become more interoperable and integrate innovations into transportation.

Furthermore, in standardization, common security standards and protocols that can enable a secure 6G transport system operation should be developed. This ranges from standardized encryption algorithms to the authentication mechanisms and access control policy.

- **Integration with existing systems (5G, Dedicated short-range communications [DSRC], and Wi-Fi 6/7):** Integration with existing technologies guarantees a seamless migration, taking advantage of sunk infrastructure investments. The incorporation should take into account the independent good performance and the smooth operation of 6G with existing communication systems like 4G and 5G. Its backward compatibility will interoperate with the legacy transportation systems, helping in the gradual transition to 6G. Moreover, the system must allow for seamless introduction of new functionalities or services without disturbing normal operation.
- **Role of international standards International Telecommunication Union [ITU], Institute of Electrical and Electronics Engineers [IEEE], Third Generation Partnership Project [3GPP] in ensuring compatibility:** International standards play a big part in ensuring the compatibility and interoperability of the different 6G-based transportation systems within different regions and countries. These requirements create the consistent foundation that enables 6G technology to be conceived and realized, which contributes to the overall harmony of 6G tech across the world and cross-border transport. Fragmentation and its associated cost may be inhibited by the lack of generally agreed-upon standards facilitating the deployment and uptake of new technologies.

5.3 Security and Privacy Architecture

Security and privacy are important issues in 6G-based transportation systems design and deployment. Vehicular ad hoc network (VANET) security standards solve most security problems in vehicular networks, but the integration of 5G and vehicular networks needs to consider design and implementation problems. Security should be built into the architecture to mitigate cyber threats and ensure the privacy of sensitive data. This involves the application of security, cryptographic, authentication, and access control techniques that will protect data sent over the network and data stored on transportation systems. As IT and networks are becoming more critical to economies and society in the age of 6G, the world's need for high-level security rises. Novel authentication, encryption, access control and communication, and malicious activity detection mechanisms can also satisfy the upper criteria of future networks.

Quantum and post-quantum cryptography will be essential to ensure the security and privacy of devices and data.

- **Quantum-safe cryptography:** More efforts are needed to deploy cryptographic and security protocols, such as lightweight cryptography, and their corresponding algorithms in order to secure confidentiality, integrity, and availability of sensitive data that will be robust against quantum computer attacks. This includes migration to cryptosystems and protocols that are quantum-resistant and resilient to the computational infeasibility of quantum computers. In addition, the introduction of trusted hardware (HW) modules is necessary to create trust to execute secure functions and protect critical data assets, thereby improving the overall resilience of the 6G-enabled transportation ecosystem against threats and vulnerabilities.
- **Secure handovers, identity management, and trust models in mobile networks:** There is a need for secure handover approaches to support secure and seamless transitions between various cells in a network. This includes adding offboarding and onboarding measures to authenticate and authorize users and devices working in the same environment without access to the network. Identity management frameworks are crucial to managing identities among diverse entities in transportation systems: vehicles, drivers, passengers, and infrastructure elements. Local training data can be used by federated learning, and the experiment can be trained locally, avoiding the disclosure of network data and safeguarding the privacy of data during the process of transmission.

5.4 Data Management and Processing

The effective operation of 6G-enabled transport relies on efficient data management and processing. Data acquisition primarily encompasses the collection of information coming from several sources, in particular from numerous sensors installed throughout the transportation system, high-resolution camera networks for internal and external traffic monitoring, and the heterogeneous device population present in vehicles and on the roadside. State-of-the-art approaches in big data analysis are used to derive valuable information and actionable intelligence from the large amounts of data that are generated in transportation. Then, the mass of data is systematically classified and processed, which is conducive to identifying the underlying patterns for drawing practical conclusions and for supporting plausible decision-making in a variety of transportation sectors, like a transport strategic plan, a parking-saturation solution, or a proactive security action. The growing needs of transportation development are increasingly important in the operation and utilization of these systems (Stylianou et al., 2019). Advanced data collection and information dissemination systems are offered, which are necessary to build the ITS of today and the future.

- **Big data analytics for transportation:** Big data analytics is essential for extracting patterns and insights from a large volume of data in the transportation system. It involves applying machine learning and statistical models to study traffic patterns, predict congestion, and ease the flow of transportation. Machine learning solutions have just started making good progress in the realm of transportation, demonstrating that they may offer higher return on investment (ROI) than traditional approaches. Through big data analysis, transportation

more strategic operations. 6G-integrated systems in aviation, enabling real-time aircraft performance monitoring, predictive services, and maintenance for critical components, could enhance operational efficiency and diminish downtime.

- **Optimal fleet management:** 6G technology is used in the management of fleets, where it facilitates fleet management systems using data on vehicle location, performance, and maintenance status in real-time to enhance routing, scheduling prioritization, and resource allocation. Vehicle control in real-time is achieved by 6G communication systems providing high-reliability and low-latency communications, to improve delivery and satisfy the CAV users. Furthermore, the incorporation of predictive analytics facilitates condition-based maintenance and minimizes equipment failure potential, thus limiting downtime and optimizing operational effectiveness.
- **Easing traffic with predictive traffic control:** 6G predicts how traffic will behave by analyzing real-time traffic, weather, and special event information to optimize traffic flow and reduce congestion. Another example is the use of mobile crowd sensing to access real-time data on vehicles through the deployment of smartphones and GPS-equipped vehicles, which can enable more dynamic monitoring of traffic situations and thus develop better dynamic traffic management strategies. Transportation agencies can reduce travel times and enhance overall network performance by predicting traffic congestion and applying proactive traffic management solutions.

6.3 Environmental Sustainability

6G technology is a key driver for achieving environmental sustainability in transportation systems by enabling the uptake of green policies, reducing emissions, and optimizing energy. When node amount grows exponentially, which makes the energy consumption of the network extremely heavy, the green internet of vehicles (IoV) comes into being as a direction to sustainable vehicle communication and networking. The integration of EVs and smart charging capabilities helps reduce greenhouse gas emissions and encourages cleaner transportation options. With the vast amount of sensors, intelligent devices, and data-driven applications, 6G networks tend to consume an enormous amount of energy, which is unsustainable and must be a design priority of 6G networks.

- **Smart routing to minimize fuel consumption:** The root cause of excess pollution is the excess fuel burned. The situation is worsened through the emission of pollutants to the surrounding environment as the fuel consumption increases. By sharing real-time information on traffic, terrain, and vehicle attributes, 6G technology contributes to more intelligent routing, which minimizes fuel burning. 6G enables intelligent transportation systems that ease traffic and congestion and minimize emissions, resulting in improvements in air quality and less environmental impact.
- **Integration with EVs and smart charging infrastructure:** The closer integration of 6G technology with EVs and smart charging infrastructure is vital for sustainable transportation and disuse of fossil fuels. Smart grid facilitates real-time energy demand and renewable power generation information, dynamic

operation of power resources on demand response, and distributed renewable power generation connected to distribution networks.

6.4 Enhanced User Experience

6G technology makes the user experience better on different kinds of transportation through the implementation of connectivity, personalized expertise, and immersive experiences. Travel time becomes a productive and enjoyable time, with high-speed internet access, real-time information updating, and interactive media options in the cabin. The transformation in network technologies can contribute to solving some of the problems regarding universal access to mobile broadband services. 6G-based transportation platforms can provide personalized services according to user preferences and requirements, e.g., personalized routing, real-time travel information, and location-based services.

- **Seamless infotainment and connectivity:** 6G technology can enable seamless infotainment and connectivity by offering high-speed internet, real-time information updates, and interactive entertainment choices for passengers. 6G allows not only a smooth connection with in-car infotainment services but also a shift from passive use to active and creative use of travel.
- **Personalized travel experiences with AR/VR and contextual services:** With the advent of newer technologies, the communications of the future will merge the virtual and physical world. 6G technology has allowed for travel experiences that are personalized in many ways (AR, VR, context-based services) that make the user journey more engaging by making content interactive and immersive. It can provide personalized curation, real-time support, and location-based guidance to enrich the overall journey experience and to form memorable experiences for passengers.

7. Challenges and Limitations

7.1 Technical Barriers

Despite its great potential in the advancement of transportation systems, there are a number of challenges and limitations to be overcome to ensure successful deployment and ubiquitous adoption of 6G technology. Those obstacles come in many forms, from technical and regulatory to security and cultural. Alleviating these challenges will depend on industry, government agencies, and research organizations working together to innovate and minimize potential risks.

- **THz propagation limitations:** The application of THz technologies in the context of the 6G technology faces a number of challenges in connection to atmospheric absorption, scattering, and increasing path-loss, which could restrict the signal coverage and reach, especially in cases of difficult meteorological conditions. Technological advancements in terms of advanced beamforming, adaptive modulation, and antenna design are to be deployed in order for the aforementioned limitations to be addressed in future systems for optimal signal propagation and environmental robustness.

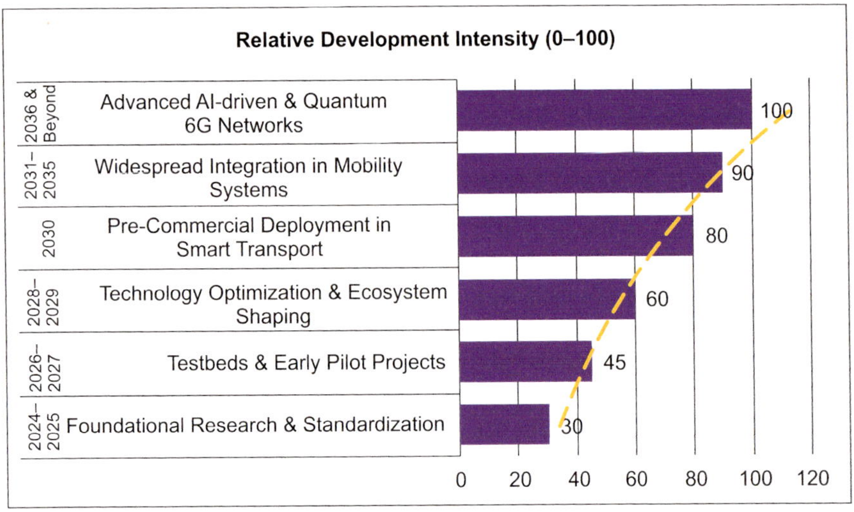

Fig. 4.8 Maturity and application range of 6G.

stakeholders, we could fully tap the potential of 6G and realize a smarter, more connected, and greener future in transportation. In this chapter, we have discussed the main elements of introducing 6G technology in transportation applications with a focus on the potential benefits, challenges, and research issues. Use cases such as UAV delivery service, holographic communication in logistics, and vehicle simulation in a digital twin illustrate the potential of 6G as a strategic transportation technology. 6G is not an incremental upgrade but a game-changer that will revolutionize how we travel, enhance efficiency, and sustainability. With 6G deployment on the horizon, technologists, transit professionals, and policymakers must work together to ensure 6G is implemented in transportation in a way that is integrated, secure, and of maximum benefit to everyone.

References

Chataut, R., M. Nankya and R. Akl. (2024). 6G networks and the AI revolution—exploring technologies, applications, and emerging challenges, *Sensors*, *24*(6), 1888. doi: 10.3390/s24061888.

Chowdhury, M.Z., M. Shahjalal, S. Ahmed and Y.M. Jang. (2020). 6G wireless communication systems: applications, requirements, technologies, challenges, and research directions, *IEEE Open Journal of the Communications Society*, *1*, 957–975. doi: 10.1109/ojcoms.2020.3010270.

Deaton, J.D., M.J. Schmitt and W.F. Jones. (2011). *U.S. Patent No. 8,078,162*, Washington, DC: U.S. Patent and Trademark Office. https://www.osti.gov/biblio/1034293.

El Faouzi, N.E. and L.A. Klein. (2016). Data fusion for ITS: techniques and research needs, *Transportation Research Procedia*, *15*, 495–512. doi: 10.1016/j.trpro.2016.06.042.

Fodor, G., J. Vinogradova, P. Hammarberg, K.K. Nagalapur, Z.T. Qi, H. Do,... and M.U. Baig. (2021). 5G new radio for automotive, rail, and air transport, *IEEE Communications Magazine*, *59*(7), 22–28. doi: 10.48550/arxiv.2101.08874.

Giordani, M., M. Polese, M. Mezzavilla, S. Rangan and M. Zorzi. (2020). Toward 6G networks: use cases and technologies, *IEEE Communications Magazine*, *58*(3), 55–61. doi: 10.48550/arxiv.1903.12216.

Islam, N., M.M. Rashid, F. Pasandideh, B. Ray, S. Moore and R. Kadel. (2021). A review of applications and communication technologies for internet of things (IoT) and unmanned aerial vehicle (UAV) based sustainable smart farming, *Sustainability*, *13*(4), 1821. doi: 10.3390/su13041821.

Javed, M.A., E.B. Hamida, A. Al-Fuqaha and B. Bhargava. (2018). Adaptive security for intelligent transport system applications, *IEEE Intelligent Transportation Systems Magazine*, *10*(2), 110–120. doi: 10.1109/mits.2018.2806636.

Jiang, H., J. Li, P. Zhao, F. Zeng, Z. Xiao and A. Iyengar. (2021). Location privacy-preserving mechanisms in location-based services: a comprehensive survey, *ACM Computing Surveys (CSUR)*, *54*(1), 1–36. https://doi.org/10.1145/342316.

Kabashkin, I. (2024). Unified aviation maintenance ecosystem on the basis of 6G technology, *Electronics*, *13*(19), 3824. doi: 10.3390/electronics13193824.

Khan, L.U., W. Saad, D. Niyato, Z. Han and C.S. Hong. (2022). Digital-twin-enabled 6G: vision, architectural trends, and future directions, *IEEE Communications Magazine*, *60*(1), 74–80. doi: 10.1109/mcom.001.21143.

Liebhart, R., M. Shafi, H. Tataria, G. Shivanandan and D. Chandramouli. (2025). Perspectives on 6G Architectures, *IEEE Wireless Communications*, *32*(1), 108–114. doi: 10.48550/arxiv.2210.03286.

Lynch, D. P., M. M. Tentzeris, V. Fusco and S. D. Asimonis. (2024). Super realized gain antenna array, *IEEE Transactions on Antennas and Propagation*. doi: 10.48550/arxiv.2309.09889.

Mach, P. and Z. Becvar. (2017). Mobile edge computing: a survey on architecture and computation offloading, *IEEE Communications Surveys & Tutorials*, *19*(3), 1628–1656. doi: 10.1109/comst.2017.2682318.

Malec, N., M. Jurgilewicz and A. Czarnecka. (2021). Litigation with Huawei 5G corp and ICT security, *Baltic Journal of Economic Studies*, *7*(2), 21–27. doi: 10.30525/2256-0742/2021-7-2-21-27.

Nguyen, V.L., R.H. Hwang, P.C. Lin, A. Vyas and V.T. Nguyen. (2022). Toward the age of intelligent vehicular networks for connected and autonomous vehicles in 6G, *IEEE Network*, *37*(3), 44–51. doi: 10.48550/arxiv.2209.01349.

Rahman, T.F., A.S. Abdalla, K. Powell, W. AlQwider and V. Marojevic. (2021). Network and physical layer attacks and countermeasures to AI-enabled 6G O-RAN. *arXiv preprint arXiv:2106.02494*. doi: 10.48550/arxiv.2106.02494.

Rasti, M., S.K. Taskou, H. Tabassum and E. Hossain. (2021). Evolution toward 6G wireless networks: a resource management perspective. *arXiv preprint arXiv:2108.06527*. doi: 10.48550/arxiv.2108.06527.

Sallahaj, D. (2024). A comparative analysis of three European airports regarding digital transformation. https://urn.fi/URN:NBN:fi:amk-2024053119113.

Stylianou, K., L. Dimitriou and M. Abdel-Aty. (2019). Big data and road safety: a comprehensive review, *Mobility Patterns, Big Data and Transport Analytics*, 297–343. doi: 10.1016/b978-0-12-812970-8.00012-9.

Wu, J., R. Li, X. An, C. Peng, Z. Liu, J. Crowcroft and H. Zhang. (2021). Toward native artificial intelligence in 6G networks: system design, architectures, and paradigms. *arXiv preprint arXiv:2103.02823*. doi: 10.48550/arxiv.2103.02823.

Wu, Q., B. Zheng, C. You, L. Zhu, K. Shen, X. Shao,... and R. Zhang. (2024). Intelligent surfaces empowered wireless network: recent advances and the road to 6G, in *Proceedings of the IEEE*. doi: 10.1109/jproc.2024.3397910.

Xiao, Y., G. Shi and M. Krunz. (2020). Towards ubiquitous AI in 6G with federated learning. *arXiv preprint arXiv:2004.13563*. doi: 10.48550/arxiv.2004.13563.

Chapter 5

Smart Mobility for All

Human-Centric Approaches to Personalized Transportation Solutions

Banu Priya Prathaban[1*], *Subashini. N. J.*[2], *Saveetha D.*[3], *Subash*[4] and *Ben Sujin Bennet*[5]

In today's fast-paced world, mobility must adapt to the individual rather than the individual adapting to mobility. Personalized mobility solutions (PMS) denote an innovative shift in how transportation systems are premeditated, accessed, and practiced. By using real-time data, artificial intelligence (AI)–powered analytics, and human-centric technologies, PMS aim to provide integrated, efficient, and wide-ranging transport skills personalized to each user's needs, preferences, and limitations. This chapter explores the fundamentals of PMS conducted via a human-centric lens, concentrating on ease, accessibility, sustainability, and impartiality. It discusses how intelligent transportation systems (ITS), ride-sharing platforms, mobility-as-a-service (MaaS), and wearable technologies are integrated to support a wide range of commuters, including daily travelers, people with disabilities, and the elderly. It focuses on a major part of behavioral information, knowledge on the present location, and prognostic modeling in considering usability and preference patterns. Additionally, the chapter covers the technical encounters, ethical implications, and privacy alarms of intensely personalized schemes. Through real-world case studies and developing trends like 6G communications, edge computing, and blockchain-based user identity administration, this chapter provides an overview of comprehensive, effective, and adjustable transportation ecosystems. By presenting human experiences as the main

[1-3] Department of Networking and Communications, School of Computing, SRM Institute of Science and Technology, SRM Nagar, Kattankulathur, Chennai, India.
[4] Department of Computing Technologies, School of Computing, SRM Institute of Science and Technology, SRM Nagar, Kattankulathur, Chennai, India.
[5] Department of Engineering, University of Technology and Applied Sciences, Nizwa, Sultanate of Oman.
https://orcid.org/0000-0003-0759-510X[1]; https://orcid.org/0009-0006-8206-8683[2]; https://orcid.org/0000-0001-6995-5583[3]; https://orcid.org/0000-0002-8570-7536[4]; https://orcid.org/0000-0001-7678-2852[5]
* Corresponding author: banuprip2@srmist.edu.in

DOI: 10.1201/9781003659822-5

key of transportation system platforms, the chapter apprehends a mobility future which is totally a smart and user personalized system.

1. Background on Personalized Mobility Solutions (PMS)

The expeditious expansion of the transportation framework in this era has shifted the world from mass transport to a human-centric, model-based environment that organizes and provides higher prioritization to individual requirements and preferences. PMS epitomizes a transformative strategy within intelligent transportation systems (ITS), exploiting enhanced technologies such as artificial intelligence (AI), the internet of things (IoT), 6G connectivity, and blockchain toward developing a personalized transportation system. Unlike conventional transportation systems, which mostly implement a standard approach, PMS emphasizes on fashioning continuous, wide-ranging, and sustainable mobility environments that adapt to varied user groups, comprising everyday passengers, persons with disabilities, and the elderly. This transformation has been the result of social factors such as rapid urbanization, the expansion of the elderly population, and the growing demand for unbiased admittance to transportation networks (Goumiri et al., 2025).

Figure 5.1 exemplifies the ways with which PMS is combined within the larger ITS frameworks. It displays the data flow from user likes and conservational inputs into AI-driven analytics, which in turn augment mobility opportunities. This theoretical illustration features the user-centric strategical method as the principal ideology of PMS. It explains the utilization of advanced technologies like 6G communication and IoT sensors to serve the diverse user needs in mobility. The user requirements comprising preferences, necessities, and constraints are entered into the ITS, beside real-time information from users and the IoT sensors. The arrangement procedures these inputs to carry optimal mobility products, such as personalized directions and manageable interfaces. The incorporation of user data and sensor inputs empowers ITS to adapt transportation resolutions to individual necessities, augmenting efficiency, expediency, and availability in transportation systems.

The incorporation of PMS with ITS permits real-time data administering, predictive analytics, and human-centric approaches, guaranteeing that flexibility resolutions are not only effective but also extremely private. By using interactive data, location intelligence, and identity protection management, PMS authorizes users to circumnavigate complex urban settings with ease and confidence. This section provides an understanding of the idea of PMS, its implications in modern transportation, the human-centric design principles that reinforce its development, and the societal drivers that demand its approval. It sets the background for an all-inclusive examination of PMS technologies, applications, and encounters in the succeeding sections.

1.1 Role of PMS in Modern Transportation System

PMS refers to transportation systems intended to adapt to the explicit requirements, favorites, and restrictions of distinct users, relying on real-time information and innovative technologies to reclaim personalized mobility involvements. Unlike

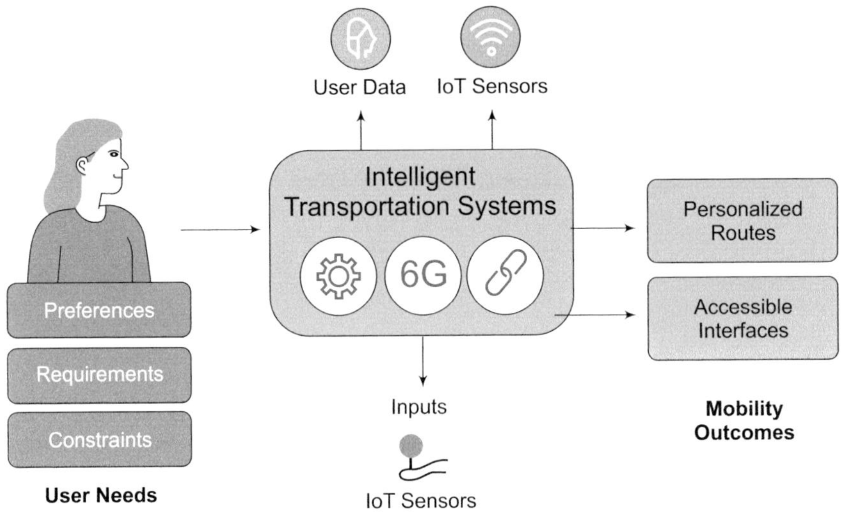

Fig. 5.1 Conceptual framework of PMS within ITS.

outdated transportation systems, which usually give importance to effectiveness, productivity, and capability for huge populations, PMS foregrounds personalization, approachability, fulfilment, and customer satisfaction. The progression of PMS can be outlined from primeval ITS progresses, like traffic signal optimization, Global Positioning System (GPS)–grounded routing, modern schemes integrating AI-motivated direction planning, mobility as a service (MaaS), and other assistive technologies for varied user groups (Oladimeji et al., 2023). For instance, PMS can augment travel directions for customers based on real-time traffic information and individual plans, while also offering reachable routing assistance for visually impaired users over voice-guided boundaries.

The purpose of PMS in current transportation networks is multilayered, tackling the rising complications of urban mobility. As cities are changing to be more thickly populated, varied user groups, ranging from everyday customers to persons with physical or cognitive infirmities, necessitate personalized resolutions. PMS participates with ITS backgrounds to influence technologies like the IoT for real-time information gathering, AI intended for extrapolative analytics, and 6G interconnectivity aimed at low-latency communication. This combination permits incessant user experiences, so realized in MaaS platforms that syndicate various transport means (e.g., buses, car-sharing, bike-sharing) into a solitary, tailored association (Smith et al., 2023). By giving importance to customer needs, necessities, and preferences, PMS transforms transportation into a self-motivated, all-encompassing network that augments mobility, making it efficient and objective.

1.2 Human-Centric Design Principles

Human-centric design is the keystone of PMS, guaranteeing that transportation systems prioritize the necessities of customers, convenience, and fulfilment. The central ideologies of human-centric strategy are to embrace convenience, inclusivity, ease, and sustainability, as depicted in Figure 5.2.

Fig. 5.2 Human-centric design principles.

Accessibility warrants that mobility resolutions are operational by everyone, comprising all individuals and even disable persons, via features such as voice-activated devices or wheelchair-manageable vehicles (Będczkowska & Grabarek, 2024). Inclusivity manages the assorted requirements of users over various demographics by delivering multilingual edges intended for non-native speakers or easy routing paths for elderly people. Comfort stresses on enlightening user involvement by means of custom-made structures, like preferred seating requirements by manipulating wide-ranging climatic periods in ride-sharing vehicles. Sustainability emphasizes eco-friendly resolutions, such as optimizing paths to reduce fuel ingestion and utilizing shared mobility services to reduce carbon emitters from the framework (Łuczyszyn, 2024). All these ideologies spark the expansion of PMS by incorporating customer response in the system design. For example, AI-motivated transportation platforms will be able to inspect the rendering of its mechanisms to distribute personalized route commendations, while IoT-aided wearables can perceive health parameters for elderly users, corroborative safety, and relaxed travelling experience. With social experiences being fundamental to mobility, recent transportation systems not only transform the efficiency level but also foster fairness and trust amongst travellers. These ideologies will be further explored in the background of advanced technologies and applications in following sections. Table 5.1 elaborates on the crucial human-centric design principles that revolve around MS and the explicit features that exploit each terminology and its relavent frameworks. It showcases how accessibility, inclusivity, comfort, and sustainability are maneuvered via practical design elements in PMS. As revealed in Table 5.1, these design principles shape the groundwork for generating mobility resolutions that encounter the varied requirements of all user groups.

1.3 Societal Drivers of PMS

Embracing PMS is ambitious in its alignment with various social trends that stress the necessity for custom-made, unbiased, and sustainable transportation schemes. Urbanization has resulted in progressively intricate mobility difficulties, with over 55

Table 5.1 Human-centric design principles vs. features.

Design Principle	*Features/Examples*
Accessibility	Wheelchair-accessible vehicles, voice-activated interfaces, big-font apps for elderly citizens
Inclusivity	Multilingual interfaces, characteristics personalized for diverse demographics
Comfort	Based on the preferences made on seating, climate control in ride-sharing, less crowded routes
Sustainability	Route optimization to decrease fuel, shared mobility, electric vehicles

percent of the global population residing in urban areas as of 2023—these numbers are expected to reach 68 percent by 2050 (Chen et al., 2022). This trend necessitates flexible, user-centric solutions to traffic congestion and varied commuter requirements. With the expansion of the aging population, elderly individuals need harmless and user-friendly transportation possibilities to sustain their independence (Remillard, et al. 2022). Moreover, the increasing prominence of mobility equity underlines the necessity to assist underserved societies, like rural areas or individuals with debilities, who are usually left out of trustworthy transportation, as shown in Figure 5.3.

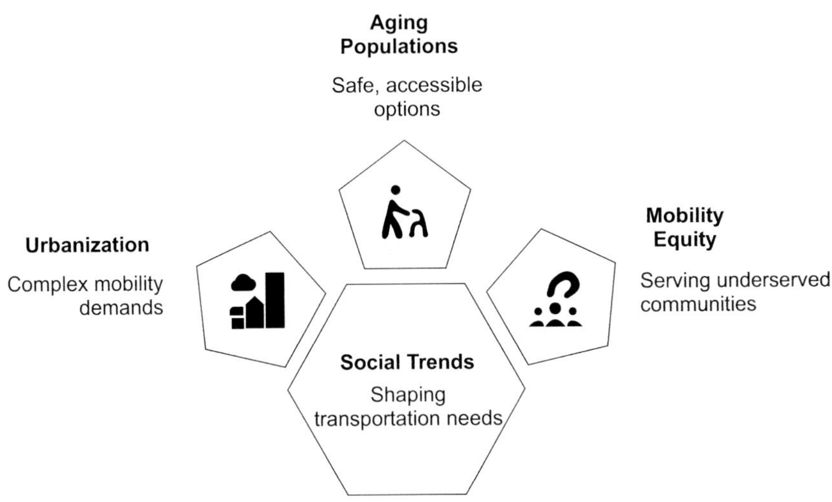

Fig. 5.3 Social trends in PMS.

The importance of PMS lies in its ability to address these challenges through tailored solutions. For example, PMS can participate in real-time traffic data to decrease travel periods for urban residents, while assistive technologies, like haptic navigation policies, can authorize visually impaired users. Additionally, PMS encourages sustainability by enhancing transport modes and dropping environmental influence, positioning itself with global goals for carbon impartiality. By tackling these social drivers, PMS not only augments separate mobility but also donates to the broader vision of comprehensive, human-centered smart cities.

2. Technological Innovations Driving Personalized Mobility

PMS depends on a collection of cutting-edge technologies to distribute personalized, effective, and all-encompassing transportation experiences. The incorporation of AI, the IoT, 6G connectivity, and blockchain technologies acts as the major support system of PMS, permitting real-time information processing, extrapolative analytics, and protected user communications (Allam & Sharifi, 2022), as shown in Figure 5.4. Each layer plays a critical role in collecting, processing, and securing data to deliver real-time personalized transportation services. These inventions discourse the varied requirements of users, from optimizing daily travels to augmenting accessibility for individuals with disabilities. This section explores the technological foundations of PMS, detailing how each component contributes to personalized, sustainable, and equitable mobility solutions. Moreover, the growth of autonomous and electric vehicles is leading to clean, safe, and reliable transport systems. These developments are shifting the focus from regularized mass transportation to smart, custom-made mobility ecosystems by spotlighting the role of behavioral data, location intelligence, and predictive modeling in understanding user patterns and optimizing mobility offerings.

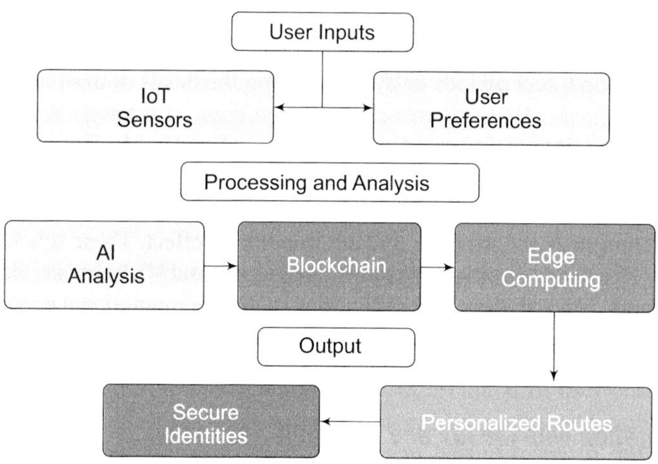

Fig. 5.4 Technological ecosystem of PMS.

The technologies are driving PMS exertion in collaboration to produce a unified, continuous, secure, and satisfied user experience. Table 5.2 summarizes the key technologies discussed in this section: AI, the IoT, 6G communication, and blockchain, along with their main applications in PMS ecosystems.

Table 5.2 provides a quick outline of how each technology provisions personalization, user-friendliness, effectiveness, and proficiency in modern transportation systems.

5.1 Technical Challenges in PMS Deployment

There are numerous technical problems when employing PMS in complicated transportation systems, such as real-time data handling, scalability, and system interoperability. For the incorporation of numerous systems, comprising public transportation systems, ride-sharing stands, and IoT-aided infrastructure, interoperability is important. However, unpredictable protocols and data presentations amongst regions and workers frequently obstruct smooth integration, resulting in fragmented user experiences. For instance, a MaaS platform might have trouble coordinating real-time data from several transit agencies, which would cause personalized route recommendations to be delayed.

Another issue is scalability since PMS must manage massive amounts of data from millions of users in urban settings. To process real-time traffic, user preferences, and environmental data, AI-driven systems need a lot of processing power, which can put a strain on the infrastructure that already exists. Applications like dynamic routing and accessibility adjustments require real-time data processing, but performance can be harmed by 5G networks' high latency or insufficient edge computing infrastructure. Although solutions like advanced edge computing nodes and 6G connectivity are being investigated, their deployment is expensive and demands a large financial commitment. Standardized protocols, strong cloud–edge architectures, and public–private cooperation are required to meet these challenges.

5.2 Privacy and Data Security Concerns

Because PMS depends on private user information like location, travel patterns, and health metrics, there are serious privacy and data security issues. The risk of data breaches or unauthorized access increases when real-time data is collected via wearables, IoT devices, and MaaS platforms. A 2024 study (Forsberg et al., 2024) showed that 70 percent of smart mobility apps do not consume robust encryption standards, making user data susceptible to probable cyberattacks. These defects diminish user confidence and hamper the acceptance of the PMS framework.

Practices like data anonymization and end-to-end encryption are utilized to decrease these risks. As discussed in section 2.4, blockchain-based identity management provides users control over who can access their data, confirming the endurance of safe data. Nevertheless, the higher computational costs and the necessity of user-friendly interfaces makes conceptualizing these resolutions at scaling difficult. Data handling becomes more difficult with regulatory frameworks such as the General Data Protection Regulation (GDPR), which enforce PMS providers to strike a balance between compliance and personalization. Defining users' safety, faith, and trust and accomplishing fair contacts differ on defending privacy without forgoing the functionality features.

5.3 Ethical Implications of Personalization

The ethical problems of algorithmic data unfairness, equitable access, and disproportionate dependence on technology are transported by the personalization that originates with PMS platforms. When AI models are accomplished on very few available datasets favored on user groups, it can give rise to disproportionate algorithmic data equivalence, which may downgrade lower-income communities or subgroups. For example, a 2022 study revealed that there is assurance in providing ride-sharing algorithms along with compact service availability in underserved neighborhoods by providing priority to areas with higher requisitions (Trigyn Technologies, 2023). Transparent AI models and a diversity of training data information given as input to the models are essential to address this bias between real-time data and model data information.

Yet another problem is providing unbiased access. PMS may unintentionally cancel travellers who are non-smartphones users or using reliable internet services, particularly in rural or lower-income zones. Mobility fairness needs universal design choices and rationally priced amenities. Users who depend a lot on technology are still in the danger of trailing their capability to leave behind any choice for themselves, which could make them less resilient to transportation arrangement catastrophes. For the stability of societal profits and personalization, ethical contexts are mandatory, such as fairness-conscious AI and comprehensive design ideologies. These issues standardize the ways with which matching human-centric values with PMS is utilized to avoid intensifying crucial social inequalities.

6. Future Trends and Opportunities

Rapid technological progressions and a growing effort on human-centric design will outline PMS's forthcoming era and overlay the customs for wide-ranging, sustainable, and operative transportation system services.

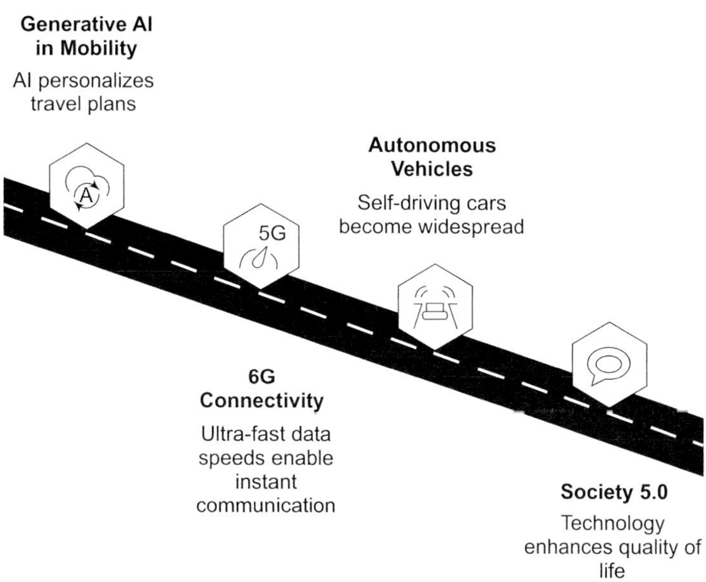

Fig. 5.8 Future trends in PMS

The competences of PMS can be improved by new changes in AI, driverless cars, and the Society 5.0 framework. This would support environmental sustainability and

of area computing include support for mobility, cognizance of place, minimum latency, and proximity to users. This paradigm is defined by its ability to provide excessive bandwidth, extremely low latency, and actual-time admission to community records, which may be leveraged by using various packages. The subsequent phase will complicate the traits of edge computing. One of the primary attributes of edge computing is its big distribution, which situates computing offerings near quit users and is tailor-made to their widely deployed computing structures (Gao et al., 2019). This arrangement facilitates rapid, massive facts analytics with stronger accuracy, as side systems can conduct actual-time analytics on a large scale. Another full-size characteristic of area computing is its low latency, which lets quit customers perform computations at the network's area near their physical region (Yuan et al., 2018). This proximity guarantees that programs sensitive to time delays can supply the most reliable computing level, effectively minimizing the latency of offering access.

Consequently, consumer level is extensively enhanced via ultra-low latency and high bandwidth. The characteristics of edge computing are shown in Figure 6.2.

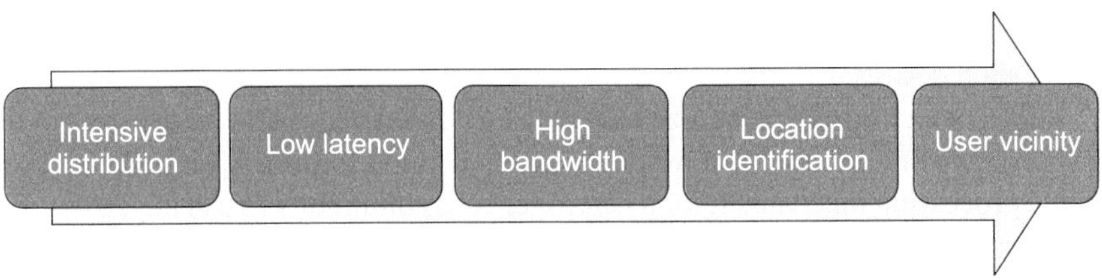

Fig. 6.2 Characteristics of EC.

2. Eiot Computing Model

Edge IoT area architecture represents a paradigm shift in how data is processed and managed in à network of related devices. At its core, IoT computing includes moving data processing and analytics toward the source of statistical technology, typically, the threshold devices themselves rather than counting on a centralized cloud infrastructure. This architectural shift addresses numerous important demanding situations, which include latency, bandwidth, and data privacy. In a traditional IoT setup, records are transmitted from side gadgets (together with sensors and actuators) to a relevant cloud server for processing (Yuan et al., 2019a). Even though it is effective, this technique can introduce significant delays because of the time required for statistical transmission over the network. Moreover, it could locate a widespread load on network bandwidth, particularly in environments with excessive information volumes or where actual processing is important. IoT side architecture mitigates these issues by deploying computational sources closer to the information source. The brink layer consists of area gadgets and facet servers. Devices often have sensors and actuators that gather and act on records locally. Aspect servers, alternatively, are more sturdy computing assets deployed at or near the information source (Yuan et al., 2019b). These servers are chargeable for jogging analytics, processing data, and making actual-time decisions without shipping data back to the cloud. This localized processing capability reduces latency and minimizes the number of records that desire to be transmitted over the community, which could notably decorate the responsiveness and performance of IoT packages. A key gain of this structure is advanced record privateness and safety. Via processing sensitive facts regionally, IoT edge computing reduces the risk of data publicity during transmission. It also allows extra granular manipulation over statistics to get entry to security rules, which is essential in environmental problems and regulatory compliance. Moreover, the IoT area structure helps with scalability and flexibility in dealing with various disbursed IoT deployments (Dong et al., 2019). It allows companies to deploy and manage resources efficiently throughout many places and scenarios, from business automation to clever towns. Integrating part computing with cloud services permits a hybrid method in which critical and time-sensitive tasks are treated at the brink, while extra full-size information evaluation and garage can still be controlled within the cloud. The IoT part structure represents a transformative technique for facts management in IoT ecosystems (Siqi et al., 2019).

By leveraging area gadgets and servers to technique information domestically, this structure addresses latency, bandwidth, and privacy issues while supplying scalability and flexibility to satisfy the demands of contemporary IoT applications. The EIOT model is shown in Figure 6.3.

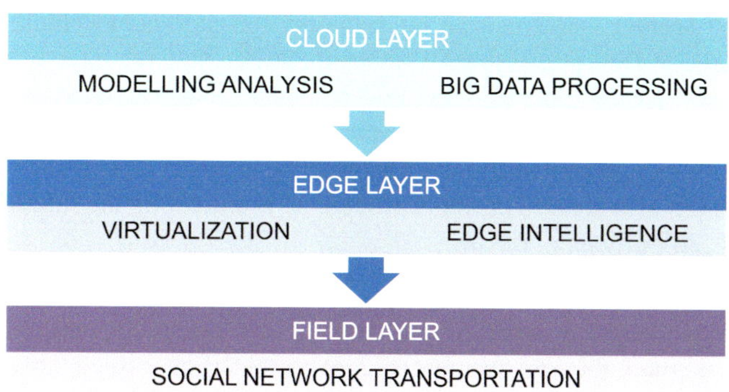

Fig. 6.3 EIOT Model.

3. The Integration of EIOT Computing

Edge computing notably complements the IoT by permitting immediate facts processing and evaluation, which is critical for programs that call for fast responses. Through parasitic gathering and processing data towards the source on the "side" of the Net, area computing reduces the latency that may occur, while statistics must travel vast stretches to centralized servers. This immediacy no longer boosts the efficiency of IoT structures; however, it guarantees that critical operations can be kept even though connectivity with critical servers is disrupted (Ma, 2019). This localized processing capability enhances machine reliability and resilience. Additionally, computing addresses the growing data demands of increasing IoT ecosystems. With the proliferation of IoT devices producing full-size quantities of records, facet computing allows us to manipulate and process this information extra efficiently. In preference to overwhelming critical servers with big information streams, area devices handle preliminary data evaluation and filtering locally (Lian, 2019). This technique reduces the volume of records transmitted over the network, alleviates community congestion, and speeds up statistics-driven selection-making approaches.

4. Sensible Programs of EIOT Computing

The synergy between IoT and edge computing is obvious throughout various sectors, demonstrating its realistic advantages.

Clever Production: In business settings, computing allows for actual time tracking and management of manufacturing approaches. Sensors and IoT devices deployed at the manufacturing unit ground accumulate facts on equipments' overall performance, environmental situations, and manufacturing metrics. Aspect computing tactics these statistics domestically, enabling on-the-spot changes to optimize operations, improve efficiency, and limit downtime. For instance, predictive maintenance may be accomplished by analyzing equipment information in actual time, potentially preventing highly priced breakdowns and improving standard manufacturing reliability.

Independent Motors: Computing is crucial in self-sufficient vehicles, allowing fast neighborhood facts processing (Wang et al., 2019). Self-reliant motors are ready with numerous sensors and cameras that constantly collect statistics about their environment. Edge computing allows these vehicles to analyze facts domestically and make cut-up 2d selections, including breaking to avoid collisions or navigating complex visitors' situations. This rapid processing is critical for the safe and effective operation of self-sufficient systems.

Smart Grids: Inside the electricity zone, smart grids use edge computing to handle and optimize electricity distribution and intake in real-time. IoT sensors and devices deployed throughout the grid accumulate facts on electricity usage, grid conditions, and delivery ranges.

Edge computing approaches these records regionally, bearing in mind actual-time modifications to stabilize supply and demand, stumble on and reply to faults, and decorate the general performance and efficiency of the electricity grid. This localized approach enables lowering power waste and enhancing grid reliability (Guo et al., 2015). The combination of EIOT computing gives a strong framework for coping with modern programs' complicated and dynamic demands, using improvements throughout various industries. The integration of EIOT computing and sensible programs is shown in Figure 6.4.

5. Challenges and Issues in ELIOTF

5.1 Facts Protection

- *Cyber Threats*: With its decentralized nature, facet computing introduces precise protection in demanding situations. Numerous cyber threats, including

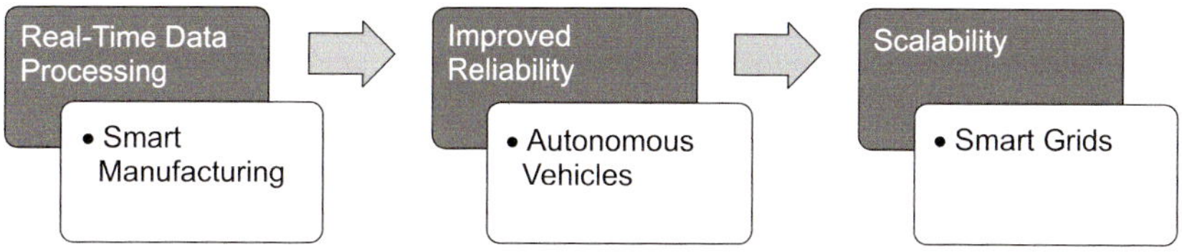

Fig. 6.4 EIOT computing and Sensible Programs.

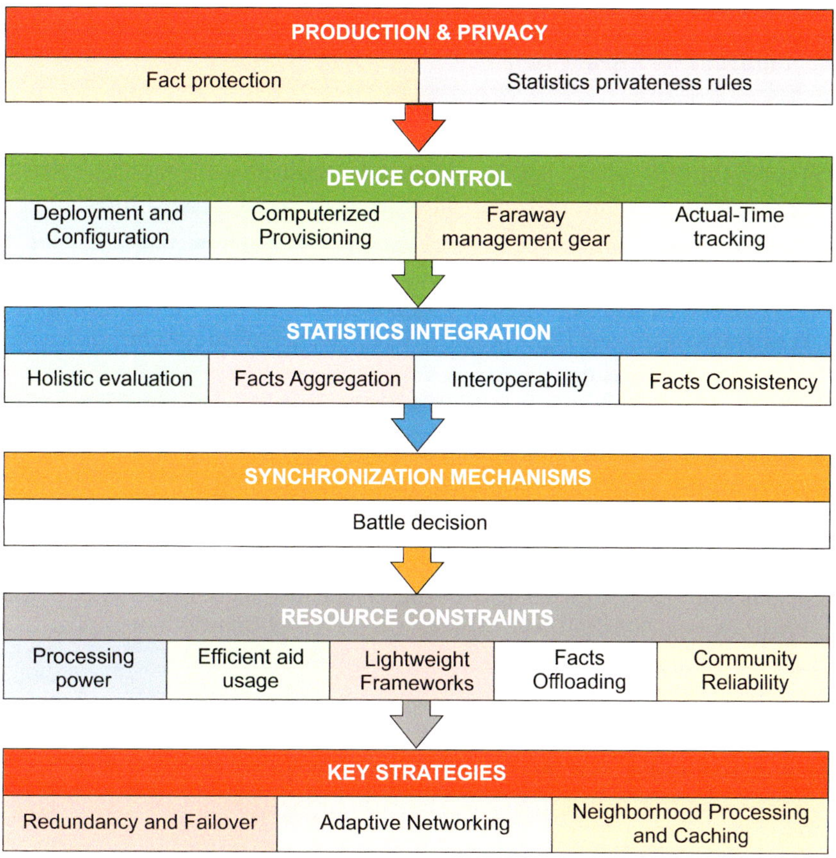

Fig. 6.5 Challenges and Issues in EIOT Computing.

unauthorized admission, information breaches, and malicious assaults, should cover information processing at the edge. Implementing sturdy security measures is important to protect touchy information. Key practices include encrypting facts in transit and relaxation to prevent unauthorized entry to advanced encryption algorithms. Comfort key control practices are essential (Wang et al., 2013). Authentication and get admission to manipulate implementing strong authentication mechanisms (e.g., multi-issue authentication) and granular get entry to control guidelines to limit information get admission to authorized customers and devices.

- *Updates and Patching*: Maintaining facet gadgets and keeping software up to date with ultra-modern security patches can protect against recognized vulnerabilities.
- *Bodily Safety*: Side gadgets are regularly deployed in less comfortable or far-off places, making them prone to bodily tampering or robbery.
- *Tamper-Resistant Hardware*: Hardware designed to face up to tampering and physical attacks should be used (Lee & Cheng, 2016). Comfortable *installation practices* ensure relaxed installation and preservation procedures to guard against physical threats as shown in Figure 6.5.

5.2 Statistics on Privacy Rules

Adhering to data privacy rules (including GDPR and CCPA) is critical for preserving legal and moral requirements in information handling. Compliance involves privateness-through-layout, i.e., incorporating information privacy ideas into the design and operation of area devices and structures (Luan & Leng, 2016).

- *Minimization*: Amassing and processing only the information necessary for functions and ensuring proper anonymization and aggregation are viable.
- *Consumer Consent and Rights*: Mechanisms can be implemented to reap consumer consent and allow people to work out their records rights, such as data admission and deletion requests.
- *Pass-Border Records Switch*: While statistics are transmitted across international borders, it should comply with varying nearby information protection laws.
- *Information Prison Necessities*: Groups must navigate these complexities by information prison necessities, familiarizing themselves with relevant guidelines in all jurisdictions where information is processed or stored.
- *Statistics Transfer Agreements*: These can be enforced by organizing suitable felony frameworks, including standard contractual clauses (SCCs) or binding company guidelines (BCRs), to make certain lawful statistics transfers.

5.3 Scalability and Control

- *Deployment and Configuration*: Dealing with many IoT gadgets and aspect nodes entail ensuring that each device is effectively deployed, configured, and maintained. This can be hard because of the size and variety of gadgets (Alwakeel, 2021).
- *Effective Strategies*: These encompass computerized provisioning and automatic systems to streamline tool setup, configuration, and onboarding processes.
- *Faraway Management Gear*: This involves leveraging remote control and tracking equipment to manipulate devices from an important location, together with skills for firmware updates and troubleshooting.
- *Tracking and Protection*: Ensuring ongoing tool functionality requires strong tracking and maintenance practices.
- *Actual-Time Tracking*: Continuously enforcing systems can reveal device performance, fitness, and connectivity.
- *Proactive Upkeep*: Predictive analytics can be used to anticipate and cope with capability problems before they impact operations.

5.4 Statistics Integration

- ***Holistic Evaluation***: Integrating statistics from edge gadgets with centralized systems is important for complete analysis and selection-making.
- ***Facts Aggregation***: Implementing effective statistics aggregation methods help combine and analyze statistics from multiple facet devices.
- ***Interoperability***: Ensuring that information formats, protocols, and standards are well suited among area devices and significant systems can facilitate seamless records trade.
- ***Facts Consistency***: Retaining consistency across facet and centralized systems is essential for accurate analysis.
- ***Synchronization Mechanisms***: Imposing robust synchronization techniques ensure that facts are as they should be and are constantly included.
- ***Battle Decision***: Strategies can be used to deal with facts, conflicts, and discrepancies that could arise throughout integration and performance optimization.

5.5 Useful Resource Constraints

- ***Processing Power and Garage***: Some devices typically have confined processing electricity and storage compared to centralized structures.
- ***Efficient Aid Usage***: Available processing energy and garage can be optimized by prioritizing important tasks and offloading non-important processing to valuable systems.
- ***Lightweight Frameworks***: Employing lightweight computing frameworks and running systems designed for resource-restrained environments can maximize device skills.
- ***Facts Offloading***: To address barriers, offloading complicated or non-urgent responsibilities to centralized servers should be avoided, even as facet gadgets take care of time-touchy operations locally.
- ***Community Reliability***: Ensuring reliable community overall performance across numerous environments is important for maintaining the effectiveness of IoT systems.

5.6 Key Strategies

- ***Redundancy and Failover***: Imposing redundant network paths and failover mechanisms can ensure continuous connectivity and performance.
- ***Adaptive Networking***: Adaptive networking strategies can be used to alter and convert community situations to maintain overall performance.
- ***Neighbourhood Processing and Caching***: Leveraging neighborhood processing and record caching allows in-area gadgets to function efficaciously despite intermittent connectivity.
- ***Neighborhood Caching***: Storing important facts regionally can reduce reliance on constant network access.
- ***Facet-Based Processing***: Performing vital computations at the brink can reduce latency and dependence on relevant structures.

Addressing these demanding situations calls for a balanced technique for security, control, and performance, ensuring that IoT and aspect computing systems are strong, scalable, and capable of meeting the needs of modern applications.

6. Applications of EIOT Computing

6.1 Smart Site Visitors' Management

Visitors' sensors and cameras reveal automobile waft, congestion, and traffic light reputation. These gadgets accumulate real-time records of site visitors' conditions, vehicle counts, and speeds. Statistics from these sensors are processed locally at site visitors' manipulated nodes (Alwarafy et al., 2020). This actual-time processing allows for adaptive visitor light management, dynamic routing, and instantaneous responses to traffic incidents. For example, side gadgets can alter visitors' signals based on contemporary traffic conditions to minimize congestion and decrease delays as shown in Figure 6.6.

Fig. 6.6 Application of EIOT.

6.2 Waste Management

Smart waste packing containers with ultrasonic sensors measure waste ranges and detect when they may be complete or need preservation. The information collected is processed locally to optimize collection routes. For example, facet devices can analyze fill degrees and determine the maximum green path for waste vehicles, reducing gasoline consumption and operational charges (Angel et al., 2021). This additionally enables the decrease in environmental effects by warding off pointless trips.

6.3 Healthcare

6.3.1 Far-Flung Affected Person Tracking

- IoT: Wearable gadgets like fitness trackers and medical video display units collect data on important symptoms, including heart price, blood pressure, and glucose ranges.
- Edge Computing: These gadgets system facts locally to discover anomalies or tendencies in real-time. For instance, if a patient's heart rate exceeds a certain threshold, the threshold device can alert healthcare providers or trigger an emergency reaction without delay. This neighborhood processing guarantees timely interventions and decreases the need for continuous record transmission.

6.3.2 Telemedicine

- IoT: Connected medical devices facilitate far-off diagnostics and patient consultations.
- Edge Computing: Area computing supports low-latency video streaming for telemedicine. Via processing records regionally, facet gadgets make certain smooth and dependable virtual consultations between patients and healthcare specialists, even in regions with confined net connectivity. Moreover, aspect computing can cope with real-time evaluation of diagnostic records for the duration of consultations.

6.4 Business IoT

6.4.1 Predictive Preservation

- IoT: Sensors on industrial machinery screen temperature, vibration, and wear-and-tear parameters.
- Edge Computing: Facts from these sensors are analyzed in real-time to determine expected capacity system screw-ups before they occur. For instance, if a sensor detects uncommon vibrations, a part tool can analyze these facts to expect whilst protection might be wished, thereby stopping unplanned downtime and reducing preservation fees.

6.4.2 Quality Control

- IoT: Cameras and sensors on meeting lines inspect merchandise for defects and ensure it meets first-class requirements.
- Edge Computing: Nearby visible and sensor data processing allows instantaneous detection of minor troubles. For example, if a camera identifies a defect, area devices can trigger corrective actions, halt the production line, or alert operators, ensuring high product satisfaction and lowering waste.

6.5 Agriculture

6.5.1 Precision Farming

- IoT: Soil moisture sensors, weather stations, and crop fitness monitors offer statistics on soil conditions, climate forecasts, and crop popularity.
- Edge Computing: Neighborhood evaluation of these facts allows precision agriculture practices. For instance, edge devices can process soil moisture records to manipulate irrigation structures more precisely, using water most effectively when needed. This optimizes water use, improves crop yield, and decreases operational costs.

6.5.2 Farm Animals Monitoring

- IoT: Wearable sensors on farm animals track health metrics, including temperature, interest degrees, and feeding behavior.
- Edge Computing: Real-time processing of these statistics at the threshold permits immediate detection of fitness issues or strange behavior. For example, if a sensor indicates a drop in activity degrees, aspect devices can alert farmers to potential health issues, allowing well-timed intervention and better cattle control.

6.6 Retail

6.6.1 Inventory Control

- IoT: RFID tags and sensors track product portions and places throughout the supply chain and in-keep.
- Edge Computing: Nearby data processing optimizes stock management by supplying real-time updates on inventory degrees and product moves. For example, area devices can alert managers when stock stages are low or when products are out of place, sparking off restocking and higher inventory manipulation.

6.6.2 Customized Shopping

- IoT: Beacons and in-keep sensors acquire records on customer behavior, options, and motion patterns.
- Edge Computing: The processing of these statistics nearby permits customized tips and promotions in actual time. For example, facet devices can analyze purchaser behavior to ship focused offers or reductions through cellular apps, improving shopping and increasing sales.

6.7 Transportation and Logistics

6.7.1 Fleet Control

- IoT: GPS and telematics devices tune automobile vicinity, speed, fuel intake, and driving force behavior.
- Edge Computing: Real-time statistics processing at the edge improves fleet control by enabling direction optimization, tracking motive force overall

performance, and managing gasoline efficiency. For example, side devices can analyze records to indicate extra green routes or offer indicators for hazardous riding practices.

6.7.2 Smart Warehousing

- IoT: Computerized structures and sensors manage inventory, music shipments, and display warehouse conditions.
- Edge Computing: Local statistics processing complements warehouse operations by automating sorting, tracking inventory in real-time, and managing logistics effectively. For example, facet gadgets can optimize garage areas by analyzing stock information and adjusting warehouse layouts dynamically.

6.8 Power Control

6.8.1 Smart Grids

- IoT: Smart meters and sensors screen power consumption, grid performance, and device repute.
- Edge Computing: Facet gadgets manage statistics domestically to optimize power distribution, hit upon outages, and manage load balancing. For example, neighborhood processing can adjust energy distribution in real-time based on on-demand styles, enhancing grid reliability and efficiency.

6.8.2 Renewable Electricity Integration

- IoT: Sensors on renewable electricity resources like solar panels and windmills monitor performance and environmental conditions.
- Edge Computing: Evaluation of nearby facts manages electricity era and garage, improving the integration of renewable resources into the grid. For example, facet devices can manipulate battery storage structures and adjust power output primarily based on actual-time weather situations and power demand.

6.9 Smart Houses

6.9.1 Domestic Automation

- IoT: Clever gadgets, such as thermostats, lights, and safety systems, join and interact to automate domestic functions.
- Edge Computing: Nearby processing allows seamless automation of home structures based on consumer choices and environmental situations. For instance, edge gadgets can modify heating and cooling primarily based on occupancy styles and time of day, enhancing comfort and energy efficiency.

6.9.2 Strength Efficiency

- IoT: Smart meters and home equipment monitor and control home strength.

- Edge Computing: Local facts analysis optimizes strength consumption by presenting insights into utilization styles and making actual-time modifications. For example, facet gadgets can routinely flip off lighting or regulate thermostat settings when a room is unoccupied, thus reducing power waste.

7. Evolution of EIOT Computing

7.1 5G Networks

The deployment of 5G networks marks a widespread jump ahead for the IoT and edge computing, as shown in Figure 6.7.

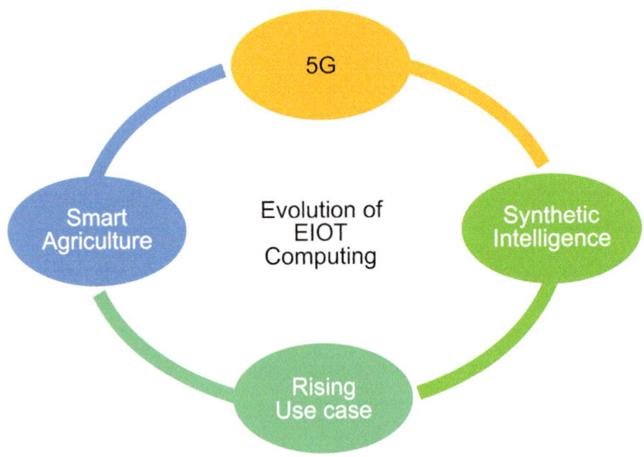

Fig. 6.7 Evolution of EIOT Computing

- *Accelerated Bandwidth*: 5G networks provide greatly accelerated record switch speeds compared to previous generations. This higher bandwidth enables IoT gadgets to quickly transmit massive records, facilitating real-time applications previously constrained by slower community speeds (Beckman & Catlett, 2020).
- *Reduced Latency*: The 5G era provides extremely low latency, which is crucial for programs requiring instant remarks and quick responses. For edge computing, statistics may be processed and acted upon almost immediately, improving the performance of packages like self-reliant automobiles and real-time far-flung monitoring.
- *More Desirable Connectivity*: 5G supports a higher density of connected devices according to a rectangular kilometer. This scalability is crucial for big-scale IoT. Deployments in smart towns and business settings, where thousands of sensors and gadgets want to operate seamlessly.
- *Community Cutting*: 5G introduces network slicing, allowing virtual networks to be tailored to precise desires. For instance, a network slice can be dedicated to critical IoT packages with stringent reliability and latency requirements, ensuring that these packages perform optimally without interference from different kinds of network traffic.

7.2 Synthetic Intelligence

AI and system learning are becoming increasingly fundamental to facet computing in the following ways.

- *Processing*: AI algorithms are being deployed without delay on facet devices, which include sensors and gateways, to carry out real-time facts processing (Bibri, 2023). This reduces the need to transmit facts to centralized cloud servers, reducing latency and bandwidth usage and enhancing privacy and security.
- *Improved Choice-Making*: By integrating AI at the brink, devices could make more informed choices without needing regular cloud conversation. For example, an AI-prepared surveillance camera can pick out and alert users to uncommon sports right now instead of sending all pictures to the cloud for evaluation.
- *Adaptive Systems*: AI at the brink permits structures to research and adapt over the years, solely on the basis of nearby records. For example, a clever thermostat can research a household's temperature preferences and regulate settings autonomously, optimizing strength utilization and luxury.
- *Predictive Preservation*: In commercial environments, AI-driven edge computing can monitor equipment at actual time, expect capability failures, and schedule renovation before a breakdown happens (Bilal et al., 2018). This method improves operational efficiency and decreases downtime.

7.3 Rising use Cases

Augmented reality (AR) and virtual reality (VR) side computing are pivotal in helping the immersive studies supplied using AR and VR. Here is how:

- *Actual-Time Rendering*: AR and VR packages require high-performance rendering of 3D graphics with minimal latency (Carvalho et al., 2021). Side computing offers vital computational energy near the consumer, essential for rendering visuals in real-time without delays that would disrupt the experience.
- *Reduced Latency*: Low latency is important in VR to maintain immersion and prevent motion sickness. Edge computing minimizes latency by processing statistics domestically, ensuring that user moves are meditated nearly instantly within the virtual environment.
- *Bandwidth Performance*: AR and VR applications regularly contain large quantities of information, consisting of high-resolution photographs and video streams. Edge computing facilitates manipulating these records more efficaciously by processing them regionally and reducing the quantity of data that needs to be sent over the community.

7.4 Smart Agriculture

IoT and area computing are revolutionizing agriculture with several superior solutions:

- *Precision Farming*: Sensors and IoT devices collect statistics on soil situations, weather styles, and crop fitness. Edge computing tactics these statistics in real-time to offer actionable insights for optimizing planting schedules, irrigation, and fertilization to accelerate crop yields and aid efficiency.

To sum up, the convergence of the IoT and edge computing represents a groundbreaking development with the ability to revolutionize how facts are processed and utilized by way of allowing actual-time facts analysis, improving consumer reviews, and fostering innovation. These technologies are poised to address complicated, demanding situations and create smarter, more linked environments. Integrating the IoT and side computing will open up new possibilities and usher transformative changes across various sectors as studies progress and technology matures.

References

Agiwal, M., Roy, A. and Saxena, N. (2016). Next Generation 5G Wireless Networks: A Comprehensive Survey. *IEEE Communications Surveys & Tutorials, 18*(3), 1617–1655.

Alwakeel, A.M. (2021). An overview of fog computing and edge computing security and privacy issues, *Sensors, 21*(24), 8226.

Alwarafy, A., K.A. Al-Thelaya, M. Abdallah, J. Schneider and M. Hamdi. (2020). A survey on security and privacy issues in edge-computing-assisted internet of things, *IEEE Internet of Things Journal, 8*(6), 4004–4022.

Angel, N.A., D. Ravindran, P.D.R. Vincent, K. Srinivasan and Y.C. Hu. (2021). Recent advances in evolving computing paradigms: cloud, edge, and fog technologies, *Sensors, 22*(1), 196.

Beckman, P. and C. Catlett (eds). (2020). *5G-Enabled Energy Innovation: advanced Wireless Networks Workshop for Science—Workshop Report*, Argonne National Laboratory (ANL), Argonne, IL, United States; Northwestern University, Evanston, IL, United States.

Bibri, S.E. (2023). Data-driven smart eco-cities of the future: an empirically informed integrated model for strategic sustainable urban development, *World Futures, 79*(7–8), 703–746.

Bilal, K., O. Khalid, A. Erbad and S.U. Khan. (2018). Potentials, trends, and prospects in edge technologies: fog, cloudlet, mobile edge, and micro data centers, *Computer Networks, 130*, 94–120.

Carvalho, G., B. Cabral, V. Pereira and J. Bernardino. (2021). Edge computing: current trends, research challenges, and future direction, *Computing, 103*, 993–1023.

Chae, B.K. (2019). The evolution of the internet of things (IoT): a computational text analysis, *Telecommunications Policy, 43*(10), 101848.

Domestics, P., et al. (2013). 5G on the horizon: key challenges for the radio-access network, *IEEE Vehicular Technology Magazine, 8*(3), 47–53.

Dong, Q., Y. Ma and J. Li. (2019). Load balancing oriented scheduling scheme in an edge computing network, *Application Research of Computer*, 1–6.

Gao, K.Z., Z.G. Cao, L. Zhang, Z.H. Chen, Y.Y. Han and Q.K. Pan. (2019). A review on swarm intelligence and evolutionary algorithms for solving flexible job shop scheduling problems, *IEEE/CAA Journal of Automatica Sinica, 6*(4), 875–887.

Guo, D., W. Wang and G. Zeng. (2015). A new cloud ware PaaS platform based on microservices architecture, *NETINFO SECURITY*, 11, 15–20.

Jia, X., Q. Feng, T. Fan and Q. Lei. (2012). RFID technology and its applications in the internet of things (IoT), in *Proceedings of the International Conference on Consumer Electronics*, pp. 1282–1285.

Khan, R., R. Zaheer and S.U. Khan. (2013). Future internet: the internet of things architecture, possible applications, and key challenges, in *Proceedings of the International Conference on Frontiers of Information Technology*. doi: 10.1109/FIT.2012.53.

Lee, D. and C.C. Cheng. (2016). Energy savings by energy management systems: a review, *Renewable and Sustainable Energy Reviews, 56*, 760–777.

Li, Y., M. Hou, H. Liu and Y. Liu. (2012). Towards a theoretical framework of strategic decision, supporting capability and information sharing under the internet of things, *Information Technology & Management, 13*(4), 205–216.

Lian, X. (2019). Research on edge computing unloading mechanism based on security management, M.S. thesis, Beijing University of Posts and Telecommunications.

Luan, H. and J. Leng. (2016). Design of energy monitoring system based on IOT,' in *Proceedings of the 2016 Chinese Control and Decision Conference (CCDC)*, Yinchuan, China, 6785–6788.

Ma, X. (2019). Research on trusted cooperative mechanism based on edge computing, M.S. thesis, Beijing University of Posts and Telecommunications.

Mo, Q., W. Song, F. Dai, L. Lin and T. Li. (2019). Development of collaborative business processes: a correctness enforcement approach, *IEEE Transactions on Services Computing*. doi: 10.1109/TSC.2019.2961346.

Siqi, D., L. Hailong and H. Lei. (2019). Task scheduling policy for mobile edge computing with user priority, *Application Research of Computers*, 1–5.

Wang, L., B. Ma and Q. Zhang. (2019). *Microservices Architecture and Practices*.

Wang, L., S. Greenberg, J. Fiegel, et al. (2013). Monitoring-based HVAC commissioning of an existing office building for energy efficiency, *Applied Energy*, *102*, 1382–1390.

Xu, X., X. Liu, Z. Xu, F. Dai and L. Qi. (2020). Trust-oriented IoT service placement for smart cities in edge computing, *IEEE Internet of Things Journal*, *7*(5), 4084–4091.

Yu, W., et al. (2018). A survey on the edge computing for the internet of things, *IEEE Access*, *6*, 6900–6919.

Yu, W., H. Xu, H. Zhang, D. Griffith and N. Golmie. (2016). Ultra-dense networks: survey of state of the art and future directions, in *Proceedings of the 25th International Conference on Computer Communication and Networks (ICCCN)*. doi: 10.1109/ICCCN.2016.7568592.

Yuan, H., J. Bi and M. Zhou. (2018). Temporal task scheduling of multiple delay—constrained applications in green hybrid cloud, *IEEE Transactions on Services Computing*. doi: 10.1109/TSC.2018.2878561, 2018.

Yuan, H., J. Bi and M. Zhou. (2019a). Multiqueue scheduling of heterogeneous tasks with bounded response time in hybrid green IaaS clouds, *IEEE Transactions on Industrial Informatics*, *15*(10), 5404–5412.

Yuan, H., J. Bi and M. Zhou. (2019b). Spatiotemporal task scheduling for heterogeneous delay-tolerant applications in distributed green data centres, *IEEE Transactions on Automation Science and Engineering*, *16*(4), 1686–1697.

CHAPTER 7

Network Infrastructure and Connectivity

Perarasi T.[1*] and *Manoj R.*[2]

The progress of the medical internet of things (MIoT) technologies has changed healthcare distribution through intelligent, mutual systems, requiring strong networks, infrastructure, and seamless connectivity. This chapter explores next-generation communication technologies such as fifth generation (5G) and beyond, high speed, ultra-wisdom, and low-oppression communication, telescargari, distance patient monitoring, and real-time diagnosis. It also highlights the supplementary use of low-power wide-area networks (LPWAN) for non-mutable but extensive monitoring needs, especially in remote or understanding areas. A comparative study between edge and cloud computing underlines trade—closes in delay, data privacy, and scalability—emphasizing hybrid architecture in the form of emerging criteria in MIOT implementation.

The chapter discusses the bandwidth and delaying requirements for various healthcare applications, expanding the harsh demands of mission-cultural medical use, and it describes periodic monitoring systems. The study of real-world case in diverse geographical regions demonstrates practical implementation, including smart ambulances, artificial intelligence (AI)–capable patient monitoring, cold-chain vaccine logistics, and detection for the elderly. The chapter ends with analytical insight into emerging technologies such as device density, power efficiency, and in AI, federated learning, and hybrid cloud-edge solutions. These developments shape a safe, scalable, and responsible ecosystem for intelligent health service services.

1. Introduction

Incessantly working toward a realizable implementation of real-time monitoring, diagnostics, treatment, and the like, the MIoT is bringing about a revolution in healthcare. An effective backbone network infrastructure thereby becomes a

[1,2] Department of Electronics and Communication Engineering, Bannari Amman Institute of Technology, Erode, Tamil Nadu, India.
Email: manojramesh2005@gmail.com
* Corresponding author: pera.sweety@gmail.com

DOI: 10.1201/9781003659822-7

transformer, making sure that the connectivity is uninterrupted, latency is within permissible limits, and data is transmitted with high reliability. This chapter presents the essentials concerning the building blocks for MIoT's network infrastructure, i.e., 5G, various LPWAN solutions, edge versus cloud paradigms, bandwidth, and latency requirements.

a. 5G and Future Connections in the MIoT Environments

The 5G technology is going to be the disruptive force in the MIoT arena, where ultra-low latency, very high bandwidth, and massive device connectivity are available (Agiwal et al., 2016; Al-Fuqaha et al., 2015; Parvez et al., 2018). These characteristics are at the very core of remote surgeries and real-time patient monitoring, speeding up the transmission of data between devices and healthcare providers.

b. Key Features of 5G in Healthcare

- **Ultra-Reliable Low Latency Communication (URLLC):** This kind of channel is for the mission-critical applications that require ultra-low responses, such as telesurgery or emergency response.
- **Enhanced Mobile Broadband (eMBB):** This channel supports high-definition video consultation and transferring huge files like magnetic resonance imaging (MRI) scans.
- **Massive Machine-Type Communications (mMTC):** This supports the concurrent connection of a very large number of medical devices.

c. Virtual-World Applications

- **Remote Patient Monitoring:** 5G facilitates uninterrupted measurement of patient vitals via wearables along with timely intervention.
- **Connected Ambulances:** In Milan, 5G-connected ambulances permit paramedics to transmit patient data during transit so that hospitals can be prepared for arrival (Li et al., 2019).
- **Telemedicine and Telesurgery:** This umbrella term potentially covers remote consultation and even robotic surgery over a high-speed and low-latency connection.

1.1 Low-Power Wide-Area Network for Medical IoT

LPWANs are intended for long-range communications at very low power and hence are suited for MIoT devices that need transaction data rarely and over long periods (Augustin et al., 2016; Centenaro, 2016; Petäjäjärvi et al., 2015; Raza et al., 2017).

a. Advantages

- Battery Life: Can run on a battery for years without requiring to be replaced.
- Coverage Area: Works both within cities and in rural settings, and in tricky environments like underground storage.
- Cost-Effective: Compared to costly cellular networks, its deployment and operations are very cheap.

b. Use Cases

- Asset Tracking: Monitoring the location and condition of medical equipment and supplies in various facilities.

- Environmental Monitoring: Measuring temperature and humidity pertinent for medical supply storage.
- Patient Compliance: Connected pill dispensers that monitor and promote adherence to medication.

1.2 Edge Computing vs. Cloud Computing in MIoT

Edge computing places the computational facility near the data source, i.e., it is closer to the data-generating point to process the data, whereas cloud computing allows the data to reside in distant servers.

- The decision between on-the-edge and off-the-edge cloud processing has major consequences on the performance, reliability, and security of MIoT deployments.

a. Edge Computing

- Near Processing: Data is processed at the source to reduce delay time (Bonomi et al., 2012).
- Privacy Preserving: The private data analyzing in edge can exactly without data send to the cloud directly.
- Reliability: Does not rely on a stable long stretching internet connection (Satyanarayanan, 2017).

b. Cloud Computing

- Elastic: Scales up or down with the load of data and processing power.
- Centrally managed minder overhead in updates and maintenance.
- Advanced Analytics: Ability to use heavy dose of computing power to dig into large volumes of data (Chen et al., 2012; Dinh et al., 2013; Mao et al., 2017) .

c. Implemented Work

- Smart Edge: The difference shows in Table 7.1 a hybrid edge and cloud computing system for diabetes prediction with low latency and higher accuracy (Zanella et al., 2014).
- Smart health IoT: Fog-supported real-time cardiac monitoring and care.

Table 7.1 Edge computing vs. cloud computing.

Feature	*Edge Computing**	*Cloud Computing***
Latency	Low	Higher
Data Privacy	High	Variable
Connectivity Needs	Minimal	Constant
Scalability	Limited	High
Maintenance	Decentralized	Centralized

Source: * Yang et al. (2014); ** Kim et al. (2017).

1.3 Bandwidth and Latency Considerations

An optimal MIoT operation depends largely upon satisfying certain bandwidth and latency constraints such that the data can be relayed in a timely manner and correctly.

a. Bandwidth Considerations

- High-Definition Imaging: Hundreds of megabytes must be transmitted for an MRI or computed tomography (CT) scan.
- Video Visits: Reliable, good quality video streams require a constant supply of bandwidth.

b. Latency Considerations

- Real-Time Monitoring: Monitoring of vital signs requires low-latency systems to quickly flag anomalies.
- Emergency Response: Real-time data transfer is required in emergencies to aid quick medical therapy.

c. Standards

- 5G URLLC: Aims for latency of 1 ms, for ultra-reliable, low-latency communications.
- LPWAN: Only suitable for non-time-critical use cases with latencies in the seconds, for support cut-off (Giordani et al., 2020).

1.4 5G and Prospective Connectivity in MIoT

The MIoT is composed of the large network of interrelated medical devices and systems which acquire, transmit, and handle health-associated data. For MIoT to be effective, the technology arc must support and have the capacity to accommodate high data throughput, URLLC, scalability to massive connectivity, and ubiquity—a set of attributes that aligns with those of 5G and beyond.

5G wireless networks will be the next major phase of mobile telecommunications and will bring us:

- Data rates of up to 10 Gbps
- Low, as in 1 millisecond, not as in being a hipster. Must be a hipster.
- Support for over 1M devices per km^2

5G specifically can be well-suited for healthcare-applications that are latency-sensitive, require connectivity of a lot of devices, and are mission-critical, due to its properties. Next-generation connectivity solutions like 6G, NTNs (including satellite and aerial platforms), and private 5th Generation networks are likely to bring these capabilities even closer to edge, particularly in remote and underserved areas.

2. Literature Review and Analysis

An increased body therapy of global research literature confirms an important role of 5G and future connectivity technologies in enhancing IoT capabilities. The following discussion underlines the major technical insights from the recent publications of scholars, which is classified by theme.

- Ultra-wisdom in medical applications: Zhang et al. (2019) examined the integration of the URLLC in telesurgery scenarios. Their test simulation, incorporating robot surgical weapons connected through 5G, maintained a delay under 2 ms, which is within the safe range for real-time response-based medical procedures. Key Tech Use: Low delay and high reliability allows life-saving tasks such as distance surgery, closed-loop insulin pumps, and defibrillator response systems to work with confidence on wireless infrastructure.

- Network slicing for healthcare priority: Sharma and Park (2020) expanded how the network slicing in 5G allows separate virtual networks (slices) for separate MIoT applications. For example, critical care telemetry can be allocated high-percentile, low-oppression slices, while inventory tracking can use a best-time slice. Technical Points: Each slice supports optimized quality of service (QOS) parameters—delay, nervousness, and reliability. Software defined networking (SDN) and network function virtualization (NFV) were obtained. Implications: Initial research can lead to immediate diagnosis, digital avatar for patient tracking and multi-sensitive immersive health education (Giordani et al., 2020).
- Edge-AI over 5G: Wang et al. (2023) model integrated AI conclusion on 5G network edge to predict the incidence of heart failure. This system used the wearable electrocardiogram (ECG) and oxygen saturation (SpO2) devices to stream data for a multi-pain edge computing (MEC) node. Outcome: The model acquires real-time classification with <5 ms delay, ensuring timely alert based on cloud delay.

3. Analytical Approach

Table 7.2 Core performance metrics for MIoT.

Metric	*Requirement for MIoT*	*5G Capability*
Latency	$\leq$ 10 ms for critical care	~1 ms (URLLC)
Data Rate	$\geq$ 1 Mbps for video streams	Up to 10 Gbps (eMBB)
Connectivity Density	10^4–10^6 devices/km^2	10^6 devices/km^2 (mMTC)
Reliability	> 99.999%	Achievable via slicing & QoS

This section presents the *quantitative evaluations and architectural considerations* that highlight the fit of 5G in the MIoT context.

Table 7.3 Quantitative evaluations and architectural considerations.

Metric	*Requirement in MIoT*	*5G Capability*	*Impact*
Latency	$\leq$ 10 ms (e.g., robotic control, emergency alerts)	~1 ms (URLLC)	Enables real-time diagnostics and surgeries
Data Rate	1–10 Mbps (telemedicine) to >1 Gbps (HD video surgery)	Up to 10 Gbps (eMBB)	Supports immersive remote diagnostics, augmented reality (AR)/virtual reality (VR) in surgery

Contd.

Table 7.3 *Contd.*

Metric	*Requirement in MIoT*	*5G Capability*	*Impact*
Device Density	10^4–10^6 per km^2 (hospital networks)	10^6 devices/km^2 (mMTC)	Scalability for dense device environments
Reliability	99.999% (ICU monitoring, infusion control)	99.999% with redundancy and slicing	Ensures continuous, fault-tolerant operation
Mobility	Ambulances, air rescue drones, mobile hospitals	Seamless handover at 500 km/h	Continuous connectivity in transit scenarios

a. Connectivity Models

- Network Slicing: A way to virtually divide the 5G network to allocate resources to certain medical services.
- Edge Federation: Federating health data processing to be closer to the patient.
- MEC: Low latency diagnostics, local AI inference enabling end-device control or device evolution.

b. Limitations Identified

- Limited 5G in rural areas and low-income countries.
- 5G deployment costs and compatibility with legacy system.
- Security and privacy in a multi-user hospital scenario.

3.1 Connectivity Model in 5G MIOT Network Slicing: Logical Separation of Physical Network Using SDN/NFV

1. Outpatient Appointment Scheduling – Maximum latency, Best-Effort Slice 1: Monitoring of ICU patients—high reliability, low delay. Slice 2: appointment scheduling—Macym Village, Best-Prayer. Slice 3: Health Data Ancillary—high bandwidth, tolerant for delay.
2. Multi-Access Edge Computing (MEC): Localizes the processing power close to data sources (eg, hospital nodes or mobile health vans). MEC nodes reduce backhaul and enable local AI decision data pre-processing and filtering increased data privacy.
3. Device-to-device (D2D) Communication: The device (e.g., wearable defibrillator and important monitors) can communicate more than 5G directly without rooting through base stations. This is useful in high-density hospital floors or emergency areas.

5G is currently limited to urban or semi-urban centers in developing countries. It has the following challenges:

- Energy Overheads: Continuous connectivity with a battery of high throw but drains in wearable medical devices until optimized with energy-skilled protocols.
- Privacy and Safety: Data division in slices and MEC nodes require strong encryption and access control mechanisms.
- Interoperability: There is a need for harmonious standards between 5G network providers and medical equipment manufacturers.

3.2 Practical Implementation and Case Study

a. **Remote telesurgery** (China Telecom, Huawei)
   - We further implement 5G-enabled robotic telesurgery over 3,000 km.
   - Result: Latency of $\leq$ 10 ms was realized, the video stream was stable, and control was sensitive.
   - Impact: Expanded surgical knowledge into isolated areas where specialist surgeons do not exist.

b. **Smart ambulance** in the EU (H2020 5G-HEART project)
   - The application: infotainment services used in cars measured on the 5.9GHz band. Implementation: The on-board unit used in cars upgraded to 5G radios running at 5.7Hz with vehicle-to-vehicle (V2V) systems and information technology (IT) infrastructure of cars and active ambulances with connected 5G routers and IoT sensors.
   - Scenario: Live video transmission of vitals to emergency room (ER) during transport.
   - Outcome: Decreased preparations for ER were cut by 23 percent.

c. **AI Monitoring for Patients** (Samsung Medical Center, Korea)
   - Architecture: Edge-AI health data monitoring through 5G parties communicating their data through URLLC to a fifth-generation cellular network.
   - Outcome: Detected early signs of SIDS with 98 percent accuracy.
   - Tools and Integration: AI, edge servers, 5G New Radio (NR).

Table 7.4 Application domains.

Domain	*5G MIoT Use Case*	*Connectivity Role*
Telemedicine	Real-time video consultations	High bandwidth, low latency
Emergency Response	Connected ambulances, drones	Mobility, fast data transmission
Chronic Care	Remote patient vitals monitoring	Device density, periodic updates
Inpatient Care	Smart beds, infusion pumps	Local edge networks, reliability
Public Health	Surveillance & outbreak tracking	Massive machine-type communication (mMTC)

3.3 Application Domain Telemedicine and Distance Consultation: The Role of 5G

Real-time HD video with minimal delay makes the doctor–patient conversation comparable to in-patient discourse, e.g., portable ultrasound in India and Italy and COVID-19 mobile units with real time environment. 5G can play a crucial role in emergency medical services 5G role, monitoring ECG, blood pressure (BP), glucose levels enroute to hospitals. It is particularly important for stroke and cardiac patients. For instance, ambulance-to-ER telemetry in South Korea reduced cardiac arrest mortality by 15 per cent. Smart hospitals and infrastructure rely on 5G role to operate smart beds, infusion pumps, and environmental sensors that connect thousands of equipment (Mumtaz et al., 2014). Central command centers can manage operations, e.g., the Samsung Medical Center (Seoul) uses 5G to coordinate lifts, Heating,

Ventilation, and Air Conditioning systems (HVACs), and emergency alerts in wards, as depicted in Figure 7.1.

5G makes possible care and chronic illness management through continuous monitoring with wearables that triggers alert on discrepancies. This can support freedom for aging population. In Japan there are 5G-connected home health kits for patients with diabetes and high blood pressure, with cloud logging and remote intervention. 5G enables preparation and monitoring of epidemics, by feeding real-time data from testing booths, portable labs, and drones for contactless delivery or spraying, For instance, Wuhan's COVID-19 response included 5G robot nurse and 5G-managed smart quarantine centers.

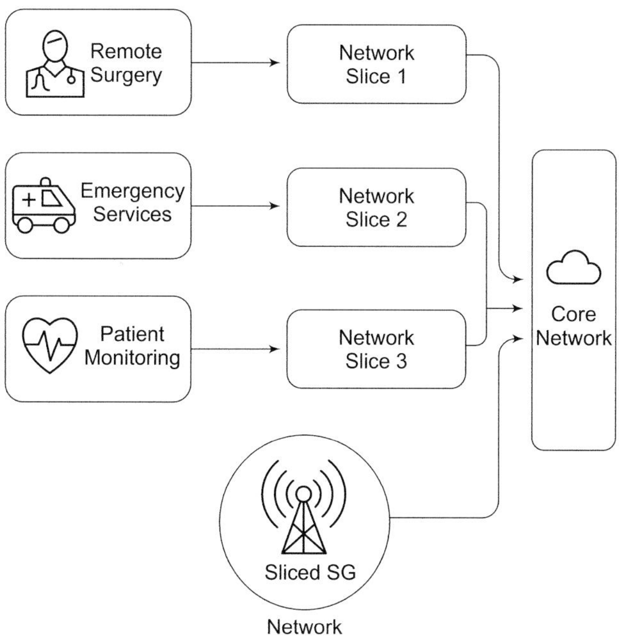

Fig. 7.1 5G Network slicing for healthcare.

3.4 Low-Power Wide-Area Networks (LPWANs) for MIoT

The MIoT ecosystem includes a diverse range of connected health equipment—such as wearables, employee and remote sensors—that produce continuous data. While high-speed networks such as 5G are essential for delayed applications, LPWANs are inevitable for battery-powered, long-term, and geographically scattered health-negative systems. LPWAN technologies such as long-range wide-area network (LoRaWAN), narrowband IoT (NB-IoT), Sigfox, and Long-Term Evolution for Machines (LTE-M) enable ultra-low power consumption and provide long distance communication, making them ideal for rural or underserved areas. Chronic disease monitoring, hospital asset and personnel tracking, and epidemiological sensing within smart city environments. These networks are usually designed for small data packets, unintentional broadcasting and live equipment usage in previous years on the same battery, thus reducing the infrastructure burden of the deployment of MIoT and the cost of maintenance.

4. Technical Overview

Table 7.5 Key characteristics of LPWANs for MIoT.

Characteristic	*LPWAN Value*	*Implication for MIoT*
Range	2–40 km (urban/rural)	Monitoring patients and environments in remote areas
Power Consumption	10–100x lower than Wi-Fi/Cellular	Long battery life (up to 10 years) for wearables
Bandwidth	0.3–50 kbps	Suitable for vital signs, alerts, metadata transmission
Device Capacity	100,000+ devices per gateway	Hospital-scale and city-wide patient/environment coverage
Cost	Low infrastructure & operating costs	Enables low-income regions to deploy smart health systems

Table 7.6 Major LPWAN technologies in MIoT.

Technology	*Frequency Band*	*Power Usage*	***Range***	***Notable Features***
LoRaWAN	Sub-GHz (868/915 MHz)	Very low	2–15 km	Unlicensed, bidirectional, adaptive data rates
NB-IoT	Licensed LTE bands	Low	1–10 km	Supports mobility, cellular-grade reliability
Sigfox	Sub-GHz ISM	Ultra low	3–50 km	Star topology, very limited payload
LTE-M	LTE Band (Cat-M1)	Moderate	1–10 km	Higher throughput than NB-IoT, better mobility support

4.1 Practical Implementations in Healthcare

Integration of LP WAN in healthcare systems is changing the distribution of medical services, especially in terms of distance monitoring, emergency warning, and operational efficiency. What follows is an overview of the actual deployment of LPWAN technologies and their practical implications in the MIoT domain (Hummen et al., 2014; Misra et al., 2016). The flow of healthcare monitoring using LPWAN and MIoT devices is shown in Figure 7.2.

a. Monitoring of Distance Patient in Rural Areas

Use of Technology: LoRaWAN

Place: Andhra Pradesh, India

Use Case: Chronic disease management in low-connectivity fields

Description: With cooperation between the Andhra Pradesh state government, Tata Trusts, and Stanford University, LoRaWAN-Saksham health monitoring stations

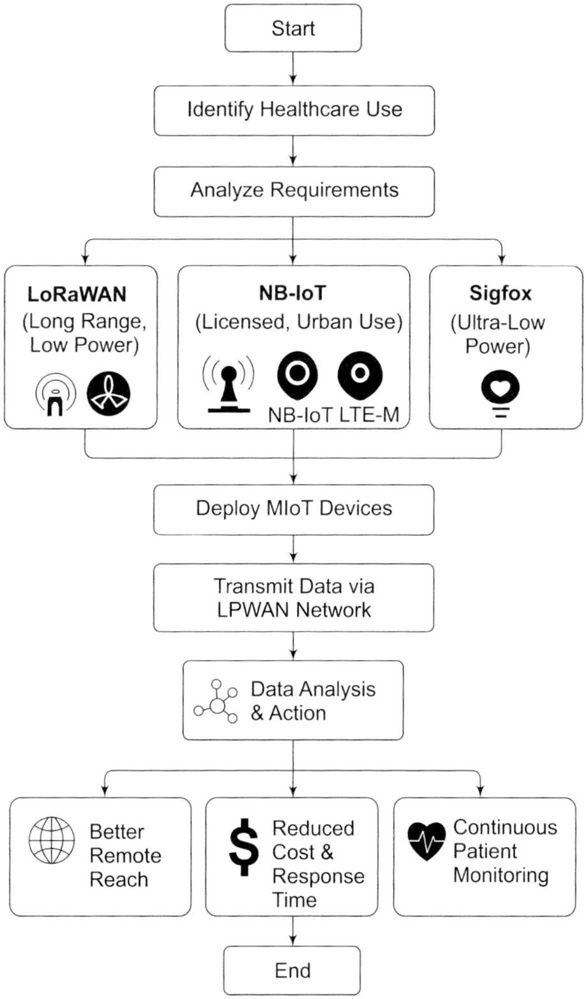

Fig. 7.2 Healthcare monitoring using LPWAN and MIoT Devices.

were deployed in rural districts (2019–2021). Patients with chronic diseases such as hypertension and diabetes were linked with Bluetooth-enabled BP monitors and glucose meters, which were connected to a LoRaWAN node in their community health center.

Functions: The devices transmit compressed data (average 2 kb) once or twice a day. Data travels to a LoRaWAN gateway climbed on telecom towers. This is then sent to the state health cloud via the internet.

Result: There was a 60 percent decrease in unnecessary hospital tours, 23 percent improvement in drug follow, and a reduction in the cost of travel for patients in remote villages.

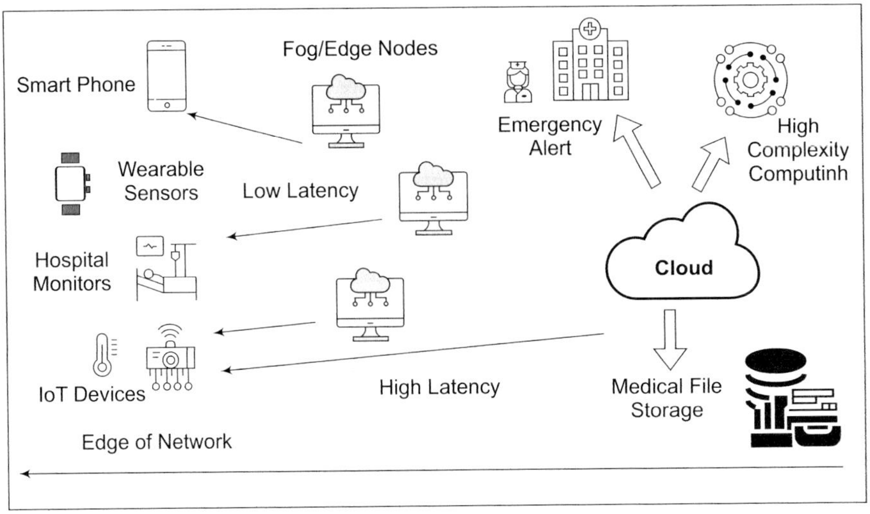

Fig. 7.4 Healthcare in IoT architecture.

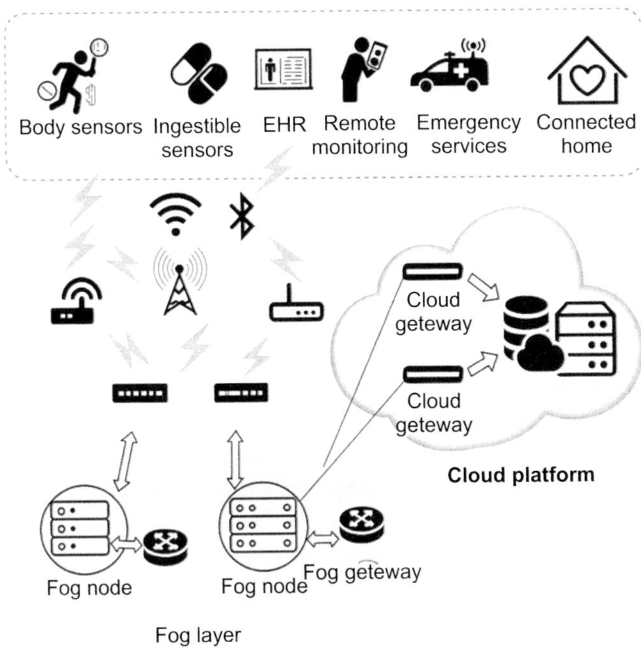

Fig. 7.5 Fog and cloud computing assisted IoT-enabled healthcare monitoring system.

alert generation, while the cloud stores long-term data, involves training of AI models, and supports regulatory compliance. For instance, edge computing can detect environmental triggers to alert an asthma patient to use an inhaler. Periodic summary data can be sent to the cloud for the doctor's review and public health tracking. The cloud architecture flow is shown in Figure 7.5.

5.1 Benefits and Challenges

Edge computing enables offline functionality (e.g., rural clinics with incredible internet), ensures immediate intervention in important events, and increases privacy and data sovereignty for patients. Cloud computing in the MIoT enables multi-device synchronization. The AI model provides centralized data lakes for development. Cross-institutional research allows for cooperation. The challenges of edge computing include calculation of limited batteries and resources, difficulties in updating or patching software, and device standardization and compatibility issues. The challenges of cloud computing include: emergency use-mammals, risk of data violations during broadcast, and infrastructure for stable internet.

5.2 Use Case

Table 7.10 Use case of different computing

Use Case	*Preferred Solution*	*Reason*
Seizure detection via brain implant	Edge	Requires sub-second response
AI-driven population health analytics	Cloud	Needs access to large-scale longitudinal data
Diabetes management via insulin pumps	Hybrid	Real-time monitoring + centralized updates
Health alerts in disaster-hit zones	Edge	Limited or no connectivity
Cloud-based hospital EMR integration	Cloud	Interoperability across systems

5.3 Future Trends

- **Federated Learning:** AI models trained on the edge using local data and then aggregated in the cloud—preserving privacy.
- **TinyML:** Machine learning models optimized for microcontrollers will enable more advanced edge-based analytics in small devices.
- **Edge AI Hardware:** New chipsets like Google's Edge TPU and NVIDIA Jetson Nano are enabling healthcare-grade AI inference on portable devices.

6. Conclusion

The MIoT represents a paradigm change in healthcare through real-time monitoring, intelligent diagnosis, and distance treatment. The incorporation of advanced communication frameworks, particularly 5G and LPWAN, underpins this development and facilitates the realization of various medical use cases aligned with critical healthcare needs. While 5G ensures high-power and low-latency connectivity for important applications such as telesurgery and emergency response, LPWAN supports long-term, low-power operations ideal for chronic care and rural deployment.

Edge computing enhances real-time accountability and data privacy by processing the information close to the source, while cloud computing facilitates large-scale analytics, historical data management, and AI model training. The hybrid models integrating both paradigms have become rapidly prevalent, providing speed, scalability and flexibility.

Through comprehensive analysis and examples from the real world, this chapter outlines the need for flexible, safe, and cost-effective network infrastructure in providing quality healthcare in different contexts. The future of the MIoT will depend on adaptive architecture by rapidly incorporating age intelligence, federated learning, and A-optimized hardware. These technologies will help remove current challenges in connectivity, power consumption, and interoperability, which can lead to a more inclusive and efficient digital health ecosystem.

References

Agiwal, M., A. Roy and N. Saxena. (2016). Next generation 5G wireless networks: a comprehensive survey, *IEEE Communications Surveys & Tutorials*, *18*(3), 1617–1655.

Al-Fuqaha, A., M. Guizani, M. Mohammadi, M. Aledhari and M. Ayyash. (2015). Internet of things: a survey on enabling technologies, protocols, and applications, *IEEE Communications Surveys & Tutorials*, *17*(4), 2347–2376.

Augustin, A., J. Yi, T. Clausen and W.M. Townsley. (2016). A study of LoRa: long range & low power networks for the internet of things, *Sensors*, *16*(9), 1466.

Bonomi, F., R. Milito, J. Zhu and S. Addepalli. (2012). Fog computing and its role in the internet of things, in *Proceedings of the Workshop on Mobile Cloud Computing*, pp. 13–16.

Centenaro, M., L. Vangelista, A. Zanella and M. Zorzi. (2016). Long-range communications in unlicensed bands: the rising stars in the IoT and smart city scenarios, *IEEE Wireless Communications*, *23*(5), 60–67.

Chen, D., H. Jin and L. Hu. (2012). Data security and privacy protection issues in cloud computing, in *IEEE International Conference on Computer Science and Electronics Engineering (ICCSEE)*.

Dinh, H.T., C. Lee, D. Niyato and P. Wang. (2013). A survey of mobile cloud computing: architecture, applications, and approaches, *Wireless Communications and Mobile Computing*, vol. *13*(18), 1587–1611.

Giordani, M., M. Polese, M. Mezzavilla, S. Rangan and M. Zorzi. (2020). Toward 6G networks: use cases and technologies, *IEEE Communications Magazine*, *58*(3), 55–61.

Granjal, J., E. Monteiro and J.S. Silva. (2015). Security for the internet of things: a survey of existing protocols and open research issues, *IEEE Communications Surveys & Tutorials*, *17*(3), 1294–1312.

Hummen, R., H. Shafagh, S. Raza, T. Voigt and K. Wehrle. (2014). Delegation-based authentication and authorization for the IP-based internet of things, in *Proceedings of the IEEE International Conference on Sensing, Communication, and Networking*.

Kim, J., A.M. Campbell and S. de Albuquerque. (2017). Wearable sensor-based mobile health monitoring system, *Sensors*, *17*(5), 1060.

Li, B., Z. Fei and Y. Zhang. (2019). UAV communications for 5G and beyond: recent advances and future trends, *IEEE Internet of Things Journal*, *6*(2), 2241–2263.

Mao, Y., C. You, J. Zhang, K. Huang and K.B. Letaief. (2017). A survey on mobile edge computing: the communication perspective, *IEEE Communications Surveys & Tutorials*, *19*(4), 2322–2358.

Misra, S., R. Maheswar and N. Saha. (2016). Security challenges and approaches in internet of things, in *Proceedings of the International Conference on Communications and Signal Processing (ICCSP)*.

Mumtaz, S., J. Rodriguez and L. Dai. (2014). *Smart Device to Smart Device Communication*, Springer.

Parvez, I., A. Rahmati, I. Guvenc, A.I. Sarwat and H. Dai. (2018). A survey on low latency towards 5G: RAN, core network and caching solutions, *IEEE Communications Surveys & Tutorials*, *20*(4), 3098–3130.

Petäjäjärvi, J., K. Mikhaylov, A. Roivainen, T. Hänninen and M. Pettissalo. (2015). On the coverage of LPWANs: range evaluation and performance analysis, in *Proceedings of the International Telecommunication Networks and Applications Conference*.

Rappaport, T.S., Sun, S., Mayzus, R., Zhao, H., Azar, Y., Wang, K., Wong, G.N., Schulz, J.K., Samimi, M. and F. Gutierrez Jr.,"Millimeter Wave Mobile Communications for 5G Cellular: It Will Work!" *IEEE Access*, vol. 1, pp. 335–349, 2013.

Rappaport, T.S. et al. (2013). Millimeter wave mobile communications for 5G cellular: it will work!, *IEEE Access*, *1*, 335–349.

Raza, U., P. Kulkarni and M. Sooriyabandara. (2017). Low power wide area networks: an overview, *IEEE Communications Surveys & Tutorials*, *19*(2), 855–873.

Satyanarayanan, M. (2017). The emergence of edge computing, *Computer*, *50*(1), 30–39.

Sharma, P. and Park, J. "A blockchain-based smart home gateway network architecture," *Human-centric Computing and Information Sciences*, vol. 10, no. 1, 2020.

Yang, G., L. Xie, M. Mantysalo, et al. (2014). A health-IoT platform based on the integration of intelligent packaging, unobtrusive bio-sensor, and intelligent medicine box, *IEEE Transactions on Industrial Informatics*, *10*(4), 2180–2191.

Zanella, A., N. Bui, A. Castellani, L. Vangelista and M. Zorzi. (2014). Internet of things for smart cities, *IEEE Internet of Things Journal*, *1*(1), 22–32.

Table 8.1 *Contd.*

Reference	*Focus Area*	*Key Contribution*	*Limitations*
Khalil et al. (2022), Ogunkan & Ogunkan (2025)	IoT & AI in smart cities	Real-time data collection and AI-based insights	Latency, centralization, data privacy issues
Yoo (2021)	Blockchain for decentralization	Secure and tamper-proof data handling and transparency	Scalability and integration challenges
Siddiqui et al. (2024), Ullah et al. (2023)	National blockchain Strategies	Real-life examples in Estonia and Dubai	Policy-dependent, domain-specific deployments
Correia et al. (2023), Harrison & Donnelly (2011)	Blockchain in energy trading	Facilitated a peer-to-peer solar energy market	Regulatory hurdles, lack of scalability
Atzori et al. (2010)	Mobility systems	Trust and transparency in car share mobility	Limited integration with legacy systems
Batty et al. (2012)	Privacy in smart cities	Frameworks for privacy-sensitive data use	Implementation complexity, user acceptance
Santos et al. (2025)	Blockchain applications review	Taxonomies and syntheses of blockchain use cases	Mostly conceptual, lacks implementation depth
Kitchin, (2013), Tyagi (2024a)	Integrated urban frameworks	Proposed holistic frameworks for future smart cities	Early-stage, theoretical models need validation

Organization of this work: The chapter explains the BIDA framework and its use in decentralized city networks. Section 2 describes how centralized computing developed and the problems it creates in current smart city settings. Section 3 covers the basics of blockchain technology, showing that its spread and unchangeable qualities fit urban infrastructure. Section 4 explores why BIDA came about, pointing to a need for trust, self-rule, and compatibility in city systems. Section 5 explains how the cloud of things (CoT) works in modern cities—it shows that connecting the CoT with blockchain builds adaptable infrastructure that grows. Section 6 addresses the problems of central systems in smart city places. It explains why moving to decentralization seems right. Section 7 shows how decentralization improves network speed, toughness, and citizen control in smart cities. Section 8 describes how blockchain and the CoT work together to create city networks that think and react better—it stresses clear data and safety. Section 9 shows real uses for BIDA in important city sectors, such as energy, transport, trash handling, pollution checks, and city rule. Section 10 looks at how BIDA and the CoT make IoT data safer, especially for city structures like smart traffic systems. Section 11 lists the issues that still face BIDA when people try to put it into practice, which include problems with size, privacy dangers, tough rules, and the absence of laws. Section 12 concludes the chapter, summing up what BIDA does for smart cities of the future and pointing out aspects where more study and teamwork are needed.

2. Centralized Computing in Today's Era

Centralized computing continues to be the primary structure of most information systems. Businesses and public services use these systems. In a centralized computing model, a single server, or a group of servers, handles data—it processes, stores, and manages this data. Client devices simply send requests or get information. People prefer this model because it is simple, affordable, and easy to run. Organizations monitor, protect, and upgrade centralized systems easily. This happens because one point controls everything. From hospital record systems to ride-sharing apps and cloud-based business software, most digital services depend on centralized designs.

Centralized computing sets up a hub-and-spoke model. The central hub, often a data center or cloud platform, performs heavy computational work. It manages databases and enforces security rules. Users connect to this central node through the internet or private networks; they interact through light devices such as phones, tablets, or office computers. The central server takes input, works on it and sends back the result. Users often do not know about the complex process—this method lessens the work on end devices and lets a central unit govern sensitive operations like data backups, access control, and software updates (Reyna et al., 2018). These are some common features of centralized computing systems

- Single point of control: A single place manages all decisions and operations.
- Governance setup: Rules and security standards apply the same way everywhere.
- Resource Utilization: Resources gather in one place, so it is easier to use them well.
- Connectivity Dependency: Users must connect to the central server to use all features.

This model has limits, which show up more as data grows in volume, complexity, and sensitivity. One major worry is the single point of failure. If the central system stops working because of a cyberattack, a power loss, or an internal error, the entire network of users can stop. Central data stores can draw hackers. That raises worries about privacy, misuse, and unauthorized viewing. People care more about data rights and trust now, especially in health and finance. As a result, many see this design as a problem, not a solution.

Despite its problems, centralized computing is still dominant because it works with old legacy systems (see Figure 8.1). Many government services, financial groups, and large businesses still use it. This occurs partly because of rules and partly because they do not want big system changes. Cities and systems become more complex and connected. The bad points of this model may become larger than the good points—this growing problem is making people look at decentralized options like blockchain. In blockchain, trust, resilience, and clarity are built into the design itself. That creates the basis for city structures that plans like BIDA envision.

Closed data access with decentralized computing grows in cities. Security along with control over data is a challenge. Centralized systems have one group that runs the data. But decentralized architectures spread control among many nodes. This

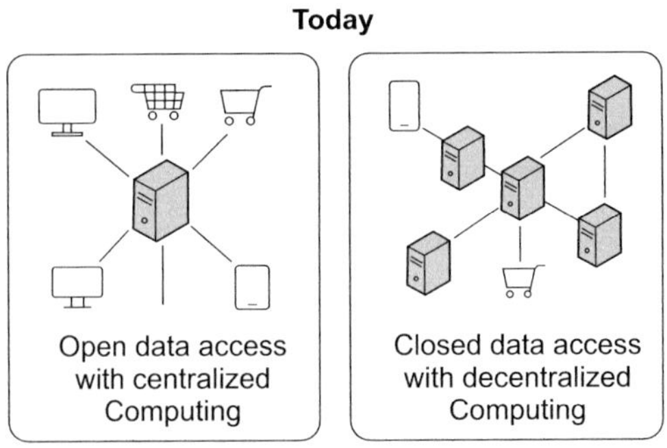

Fig. 8.1 Existing cloud of things infrastructure.

lowers the chance of data leaks and censorship. In cities, this model lets people, local leaders, and separate systems handle their own data, while still working inside a larger connected structure. Smart traffic systems, decentralized energy grids, and local health monitors can all work alone (Khan et al., 2020). Yet they also work with others, which keeps them steady and less reliant on a central power. This way of doing things supports data control promotes new ideas because decisions happen locally and people see more.

3. Blockchain Technology

Blockchain technology first started to gain popularity in 2008. It became the base for Bitcoin and showed a secure, decentralized, and tamper-resistant way to record digital transactions. The core of blockchain is a distributed ledger. A network of nodes holds this ledger, with each node verifying and storing data through cryptographic ways like hash functions in addition to Merkle trees. Proof of work (PoW) algorithm is used, and blockchain secures data so that it cannot be changed after it is recorded, unless the network agrees. Cryptocurrency was first major use of blockchain, but as the structure of blockchain grew, it became a transforming tool for many sectors, especially in modern city systems and smart city uses.

In cities, large amounts of data come from sensors, IoT devices, mobility platforms, and utility networks. Trust, security, and transparency become important in this context. Blockchain provides a reliable trust model by creating a permanent record of transactions and activities. This record is verifiable and can be checked. For example, data from smart traffic systems or energy grids can be stored safely on a blockchain ledger—this stops unauthorized changes and lets all people involved, like citizens, service providers, or regulators, get consistent information. This distributed way of operation removes the need for one central authority or a single point of failure, which matters for building durable urban infrastructure that serves citizens. A main benefit of blockchain is how it can be put to use (Conoscenti et al., 2016). There are three types of blockchains:

- Public blockchains: Anyone can join the network. This makes it good for open data applications like land title records or tracking public spending.
- Private blockchains: One organization controls these, making them suitable for internal government processes or business applications.

- Consortium blockchains: A group of trusted organizations share control. This works well for cooperation between agencies, for example, in waste management logistics or emergency services.

Beyond its structure, blockchain uses asymmetric encryption with public and private keys. Every person in the network has a private key, which only they know, and a public key, which they share openly. These keys sign and check transactions, which makes data private, whole, and real. In smart city systems, this means that a citizen can interact with city services securely. For instance, paying utility bills, verifying digital identity, or getting health records happens without revealing private information to others who should not see it.

Every transaction on the blockchain gets a timestamp. It also includes other data like the identities (or pseudonymous addresses) of participants, the transaction value, and other relevant information. The network groups these transactions into blocks and checks them through consensus mechanisms. These could be PoW, proof of stake (PoS), or newer, energy saving models. For smart cities, consensus makes sure that updates to data, such as real-time transit status or water usage analysis, are correct, synchronized, and free from tampering. If someone tries to change any transaction, the rest of the network detects the difference and rejects the change, which keeps the ledger sound (Sharma & Park, 2018).

In essence, blockchain offers a strong layer of trust and decentralization in city systems, supporting secure data exchange, clear governance, and effective service delivery. Its unchanging nature and cryptographic sealing make it hard to tamper with data. Moreover, its decentralized design improves how well systems hold up. Blockchain helps create smarter, more inclusive, and more trustworthy cities. This includes managing decentralized energy grids and verifying digital identities. It also helps with clear urban planning and public service accountability. As urban problems grow, putting blockchain into the digital structure of cities will be key to growing sustainability, fairness, and new ideas. Today's cities quickly change into intricate cyber-physical systems. Information from millions of sensors, devices, and services support everything from traffic control and energy management to public safety and trash removal. As this digital change speeds up, cities confront important problems concerning data honesty, how well systems work together, and public confidence. Old urban data systems usually stay separate, with a central group managing them, and they can fail from one point or from someone changing them (Kang et al., 2017). When various groups—city governments, private sellers, citizens, and self-operating systems—work together through common digital structures, it gets harder to hold people accountable and show how things work. These problems show that a basic data structure is needed that is safe, shows if anyone changes it, and can be managed together by different groups.

BIDA solves this problem by suggesting a blockchain structure. This structure helps decentralized urban systems. It differs from general blockchain solutions because BIDA considers the specific needs of smart cities—it supports sensor data that comes in quickly, helps semi-trusted groups agree well, and works with IoT in addition to AI services. BIDA wants to protect urban data from changes and unauthorized access. It also helps new kinds of shared work among decentralized groups. For example, BIDA lets several city departments, outside contractors, and people living in the city work together on a shared, checkable record, without needing a central group to oversee them. BIDA helps cities get a trust structure that grows, withstands problems, and runs fair, accountable, and lasting digital urban services.

4. Motivation

Today, with BIDA, we aim to redefine urban living by harnessing cutting-edge blockchain technology. The motivation behind this initiative stems from the urgent need to address contemporary urban challenges, such as overpopulation, resource scarcity, inefficient governance, and economic inequality, while simultaneously creating sustainable, inclusive, and smart urban environments tailored to modern lifestyles.

Traditional city models often struggle with centralized systems, which frequently result in bureaucratic delays, lack of transparency, and inequitable distribution of resources. At its core, BIDA seeks to decentralize control through blockchain technology, enabling secure, transparent, and efficient management of city resources and services. This decentralization empowers residents by granting them direct participation in governance and decision-making processes, thereby fostering accountability and trust within communities.

To ensure affordability, BIDA integrates blockchain with innovative urban planning and resource management strategies. This integration significantly reduces operational costs and leverages peer-to-peer networks to optimize critical services such as energy, water, housing, and public infrastructure. Consequently, quality urban living becomes accessible to a broader population, helping to dismantle the social and economic barriers that typically marginalize vulnerable groups.

Moreover, BIDA addresses the pressing environmental imperative by promoting eco-friendly infrastructure and sustainable resource cycles powered by transparent blockchain protocols. This approach not only reduces the carbon footprint of urban areas but also fosters resilient cities capable of adapting to future challenges. For the modern generation—which demands flexibility, digital connectivity, and ethical governance—BIDA offers a blueprint that aligns urban life with these values. Its driving force is to connect people, technology, and sustainability within a decentralized framework that reimagines cities as vibrant, affordable, equitable, and forward-looking hubs of innovation and community.

5. Cloud of Things and Its Role in Urban Systems

The CoT appears as the natural growth of cloud computing and the IoT, both of which are evolving technology. The IoT links devices like sensors, vehicles, and appliances to the internet. Cloud computing, in contrast, gives the structure to work with—it studies and stores the large amounts of data that devices produce. The CoT joins these two abilities. This lets cities change data services as they grow, store large sets of data in a central place or near the edge, and get facts at once with little equipment on site—this connection matters much more in smart cities.

Urban systems today depend on a constant flow of data to work well. From checking air quality to adjusting traffic lights or managing public transport, smart city tasks require platforms that can handle changing real-time data from many points. The CoT provides this because it offers central and flexible computing power—these resources can grow when many people use them and shrink when few people do. This flexibility is important for city offices; they need to use resources well and make public services respond quickly. Another main benefit of the CoT in smart cities is its usefulness for central thought and study. For example, a city's waste system can collect data from bin sensors in different parts. The CoT gathers and works with this data to find out how full bins get to better plan better collection routes and assess needs (Dorri et al., 2017). The CoT also helps with predicting repairs in utilities—it constantly checks how well equipment works in power grids or water pipes. These abilities help cities to not only work better but also act more ahead of time in how they manage things. The CoT structure helps different city systems work together and connect. In many old city structures, data systems stay separate. Each department keeps its own tools and data. The CoT offers a place to bring these separate streams into one setting; this allows data sharing among departments like transport, public safety, and energy. This connection matters for systems that cover many areas. One example is emergency response systems that bring together police, ambulance, and traffic units.

Gathering more data in CoT environments also brings serious worries about data safety, privacy, and who owns the data. City data, much of which concerns what citizens do, who they are, and their habits, is very private. When central cloud groups control this data, questions about who owns it, who can see it, and how people use it become harder to answer. Cloud platforms do offer basic safety checks. But they often do not give the kind of openness and trust that public city systems need. Decentralized models, such as blockchain, are being thought of as helpful additions. In the end, the CoT helps smart urban systems (Li et al., 2020). It allows city services to be more useful, quick to react, and focused on citizens. Its ability to combine different data inputs, power real time studies, and support adaptable systems makes it a base for how cities will plan and manage in the future. But for the CoT to deliver all it promises, people must put it in place with a clear understanding of privacy, trust, and lasting strength. Decentralized systems, like BIDA, aim to add to these ideas. When the CoT and decentralized blockchain structures come together, they offer a good way forward. They mix the best of central use with trust.

6. Limitations of Centralized Architecture in Urban Systems

Centralized computing has a history of serving many digital services, even those in urban areas. However, it has become more complex and produces much data. The limits of centralized setups appear more often. The problem stems from relying on one or a few data centers to handle the data flow. This central point creates delays, especially as smart city devices increase quickly. Much data from traffic sensors, cameras, transit, and environmental monitors often fills the central systems or causes delays—this stops fast action. A main issue with central systems in urban areas is their weakness to single failures. If a main server stops working because of technical issues, cyberattacks, or natural problems, it can halt city services. That poses a particular concern in key infrastructure, like emergency systems, water plants, or traffic light networks (Zhang & Wen, 2017). A failure in one central spot can spread, which shows a big weakness in system strength; this weakness harms the reliability and toughness expected in city services, especially when people's safety faces risk. Data privacy and how it is used also cause major concern in central computing. Smart cities gather much sensitive data daily, including personal identification, travel paths, and behavior details. When such data collects on central servers, often run by private firms, questions arise about who owns the data. People also ask how the data is used and if citizens can see or control these actions. Centralization gathers not only data but also power. This may allow misuse, observation, or unanswerable decisions by a small group.

Central setups struggle to adapt and grow with fast-changing city needs. As cities expand and digital services grow, changing or setting up a central system can take time, increase costs, and cause problems. New devices, programs, and data types must fit into an already complex system—this often requires much technical work. Such stiffness limits a city's ability to quickly adjust to new problems or technologies. That proves especially difficult where quick action and new ideas matter much. Delays and data transfer limits are other technical problems of central computing. Real-time programs, such as self-driving public transport, changing traffic plans, or remote medical care, need very little delay to work safely and well. Sending data between edge devices and a central server can cause delays that are too long in these urgent situations. As urban sensors and devices grow in number, central systems often lack enough data transfer ability; this slows performance as well as lowers service quality for users (Salman et al., 2019).

Central systems can accidentally worsen unfairness in digital access and service spread. City areas with poor connections or little access to data networks may receive less service from central models because their data is harder to send and combine. This can lead to differences in how smart services arrive, strengthening existing social and spatial divisions within cities. For truly open and strong urban digital infrastructure, cities should move past old central models and look at spread-out or mixed models, which better suit the complexity as well as variety of city life.

7. Role of Decentralization in an Efficient Network

Decentralization is key to shaping helpful, sturdy networks for clever, growing cities. City populations grow, and frameworks get more complex. Centralized ways often struggle to grow. Single points of failure, and shortcomings can break necessary services—these cover traffic flow, trash removal, power supply, and public welfare. Decentralized networks, however, let operations, choices, and services pass to local places (Logota et al., 2010). This makes city systems change quickly and answer actual needs. One direct benefit of decentralization in clever cities is better toughness. When control splits among many spots or groups, systems become less open to breaks or cyberattacks. For example, if a central power grid fails, the whole city can lose power. But a decentralized microgrid way can close off faults and keep power local, and, this way, keep services running and build lasting firmness into important structures.

City systems must handle much data from sensors, cars, buildings, and people. Centralized computing can cause delays because of network crowding as well as process limits. By moving computing tasks to the edge, closer to where data comes, clever cities can process facts quickly and more helpfully. For instance, traffic lights can change right away based on local car movement—they do not have to wait for central approval. That lessens jams and bad air quality. Decentralization also promotes new ideas and a broader reach. In a central setup, choices and updates often need coordination through official or private paths; this slows down new ideas. A decentralized way supports testing at the local spot. Groups, or even single buildings, can try new tech or services. They do not have to wait for city-wide use. By letting more people join directly in services, rule, or data handling, cities can include more and fit the needs of many people.

Data control as well as privacy also work better with decentralization. People and groups keep hold of their facts—they do not send all personal or group data to a central agency. This way fits with privacy tech like blockchain and zero knowledge proofs. It sees that private data, like health or travel paths, can help the public good without hurting private rights. It also builds public trust, which matters for clever city plans to begin (Kundu, 2019). Decentralization is not just a tech choice. It is a plan idea for the cities of the future. By splitting power, duty, and thought across the city plan, decentralized networks build cities that answer more, work better, stand firmer, and treat everyone fairly. City problems keep changing, especially with climate change and population growth. Decentralization shows a way forward that can grow also last—it makes sure clever cities stay not only tech advanced but also focused on people. Figure 8.2 illustrates several application of blockchain integration.

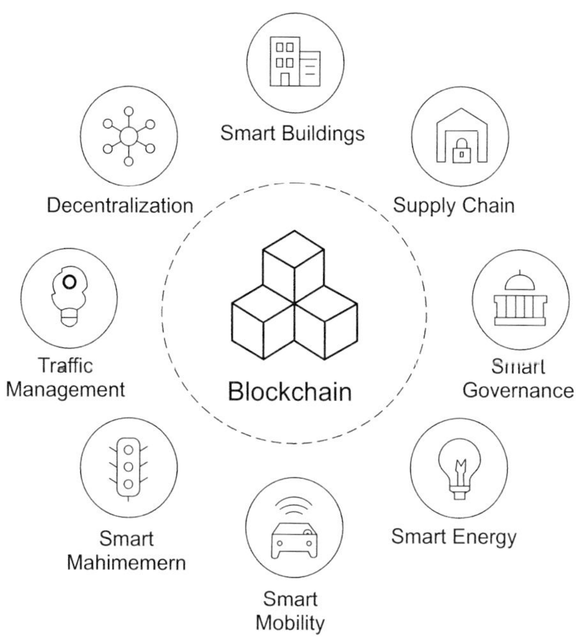

Fig. 8.2 Blockchain integration in growing cities.

7.1 Proposed Conceptual Framework

BIDA is a converged platform that leverages the strengths of the IoT, edge computing, and digital twins to improve infrastructure management. BIDA provides real-time data and the associated state of the connected infrastructure—this ensures the integrity of the data and secure access to it. Edge computing allows the connected infrastructure be less reliant on centralized data centers, provides better privacy (with regard to personal identifiable information), and time-sensitive applications like emergency response and traffic optimization. BIDA also supports a city-scale digital twin environment that mimics real-world entities, where infrastructure performance can be monitored, anomaly detection conducted, and future scenarios simulated. Digital twins are maintained with real-time data flows and blockchain updates using data from the IoT. The BIDA ecosystem is participatory in nature, involving citizens, governments, and service providers. Citizens own their data by using decentralized identity mechanisms and have a possibile role in governance. Governments establish policies and enforce infrastructure, while service providers provide utilities and digital services. Smart contracts ensure transparency and fair service. Further, both private and public entities can interact with the blockchain to log service activities, processes payments, and manages access rights. BIDA provides a new way of governing data using blockchain-based data governance technology that is embedded with self-sovereign identity (SSI) and decentralized data ownership principles. Every citizen has their own identity linked to a memory (think blockchain wallet) and can securely access a service through a digital interface, all the while being assured of their privacy and outright ownership of their data. All data is encrypted and recorded in a distributed ledger, while access to the data is controlled by smart contracts. Most importantly, the digital identity (DID) allows citizens to authenticate themselves across multiple services, all without fear of theft of their identity from gypsy hackers, not to mention no one spying on their access to a service. The services are completed through a smart contract, which adds a layer of trust while reducing costs.

8. Blockchain and CoT Integration for Smarter Urban Cities Network

A mix of blockchain and CoT matters much for cities in the future. When cities depend more on data, the amount of facts that sensors, devices, and services create grows fast. CoT systems let cities handle much data quickly. This data rise brings worries about safety, ownership, and trust. With its public record and open rules, the CoT offers an effective way to fix these issues. When they work together, CoT and blockchain can make cities safer, clearer, and more about the people.

In today's cities, many systems, such as traffic control, power lines, trash pickup, and help for urgent needs, often work apart. The CoT pulls these different data flows together, offering central checks and rule. But putting such large amounts of private data in one place makes city setups weak to attacks and single points of failure (Alam & Salah, 2019). Blockchain helps here by spreading out control and record keeping. Each part on the network checks deals, so it becomes very hard to change data without everyone agreeing. This builds a tough digital base for cities to use. Another change that blockchain-CoT brings lies in data origins and faith. City systems rely much on sensor data, from air quality readers to cameras, and faith in this data counts.

Blockchain can make sure that once a data packet goes in, no one can change or erase it later—this gives city leaders, planners, and citizens belief in the data's truth. For instance, pollution data checked on a blockchain becomes true for both citizens and green groups, paving a path for quicker and fairer rules.

Smart contracts, which are rules stored and run on the blockchain, add more uses to city systems; they can automate actions such as traffic light changes based on how many cars are on the road or by charging for high gas cars. In CoT-ready cities, sensor data starts these actions, and the blockchain runs them openly. This reduces the need for people to do things manually or for central checks. This not only makes things work better but also lessens chances for cheating, as the contract rules are clear and the network can enforce them. People who live in cities also gain much from blockchain-CoT. In many cities, a lack of clear rules and safe digital identity prevents real involvement. By using blockchain for digital identity and safe voting, cities can let people participate in decisions about local matters. Along with the CoT's fast feedback mechanisms, such as reporting holes in the road or asking for trash pickup, this creates a loop of steady joining. People then do not just use services, but can help shape their city.

From a view of lasting good, blockchain-CoT allows better checking and handling of goods. Smart grids can track power use across areas and encourage less use during busy times through blockchain tokens or credits. Similarly, water use and trash systems can work best by using blockchain to make vendors and users answerable. This close checking and checked data sharing brings openness and pushes long-term goals for lasting good.

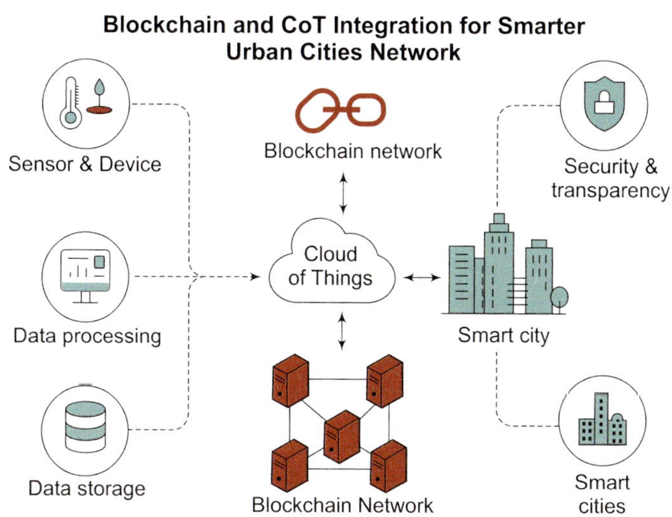

Fig. 8.3 Blockchain and CoT integration overview.

Joining with other systems is another area where blockchain helps CoT systems (see Figure 8.3). In a conventional city, many sellers and groups manage different parts of city setups. Blockchain can act as a fair, open place for data sharing, making sure that no one group controls the system. Whether public transport systems link with private ride share networks, or hospitals share patient data with urgent help groups, a single blockchain layer brings smooth work together without hurting data truth or safety.

In essence, the blend of blockchain and CoT holds much promise for shaping how cities live in the future—this mix can turn cities into places that change, are open, are inclusive, and where data flows safely and fairly among all groups. By building faith, letting things run themselves, and keeping citizen rights safe, blockchain-CoT systems are set to become the digital backbone of future cities. For city planners, tech people, and rule makers, now is the time to put money into this blend to show its changing power.

9. Case Study: Scenario Analysis of BIDA Adoption in NeoCity

In order to demonstrate the real world applicability of the BIDA framework, this section proposes a scenario based on real-world problems and future visions of a mid-sized urban center—NeoCity. NeoCity is facing the real-time challenges of urbanization, increased demands of citizens, increased pressure for infrastructure, and digital silos. NeoCity engages in a phased approach to systematically adopt the BIDA model to become a completely decentralized and blockchain-based urban environment. BIDA is implemented in NeoCity in four stages, all of which involve the active collaboration of various stakeholders such as the city government, technology suppliers, civil society, universities, local businesses, and citizens.

Each functionality under BIDA's key components has been discussed in detail.

9.1 Smart Governance

Governance is usually the most complicated and contentious function of urban systems, and it is often based on centralized control, hierarchical decision-making, and opaque public records. BIDA introduces a smart governance framework that uses blockchain-based decentralized ledgers to secure, verify, and include citizens in public processes. An example of governance can be in the form of blockchain-enabled electronic voting, which produces immutable and secure municipal elections, community decision-making, and participatory budgeting by storing votes on the ledger and making the counting transparent. Likewise, public records, like landownership, permit, or licenses, may be housed on the blockchain, so as to be tampered with and/or manipulated by a bureaucratic system.

9.2 Smart Economy

The economic part of BIDA is focused on supporting the creation of local economic ecosystems that utilize blockchain to create the foundational transaction infrastructure. A significant piece of this is implementing a city-based cryptocurrency or token economy that allows residents, businesses, and governments to conduct transactions that do not rely on cash, but rather on digital transaction technologies. Importantly, these tokens will enable a local economy focused on cashless, secure, and verifiable transactions between local participants (e.g. residents, businesses, local government) for a range of purposes, including public transport, utility bills, local buying/selling, or rewards for local civic participation events.

9.3 Smart Mobility

Urban transport continues to be a problem for cities in the modern context, primarily by way of congestion, inefficiencies, and the convention of fractured or un-integrated services. BIDA's smart mobility framework utilizes blockchain as a means to decentralize data exchange and coordination between transport systems. A decentralized ride-sharing service using blockchain can allow drivers and passengers to make direct connections to negotiate the ride price using smart contracts. Furthermore, passengers can validate their rides and make payment directly, without the need for any intermediary. This service structure creates absolute transparency, and the blockchain provides security. Real-time data exchange of traffic data from autonomous vehicles, IoT-enabled infrastructure (e.g. smart traffic signals), and city administrators can all be logged and processed on the blockchain. Information can then be disseminated and be tamper-proof. This will optimize the flow of traffic, allow for travelers to optimize their routes, and allow for enhanced responses during emergencies.

9.4 Smart Health

Urban health systems have become very siloed and have major issues with interoperability. BIDA utilizes a decentralized health data system where patient records are stored on distributed ledgers (in encrypted format), enabling secure access by healthcare providers, emergency responders, or AI diagnostic tools, all with patient consent managed through smart contracts. This model works to improve the portability, availability, and integrity of medical record systems. The benefits are that individuals have more control over their health data, emergency services and hospitals are able to inform and communicate with each other in a secure and real-time manner, and there are much better connections between hospitals, clinics, insurers, and public health figures. It is also possible to establish decentralized health networks where verified health professionals provide services and consultations transparently through blockchain-registered credentials and job logs.

9.5 Smart Energy

Traditional energy consumption and generation is a centralized ecosystem, which produces inefficiencies, dependence on utility companies, and restricts the potential for use of renewables. BIDA enables peer-to-peer (P2P) energy trading in microgrids that are powered by blockchain technology. This will allow households with solar panels or other renewable energy sources to sell their excess energy directly to neighbors through smart contracts, dynamically setting the price on energy by tracking supply and demand. The public and secure blockchain will ensure that every trade is recorded, and brand provenance and carbon credits can be tied to its point of origin. It will also create a decentralized grid, where energy consumption and flows can be logged, and maintenance can be determined and optimized separately, allowing for greater sustainability and resilience

9.6 Smart Contracts for Service Automation

A common element in all of these applications is smart contracts—self-executing scripts that operate with conditional logic that can allow urban services to be performed in many ways automatically. In the case of smart contracts for welfare payments, funds may be paid out when the eligibility check has been satisfied. For emergency responses, smart contracts may activate pre-established protocols based on real-time sensor readings. In the case of predictive analytics, a smart contract could estimate repairs and maintenance based on conditions in a digital twin. In addition to eliminating third-party intermediaries, smart contracts can create increased efficiencies by providing independent service delivery upon event trigger. For example, a smart parking contract could determine when a vehicle occupies a parking space, initiate billing based on occupancy duration, and release the parking space in the ledger when the vehicle exits, all without a human involvement.

10. Research and Technology Development of BIDA in Growing Cities

The main focus is to understand the way forward with BIDA by leveraging blockchain and CoTs to make a smarter and more efficient environment (Xu et al., 2020; Tyagi, 2024b; Tyagi, Tyagi, 2023a; Tyagi, 2023c; Tyagi, 2023d; Tyagi, 2024a; Tyagi, 2023b; Tyagi, 2021; Tyagi, 2020; Tyagi, 2022). We will go through some potential applications where this could make a major impact in the near future.

10.1 Energy and Utilities

Improving energy and utilities networks poses a central problem for cities that grow quickly. Usual systems that have a center often show faults, such as low output, high costs to upkeep them, and a weakness to breaks in service; they also struggle to add sources of power that renew. When cities join BIDA with the CoT, they change their energy and utilities networks—these become more nimble, lasting, centered on people. BIDA offers a clear, open, and spread-out plan for handling assets of infrastructure. The CoT provides the instant connection and processing power needed for smart energy work. At the core of this change is the ability of the CoT to gather and look at data from devices spread out across the city (Wong et al., 2020). These devices include smart meters, sensors, and systems that store energy. Such devices watch how much energy people use, how much demand changes, and how well equipment works at that moment. When these devices join with BIDA, people do not just examine this data. They also record it permanently and share it across a distributed ledger, which means many groups, like utility companies, people who set rules, and even regular citizens, get to see confirmed data on energy use—they do not need to depend on one main power.

This joining of systems has practical benefits for both the supply and the demand parts of energy infrastructure. For instance, spread-out energy sources, like solar panels on roofs, small community power grids, and home batteries, fit better into the city's energy mix. With BIDA, every spot in the network, such as a home with solar panels, acts as both a maker and a user. People call such a spot a "prosumer." Smart agreements on the blockchain automatically guide energy trades between groups. This makes it easier for people to trade energy directly with each other, without needing go-betweens. This improves how well energy systems cope with trouble and also lets citizens take part directly in the energy economy. For utility companies and city planners, BIDA and the CoT significantly improve how well things run.

Infrastructure like water pipes, electric lines, and gas systems fit with smart sensors; these sensors find issues like leaks, too much load, or breaks in real time. The CoT lets maintenance groups know about these issues right away. The blockchain part makes sure that records of upkeep, costs, and performance data are open and no one can change them, which means faster fixes, less waste, and better long-term plans based on confirmed old data.

From the point of view of lasting systems, joining the CoT and BIDA helps the grid act in a smarter way and change more easily. For example, during times of high demand, smart systems move energy from parts of the grid that people do not use much or take from stored renewable sources. Incentives based on blockchain, which BIDA manages, pay consumers who shift their energy use to times when demand is low or who put extra energy back into the grid (Sun et al., 2016). This makes an energy system that is more even and cares for the Earth; this system also matches goals for climate and urban sustainability.

BIDA and the CoT form a useful team for rethinking energy and utilities in smart cities. They build a system that is not only good with technology but also spread out, productive, and able to recover. By spreading out control and putting smart features into infrastructure, cities better meet the demands of expansion; they also lower waste, support lasting systems, and build connections between providers and citizens that people trust more. This new way of arranging city energy systems is necessary for cities of the future to stay livable and care for the environment over time.

10.2 Transportation and Mobility

Building transportation infrastructure helps the growth of smart cities. As urban populations rise, cities face more pressure to supply safe, workable, and lasting mobility choices. Old systems often break up, depend on central control, and respond slowly to current conditions. But the connection of BIDA with the CoT offers a helpful answer. This connection helps with collecting data in real time, making decisions without central control, and openly working with different groups in urban mobility systems.

The CoT holds a basic role by connecting and managing many devices that relate to transportation, which include GPS-enabled buses, smart traffic lights, ride-sharing platforms, electric vehicles (EVs) charging stations, and self-driving car systems. The devices that connect always produce useful data about traffic, car use, weather, and how people travel. With support from cloud infrastructure, workers can process and look at this data at once to find areas with many cars, adjust routes, and coordinate traffic lights better. To get a truly reliable and connected mobility system, BIDA is important. Through BIDA, makes sure that data shared across different systems and groups—public transit, private transport, and infrastructure firms—is safe, can be checked, and does not change (Enescu et al., 2022). For instance, ride-sharing firms and public buses can share current location and open seat data on a shared blockchain platform, which allows city planners and navigation apps to offer users the best routes with many modes of travel. Smart contracts can also put shared money among groups automatically or make people follow urban mobility rules without needing a central judge. One important use of BIDA and the CoT in transportation involves making pricing as well as access control change. Through smart contracts backed by blockchain, workers can change tolls, parking fees, and congestion charges based on current demand and how people use things. CoT sensors put in road structures, vehicles feed this data into the system, and the blockchain makes sure that pricing rules are clear and cannot be changed. People can track how charges apply and how the money goes to road upkeep or projects for lasting use, which improves trust in government systems.

Another useful point involves the progress of electric as well as self-driving vehicles. The CoT allows for real-time cooperation between vehicles and structures, such as charging stations, and BIDA handles safe identity management and payment for vehicles as self-governing agents. For example, a self-driving EV could agree on charging terms, hold a station, and handle payment through a smart contract—all without a person. Car use and repair records put on the blockchain help improve safety and cut down on fraud in car ownership or leasing markets. In short, the connection of BIDA with the CoT changes urban mobility from a system that reacts and has central control into one that adjusts, is clear, and focuses on the user—this mix lets cities fit more commuters while reducing traffic, pollution, and wasted effort. By using real-time connections and ways to trust without central control, transportation systems can become more open, work better, and respond to citizen needs. BIDA and the CoT provide the digital base needed for the flexible and lasting mobility systems that smart cities require.

10.3 Waste Management and Environmental Services

Waste operations and environmental care are central to a city's health, its ability to last, and its pleasantness. But old systems often have problems with bad work, separate data, and no immediate view. The combination of BIDA and the CoT offers cities a chance to update their waste and environmental service setups. When cities join decentralized trust with connected sensor systems, they can build clear, quick, and data-driven methods that cut waste and lessen harm to the environment. The CoT brings understanding and real-time monitoring to waste management (Laouar et al., 2019). Smart bins that have fill-level sensors, GPS tracking for collection trucks, and environmental sensors for air and soil quality can gather data across the city continuously. Cloud platforms collect and study this information to find the best waste pickup paths, predict when bins will overflow, and find places where people dump things illegally or cause pollution. This helps use resources better, lower costs for running things, and react faster to environmental problems. This data system needs a safe and sure base, and BIDA holds a major role here.

BIDA keeps data from the CoT on a decentralized record, so it stays the same and you can check it—this matters when many groups work together, like city offices, private waste businesses, recycling firms, and environmental groups. A blockchain record lets these people look at and check data without needing a main leader. For example, smart contracts can record when and where waste was picked up, and pay contractors only when sensor data proves the work has been done. This improves accountability and the performance of services. Recycling methods also gain much from this integration. CoT sensors can follow recyclable items through the supply chain, while BIDA can keep an open record of each batch of waste. People or businesses who sort waste correctly can get paid through a blockchain-based reward system, with digital tokens or credits sent automatically by smart contracts; this causes more people to take part in recycling plans and builds a clear way to follow how materials move in circular economy models, which always reuse or repurpose waste.

Environmental services, like watching air and water quality, finding noise pollution, and handling city trees, can also get better with this mix. CoT sensors placed in neighborhoods and green areas collect facts on pollution levels, weather patterns, and nature signs. This data, once put on the blockchain through BIDA, becomes part of a lasting public record. Policymakers, researchers, and the public can use it to ask for better environmental ways or to start automatic cleanup efforts through decentralized groups of repair teams or drones. In essence, combining BIDA with the CoT in waste and environmental services lets cities move past just reacting and toward methods that foresee and understand. By letting go of central control and making sure data is correct, cities can help different groups work together while holding each to shared goals for lasting health. With immediate views of waste and environmental situations, and the trust blockchain provides, smart cities can build greener, cleaner, and more lasting communities for those alive now and also in the future.

10.4 Governance and Civic Participation

Governance and civic involvement underpin a well-running smart city. Traditional governance often faces issues such as sluggish administration, little openness, little citizen participation, and slow replies. Combining BIDA with the CoT makes a more open, quick, and shared model of urban governance—these technologies help governments move from top-down systems to shared frameworks that involve residents directly in the democratic process. The CoT offers the setup for real-time data gathering across a city. This includes information from public services, utilities, infrastructure status, and citizen reports. For instance, smart kiosks, mobile applications, and connected devices in public areas let residents report potholes, noise issues, or safety worries directly to city agencies—this data, gathered all the time and on its own, becomes the base for responsive governance. The CoT ensures that citizens' concerns are recorded correctly and quickly, forming a digital link between people and their local governments.

BIDA plays a large part in making sure the data and exchanges from CoT remain safe, provable, and unchanged. Each report or civic interaction can be put on a decentralized blockchain ledger, where it gets a time stamp, stays anonymous if needed, and becomes open for public checks (Ahram et al., 2017). This builds trust in the system, especially in areas where people doubt the government. It creates a lasting public record of civic actions as well as government replies, which people can review, analyze, and improve over time. Civic participation can also change with smart voting systems built on the blockchain. BIDA allows for safe, open digital voting for local votes, community choices, or shared budgeting. Through digital identity systems tied to blockchain, citizens vote from their devices without fear of cheating. Smart contracts can tally the votes and release results right away, cutting down delays and building faith in the fairness of the democratic process. Such systems raise voter turnout by removing difficulties with voting.

BIDA and CoT help make fairer decisions by creating open platforms where citizens submit ideas, vote on city plans, or see live data about city conditions and government work. CoT sensors in public transport, utilities, and environmental zones send data to public sites that show citizens how their neighborhoods operate. With blockchain's unchanging nature, this creates a base for citizen-led governance that uses data and involves teamwork, instead of being passive or just reacting. Joining BIDA with the CoT can change civic engagement from an occasional event to a constant, connected experience. Citizens no longer wait for election times or official town meetings to speak their minds or join in city growth. They become active participants in real-time city governance; they get the tools to add to, watch, and even help design policies. By combining decentralized infrastructure with sensing everywhere, smart cities become not only more effective but also more democratic, welcoming, and attentive to the people they serve.

10.5 IoT Data Security

Smart cities now depend on internet-connected tools, like smart traffic lights, to make city operations better and safer. The many connections, however, create weak points, especially for data safety. BIDA, a data program that uses AI and the CoT, a cloud system for IoT tools, changes how cities keep data safe. It offers a firm plan to find, stop, and act on safety threats at once.

BIDA collects data and uses AI to watch internet-connected tools all the time. For smart traffic lights, BIDA studies large amounts of data, like traffic patterns, sensor states, and log entries—it looks for unusual things or actions that may show a cyberattack or a system problem. BIDA uses machine learning to tell the difference between normal changes in how things work and possible safety breaches (Lin & Liao, 2017). City managers can then act fast before problems grow. The CoT works with BIDA by offering a safe, large platform to handle and link many internet-connected tools across the city. Its cloud setup makes sure that data from smart traffic lights stays coded during transfer and stays safe, which lowers the chance of someone taking it or changing it. The CoT also handles updates and fixes, which matters for internet-connected tools that often lack strong safety parts. Regular updates to tool software through a central cloud system reduce weak points and help keep the whole system sound.

A big problem for smart cities is handling many tools and making sure safety rules apply to all of them. BIDA works with the CoT to provide a single view where city staff can check tool health, review logs, and see safety warnings all in one place. This central view helps them make wise choices and act quickly when things go wrong. For example, if someone tries to get into a traffic light control unit, safety teams can disconnect that tool from far away, stopping the threat from spreading to other linked systems. These technologies also support better ways to control who can use things, which ensures that only authorized people and systems can work with important IoT parts. BIDA uses AI to check identities. The CoT uses safe ways to check who logs in (Lin & Liao, 2017). That stops people who do not have permission from getting in, whether they are from inside or outside; this layered safety plan is important in smart cities. Physical city parts often meet digital networks, which increases the ways people can attack the systems.

BIDA and the CoT together offer a strong answer to the difficult job of keeping internet-connected systems safe, such as smart traffic lights in smart cities. By putting together smart data study, central tool management, and good coding and access controls, cities can become better at handling cyber threats. This keeps important city systems safe and also makes services for citizens more secure and steady—this opens the way for truly smart and safe city environments.

11. Open Issues in BIDA Systems

As cities connect more digitally, the idea of BIDA gains ground. BIDA pictures city structures—like transport, utilities, governance, and services—running on decentralized blockchain systems. This design offers benefits, such as openness, unchanging data, and citizens' participation. But it also introduces many problems that cities must settle before they can use it fully. One of the biggest worries for BIDA is how it will grow. Especially public ones, generally operate slowly and use much computer power. City systems create huge amounts of data—this data comes from traffic sensors, smart grids, public records, and money transfers. Current blockchain systems may struggle to manage such data loads at once. Without better blockchain growth, BIDA systems may become slow or too costly because of transaction fees and energy use.

Data privacy and remaining unknown cause another hard problem. Blockchain's openness seems helpful, but it can clash with what citizens expect for their private data. City data often holds sensitive details, like people's names, how they move, and their health records. Storing or referencing such data on ledgers that do not change needs advanced ways to keep things private, which include zero knowledge proofs or storage models that sit outside the main chain (Kim & Laskowski, 2018). Many of these ways are still in early stages and cities do not use them widely.

A third problem deals with working together. City systems span many areas, laws, and companies. For a BIDA system to work well, different blockchain systems and older computer systems must talk and share data safely. The blockchain world remains broken, with different ways to communicate and different rules. Without common ways to work together, the idea of a truly connected decentralized city will stay separate and not work well. Who runs things is another big concern in BIDA deployments. Who owns the data? Who controls the smart contracts that automate public services? What happens if a blockchain-based choice—say, for traffic or city plans—is wrong or disputed? In decentralized systems, a central power to solve disagreements often does not exist. Designing fair ways to run things that can change to fit local laws as well as citizen thoughts remains an open and hard task. Security problems grow in decentralized city systems. Smart contracts, while helpful, often contain bugs and holes if people do not write and check them carefully. In a BIDA setting, a smart contract that gets attacked could stop important services, like water supply or public transport. The ways systems agree must be strong enough to stop bad actors from taking over the system, especially when people use public blockchains.

Then there is the problem of public acceptance and knowing how to use digital tools. Using blockchain-based services assumes city dwellers and leaders feel comfortable with technology. Digital inequality still exists, even in modern city settings. If people do not design BIDA systems with easy use and teaching in mind, they risk pushing away people who need help and creating a new digital gap in access to public services (Ferrag et al., 2019; Xie et al., 2019). The rules and laws for BIDA are still not fully formed. Many countries lack clear laws about blockchain data, smart contracts, and digital identities. Questions about who is responsible, who owns ideas, and following rules are big. Until strong rules come up, many cities may wait to go beyond test projects; they fear legal uncertainty or problems later.

Table 8.2 Taxonomy of blockchain urban system applications.

Domain	*Focus Area*	*Blockchain Platform*	*Contribution*	*Limitations*
Smart Cities	Decentralized Data Auditing	–	Introduces a decentralized cloud-based data auditing system for urban applications.	Lacks smart contract implementation and security assessment.
Authorization Services	Access Authorization & Delegation	Ethereum	Proposes a blockchain-based scheme for authorization in smart city IoT environments.	Does not address user privacy concerns.
Smart City Security	Architecture Security	Ethereum	Develops a secure smart city platform leveraging blockchain for improved protection.	Ignores access control mechanisms for cloud data storage.
Sharing Economy	Secure Service Framework	Ethereum & Hyperledger	Establishes a blockchain system to secure peer-to-peer services in urban sharing models.	Data privacy considerations remain unexplored.
Smart Homes	Privacy & Security Framework	–	Outlines a privacy-focused architecture for smart home systems using BIoT: Blockchain Internet of Things	Does not include actual blockchain integration.
Smart Homes	Security-Oriented Design	–	Presents a smart home design with implemented blockchain-based security mechanisms.	Includes practical blockchain application.

Table 8.2 disucusses the taxonomy of blockchain urban system applications with benefits and limitations.

12. Conclusion

The integration of blockchain technology within decentralized urban city networks represents a transformative step toward smarter, more secure, and resilient urban environments. By decentralizing data management and enabling transparent, tamper-proof interactions, blockchain adds a foundational layer of trust to city infrastructures. Whether applied to traffic systems, energy grids, or public services, BIDA can address key challenges related to data integrity, service automation, and stakeholder accountability. However, the implementation of BIDA is not without its hurdles. Issues such as scalability, privacy concerns, interoperability between different systems, and the need for robust governance mechanisms still require thorough research and practical solutions. The complexity of integrating blockchain with existing urban infrastructures and IoT frameworks means that cities must tread carefully, balancing innovation with operational feasibility and legal compliance. Moreover, public trust and digital inclusivity will be critical to the success of BIDA systems. Without proper user education, equitable access, and transparent governance, the risk of creating new forms of digital inequality or public resistance increases. Therefore, stakeholders—from policymakers and technologists to local communities—must collaborate to ensure that BIDA enhances civic life rather than complicating it. Ultimately, BIDA holds tremendous potential in reshaping how modern cities function, enabling decentralized decision-making and more secure digital ecosystems. While challenges remain, continued research, experimentation, and policy development will be key to unlocking the full benefits of blockchain-powered decentralized cities. With the right strategies in place, BIDA could become a cornerstone of the future smart city paradigm.

Conflict of Interest

The authors declare that they do not have any conflict of interest with respect to the publication of this research work.

References

Ahram, T., A. Sargolzaei, M. Sargolzaei, J. Daniels and B. Amaba. (2017). Blockchain technology innovations, in *2017 IEEE Technology & Engineering Management Conference (TEMSCON)*, pp. 137–141. https://doi.org/10.1109/TEMSCON.2017.7998367.

Alam, T. and K. Salah. (2019). IoT security: review, blockchain solutions, and open challenges, *Future Generation Computer Systems*, *82*, 395–411. https://doi.org/10.1016/j.future.2017.11.022.

Alladi, T., V. Chamola, R.M. Parizi and K.K.R. Choo. (2019). Blockchain applications for Industry 4.0 and industrial IoT: a review, *IEEE Access*, *7*, 93788–93804. https://doi.org/10.1109/ACCESS.2019.2929361.

Atzori, L., A. Iera and G. Morabito. (2010). The internet of things: a survey, *Computer Networks*, *54*(15), 2787–2805. https://doi.org/10.1016/j.comnet.2010.05.010.

Batty, M., K.W. Axhausen, F. Giannotti, A. Pozdnoukhov, A. Bazzani, M. Wachowicz, G. Ouzounis and Y. Portugali. (2012). Smart cities of the future, *The European Physical Journal Special Topics*, *214*(1), 481–518. https://doi.org/10.1140/epjst/e2012-01703-3.

Christidis, K. and M. Devetsikiotis. (2016). Blockchains and smart contracts for the internet of things, *IEEE Access*, *4*, 2292–2303. https://doi.org/10.1109/ACCESS.2016.2566339.

Conoscenti, M., A. Vetrò and J.C. De Martin. (2016). Blockchain for the internet of things: a systematic literature review, in *Proceedings of the 2016 IEEE/ACS 13th International Conference on Computer Systems and Applications (AICCSA)*, pp. 1–6. https://doi.org/10.1109/AICCSA.2016.7945805.

Correia, D., J.L. Marques and L. Teixeira. (2023). Assessing and ranking EU cities based on the development phase of the smart city concept, *Sustainability*, *15*(18), 13675. https://doi.org/10.3390/su151813675.

Dorri, A., S.S. Kanhere and R. Jurdak. (2017). Towards an optimized blockchain for IoT, in *Proceedings of the Second International Conference on Internet-of-Things Design and Implementation (IoTDI '17)*, pp. 173–178. https://doi.org/10.1145/3054977.3055003.

Enescu, F.M., F.G. Birleanu, M.S. Raboaca, N. Bizon and P. Thounthong. (2022). A review of the public transport services based on the blockchain technology, *Sustainability*, *14*(20), 13027. https://doi.org/10.3390/su142013027.

Ferrag, M.A., L. Maglaras, A. Derhab and H. Janicke. (2019). Blockchain technologies for the internet of things: research issues and challenges, *IEEE Internet of Things Journal*, *6*(2), 2188–2204. https://doi.org/10.1109/JIOT.2018.2882794.

Harrison, C. and I.A. Donnelly. (2011). A theory of smart cities, in *ISSS Proceedings of the 55th Annual Meeting*. https://journals.isss.org/index.php/proceedings55th/article/view/1703.

Kang, J., R. Yu, X. Huang, S. Maharjan, Y. Zhang and E. Hossain. (2017). Enabling localized peer-to-peer electricity trading among plug-in hybrid electric vehicles using consortium blockchains, *IEEE Transactions on Industrial Informatics*, *13*(6), 3154–3164. https://doi.org/10.1109/TII.2017.2709784.

Khalil, U., O.A. Malik, M. Uddin and C.-L. Chen. (2022). A comparative analysis on blockchain versus centralized authentication architectures for IoT-enabled smart devices in smart cities: a comprehensive review, recent advances, and future research directions, *Sensors*, *22*(14), 5168. https://doi.org/10.3390/s22145168.

Khan, M.A., K. Salah, N. Crespi and M.H. Rehman. (2020). Blockchain-based smart grid architecture for energy trading with charging stations, *Internet of Things*, *11*, 100237. https://doi.org/10.1016/j.iot.2020.100237.

Kim, H.M. and M. Laskowski. (2018). Toward an ontology-driven blockchain design for supply-chain provenance, *Intelligent Systems in Accounting, Finance and Management*, *25*(1), 18–27. https://doi.org/10.1002/isaf.1424.

Kitchin, R. (2013). The real-time city? Big data and smart urbanism, *GeoJournal*, *79*(1), 1–14. https://doi.org/10.1007/s10708-013-9516-8.

Kundu, D. (2019). Blockchain and trust in a smart city, *Environment and Urbanization ASIA*, *10*(1), 31–43. https://doi.org/10.1177/0975425319832392.

Laouar, M.R., Z.T. Hamad and S. Eom. (2019). *Towards blockchain-based urban planning: application for waste collection management*, in *Proceedings of the 9th International Conference on Information Systems and Technologies (ICIST 2019)*, pp. 1–6. ACM. https://doi.org/10.1145/3361570.3361604.

Li, X., P. Jiang, T. Chen, X. Luo and Q. Wen. (2020). A survey on the security of blockchain systems, *Future Generation Computer Systems*, *107*, 841–853. https://doi.org/10.1016/j.future.2017.08.020.

Lin, I.C. and T.C. Liao. (2017). A survey of blockchain security issues and challenges, *International Journal of Network Security*, *19*(5), 653–659. https://doi.org/10.6633/IJNS.201709.19(5).01.

Logota, E., A. Neto and S. Sargento. (2010). A new strategy for efficient decentralized network control, in *Proceedings of the 2010 IEEE Global Telecommunications Conference (GLOBECOM 2010)*, pp. 1–6. https://doi.org/10.1109/GLOCOM.2010.5683684.

Ogunkan, D.V. and S.K. Ogunkan. (2025). Exploring big data applications in sustainable urban infrastructure: a review, *Urban Governance*, advance online publication. https://doi.org/10.1016/j.ugj.2025.02.003.

Panarello, A., N. Tapas, G. Merlino, F. Longo and A. Puliafito. (2018). Blockchain and IoT integration: a systematic survey, *Sensors*, *18*(8), 2575. https://doi.org/10.3390/s18082575.

Reyna, A., C. Martín, J. Chen, E. Soler and M. Díaz. (2018). On blockchain and its integration with IoT: challenges and opportunities, *Future Generation Computer Systems*, *88*, 173–190. https://doi.org/10.1016/j.future.2018.05.046.

Salman, T., M. Zolanvari, A. Erbad, R. Jain and M. Samaka. (2019). Security services using blockchains: a state of the art survey, *IEEE Communications Surveys & Tutorials*, *21*(1), 858–880. https://doi.org/10.1109/COMST.2018.2863956.

Santos, J.P.F.D., C.A. De Matos and A. Groznik. (2025). The role of artificial intelligence in smart city systems usage: drivers, barriers, and behavioural outcomes, *Technology in Society*, *74*, 102867. https://doi.org/10.1016/j.techsoc.2025.102867.

Sharma, P. K. and J.H. Park. (2018). Blockchain-based hybrid network architecture for the smart city, *Future Generation Computer Systems*, *86*, 650–655. https://doi.org/10.1016/j.future.2018.04.051.

Siddiqui, S., S. Hameed, S.A. Shah, J. Arshad, Y. Ahmed and D. Draheim. (2024). A smart-contract-based adaptive security governance architecture for smart city service interoperations, *Sustainable Cities and Society*, *113*, 105717. https://doi.org/10.1016/j.scs.2024.105717.

Sun, J., J. Yan and K.Z. Zhang. (2016). Blockchain-based sharing services: what blockchain technology can contribute to smart cities, *Financial Innovation*, *2*(1), 1–9. https://doi.org/10.1186/s40854-016-0040-y.

Tyagi, A.K. (2020). Prediction models, in *Handbook of Research on Disease Prediction through Data Analytics and Machine Learning*, IGI Global, pp. 50–69. https://doi.org/10.4018/978-1-7998-2742-9.ch004.

Tyagi, A.K. (2021). Analysis of security and privacy aspects of blockchain technologies from smart era's perspective: the challenges and a way forward, in *Recent Trends in Blockchain for Information Systems Security and Privacy*, CRC Press, pp. 1–17. https://doi.org/10.1201/9781003139737.

Tyagi, A.K. (2022). SecVT: Securing the vehicles of tomorrow using blockchain technology, in *Proceedings of the International Symposium on Artificial Intelligence (ISAI 2022)*, Springer, pp. 141–152. https://doi.org/10.1007/978-3-031-22485-0_11.

Tyagi, A.K. (2023a). Blockchain–artificial intelligence-based secured solutions for smart environment, in *Digital Twin and Blockchain for Smart Cities* (Chapter 23), Wiley. https://doi.org/10.1002/9781394303564.ch23.

Tyagi, A.K. (2023b). Blockchain-enabled internet of things (IoTs) platforms for IoT-based healthcare and biomedical sectors, in *Artificial Intelligence-Enabled Blockchain Technology and Digital Twin for Smart Hospitals* (Chapter 10), Wiley. https://doi.org/10.1002/9781394287420.ch10.

Tyagi, A.K. (2023c). Decentralized everything: a practical use of blockchain technology in future applications, in *Distributed Computing to Blockchain: Architecture, Technology, and Applications*, Elsevier, pp. 19–38. https://doi.org/10.1016/B978-0-323-96146-2.00010-3.

Tyagi, A.K. (2023d). Sensors and digital twin application in healthcare facilities management, in *Artificial Intelligence-Enabled Blockchain Technology and Digital Twin for Smart Hospitals* (Chapter 19), Wiley. https://doi.org/10.1002/9781394287420.ch19.

Tyagi, A.K. (2024a). Dew computing: state of the art, opportunities, and research challenges, in *Machine Learning Algorithms Using Scikit and TensorFlow Environments* (Chapter 17), IGI Global. https://doi.org/10.4018/978-1-6684-8531-6.ch017.

Tyagi, A.K. (2024b). Transformative effects of ChatGPT on modern era of education and society: from society and industry's perspective, in *Machine Learning Algorithms Using Scikit and TensorFlow Environments* (Chapter 19), IGI Global. https://doi.org/10.4018/978-1-6684-8531-6.ch019.

Ullah, Z., M. Naeem, A. Coronato, P. Ribino and G. De Pietro. (2023). Blockchain applications in sustainable smart cities, *Sustainable Cities and Society*, *97*, 104697. https://doi.org/10.1016/j.scs.2023.104697.

Wong, P.F., F.C. Chia, M.S. Kiu and E.C.W. Lou. (2020). The potential of integrating blockchain technology into smart sustainable city development, in *IOP Conference Series: Earth and Environmental Science*, *463*(1), 012020, IOP Publishing. https://doi.org/10.1088/1755-1315/463/1/012020.

Xie, J., H. Tang, T. Huang, F.R. Yu, R. Xie, J. Liu and Y. Liu. (2019). A survey of blockchain technology applied to smart cities: research issues and challenges, *IEEE Communications Surveys & Tutorials*, *21*(3), 2794–2830. https://doi.org/10.1109/COMST.2019.2899617.

Xu, R., S.Y. Nikouei, D. Nagothu, A. Fitwi and Y. Chen. (2020). BlendSPS: a blockchain-enabled decentralized smart public safety system, *Smart Cities*, *3*(3), 928–951. https://doi.org/10.3390/smartcities3030047.

Yoo, Y. (2021). Toward sustainable governance: strategic analysis of the Smart City Seoul Portal in Korea, *Sustainability*, *13*(11), 5886. https://doi.org/10.3390/su13115886.

Zhang, Y. and J. Wen. (2017). The IoT electric business model: using blockchain technology for the internet of things, *Peer-to-Peer Networking and Applications*, *10*(4), 983–994. https://doi.org/10.1007/s12083-016-0456-1.

Zhu, X., Z. Chen, T. Cheng, C. Yang, D. Wu, Y. Wu and H. Wang. (2024). Blockchain for urban governance: enhancing trust in smart city systems with advanced techniques, *Sustainable Cities and Society*, *108*, 105438. https://doi.org/10.1016/j.scs.2024.105438.

CHAPTER 9

Performance Metrics and Evaluation

R. Felista Sugirtha Lizy[1*] and *I. Gusti Bagus Yosia Wiryakusuma*[2]

The last few decades have been a time of increasing smart mobility and digital transformation. As a result, human-centric intelligent transportation systems (HC-ITS) are emerging that address both safety and human-level experience. This chapter presents an extensive discussion on the performance metrics and evaluation frameworks applied to design, implement, and advance HC-ITS. It dissects foundational concepts, relevant technologies, real-world case studies, and the growing challenges facing the transportation system in the context of human behavior and societal dynamics. It provides a detailed analysis of qualitative and quantitative metrics including safety, accessibility, reliability, user satisfaction, and equity, and presents a series of qualitative and quantitative metrics categorized as social and environmental sustainability. This discussion highlights directions for the future of human-centered, adaptive, and intelligent mobility systems based on people and technology and in alignment with ethical considerations.

1. Introduction

The transport sector is experiencing a paradigm shift as advanced technologies are integrated into the sector, while there is also an increased emphasis on human-centered design. Intelligent transportation systems (ITS) have historically focused on operating a transport agency with efficiency and managing traffic flows. ITS and the Federal Highway Administration is evolving towards human-centric intelligent transportation systems (HC-ITS) with essential elements of user behaviors, options, preferences, safety, and inclusivity. This chapter introduces the essential elements and the rationale for developing performance metrics for HC-ITS in mobility (how people move) systems in urban and rural areas. It will discuss the ways in which emerging technologies such as artificial intelligence (AI), the internet of things (IoT), and edge

[1] Assistant Professor, Department of Computer Science, A.P.C. Mahalaxmi College for Women, Thoothukudi, Tamil Nadu, India. https://orcid.org/0000-0003-3648-0205

[2] Assistant Professor, Department of International Business Management, Universitas Ciputra. https://orcid.org/0000-0001-5845-8661

* Corresponding author: 21felistaa@gmail.com

DOI: 10.1201/9781003659822-9

computing emphasize how human-centered decision-making and adaptable systems can be implemented.

2. Literature Survey

Table 9.1 Performance metrics and evaluation for HC-ITS.

Topical Cluster	*Key References*	*Core Contributions*	*Evaluation/ Metric Focus*	*Identified Gaps*
Foundational reviews & enabling technologies	Al-Khateeb & Johari (2023); Papadimitratos et al. (2021); Zhou & Xie (2020)	Consolidate the concept of HC-ITS, map sensors/edge–cloud/ AI stack, outline vehicle-to-everything (V2X) security & mobility-management layers	Qualitative maturity grids, taxonomy of user-centric functions	Need for a common *human-impact score* that mixes safety, equity, and trust across modes
Autonomous-vehicle control & human interaction	Amini et al. (2022); Rasouli & Tsotsos (2020); Shladover (2020)	Survey deep-learning control pipelines, pedestrian-aware AV–Autonomous Vehicle behavior, CAV–Connected and Autonomous Vehicle operating-domain definitions	Benchmark suites for perception/planning latency, disengagement-rate, near-miss density; driver–pedestrian interaction metrics	Lack of standardized "near-miss" ground-truth sets and cross-fleet reproducibility
Human-in-the-loop evaluation	Chen et al. (2022); NIST (2022)	Propose closed-loop testbeds where human drivers override or label AV actions; NIST offers task-load, situational- awareness & trust questionnaires	Human-state probes (eye-tracking, NASA-Task Load Index [TLX), intervention frequency, explainability score	Scaling human in the loop (HITL) experiments beyond controlled labs; privacy-preserving logging of biometric data
Sustainability & multi-criteria decision-making	García et al. (2023); UNECE (2023) policy; MCDA–Multi-Criteria Decision Analysis-primer	Apply AHP–Analytic Hierarchy Process & TOPSIS–Technique for Order Preference by Similarity to Ideal Solution to rank BRT–Bus Rapid Transit, EV–Electric Vehicle-bus, LRT–Light Rail Transit alternatives on cost–CO_2–accessibility; UNECE sets equity & Sustainable Development Goals (SDGs) alignment targets	Weighted composite "sustainability index" normalized to local baselines	Subjective weight setting; weak coupling to real-time emission sensors

Contd.

Table 9.1 *Contd.*

Topical Cluster	*Key References*	*Core Contributions*	*Evaluation/ Metric Focus*	*Identified Gaps*
Simulation & digital-twin tools	Sliwa et al. (2021); Törngren & Selberg (2022)	Catalogue SUMO, CARLA, MATSim and their human-factor plug-ins; outline cyber-physical digital-twin loop for continuous safety-case evidence	Coverage metrics (km, scenario diversity), simulation-to-real gap estimators	Need unified interoperability schema so traffic-, energy-, and crowd-sims share states
Ethics, data-governance, & privacy	Raza et al. (2022); WEF (2021)	Map attack surfaces (location tracking, adversarial spoofing) and propose privacy-by-design checklists; WEF offers AI-ethics principles for transport	Compliance scorecards vs General Data Protection Regulation (GDPR), differential-privacy budget tracking	Limited empirical studies on user acceptance of privacy trade-offs

3. The Basics of HC-ITS

HC-ITS are hybrid transportation systems that combine smart technology with a deliberate emphasis on people, their needs, behavior, and safety—such as AI, IoT, and analytics. While traditional ITS are focused on the infrastructure to improve system efficiency and traffic flow, HC-ITS are designed to increase human usability. For HC-ITS, usability, accessibility, safety, trust, and inclusivity—in both the system design and operation—are of paramount importance (see Figure 9.1).

Core Principles of HC-ITS

1. **Human-Centric Design:** Systems are designed and built based on the user's needs, limitations, and expectations with the goal of prioritizing usability and adoption across generations.
2. **Real-Time Responsiveness:** HC-ITS can help make responsive decisions in real-time, using data from sensors, vehicles, and users to help improve a situation (e.g., adaptive signal control, alerts for pedestrians, and rerouting of automobiles or transit vehicles from an emergency).
3. **Safety and Behavior Change:** The interests are to mitigate human error, create pedestrian and vehicle safety, and learn about human behavior using AI and machine learning capabilities.
4. **Multimodal and Inclusive Mobility:** There are several mobility (walking, cycling, public transit, autonomous vehicles) pathways to remove access barriers or enable equitable treatment of users, particularly seniors and persons with disabilities.
5. **Connected and Intelligent Infrastructure:** The roads, intersections, and transit/transportation systems include devices that are intelligent, providing a channel of communication between vehicles and users, making an open, intelligent mobility ecosystem.

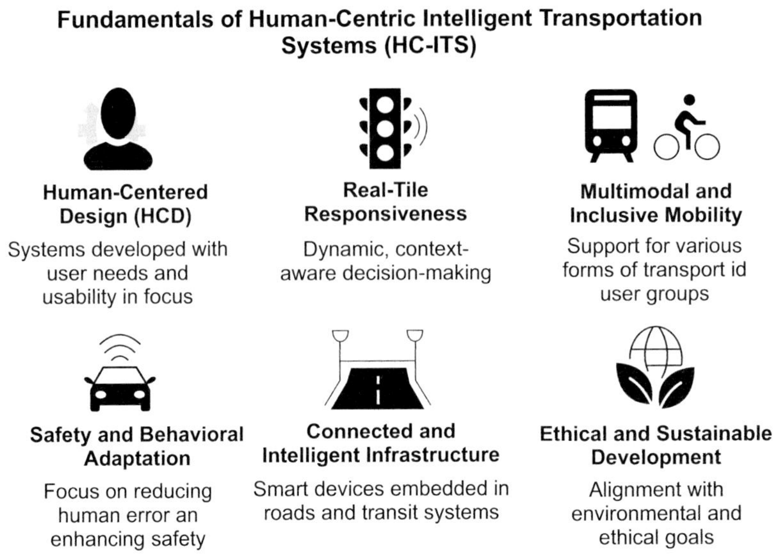

Fig. 9.1 Fundamentals of HC-ITS.

6. **Ethical and Sustainable Development:** HC-ITS aim to reconcile and harmonize transportation solutions with environmental outcomes, ethically treating and using data and practices for long-term sustainability for the community.

Table 9.2 lays down the fundamentals of HC-ITS.

Table 9.2 Fundamentals of HC-ITS.

Fundamental Aspect	*Description*	*Key Benefits*	*Technological Enablers*
Human-Centered Design Principles	Designs transportation systems based on human capabilities, limitations, and behaviors.	Improves usability, safety, and user satisfaction	HCI–Human–Computer Interaction, UI–User Interface/ UX–User Experience design, participatory design
Real-Time Responsiveness	Systems respond to dynamic road and user conditions with minimal delay.	Traffic Fluidity and Safety Enhancement	AI and IoT sensors and edge computing
Safety and Behavior Adaptation	Concerning human errors, it studies those errors in terms of cognitive load, reaction time, and decision-making processes.	Also helps to look at accident reduction and hence builds trust in the system	Cognitive models, machine learning, biometric data

Contd.

Table 9.2 *Contd.*

Fundamental Aspect	*Description*	*Key Benefits*	*Technological Enablers*
Multimodal and Inclusive Mobility	Ensures transportation modes and modalities for all users, including the elderly and disabled.	Ensures free equity and reduces dependency on private vehicles	Mobility-as-a-service (MaaS), wayfinding apps
Connected and Intelligent Infrastructure	Communicates and reacts to road circumstance and presence of users in real-time with roadside units, traffic signals, and vehicles.	Enables smooth, safe, and efficient traffic flow	V2X communication, smart grids, cloud platforms
Ethical and Sustainable Development	Values such as fairness, transparency, and environmental responsibility are embedded in both system design and evaluation.	Binding transportation in terms of societal goals and environment	Green tech, AI ethics frameworks, data governance

3.1 Definition and Evolution of HC-ITS

HC-ITS represent modern transportation solutions that holistically place human users—drivers, passengers, pedestrians, and vulnerable road users—as central to the design, operation, and evaluation of transportation systems. Traditional intelligent transportation systems primarily valued traffic flow and operational efficiency of infrastructure. HC-ITS emphasizes safety, accessibility, usability, and user satisfaction.

An evolution path could be as follows:

- Traditional ITS: Infrastructure based, i.e., traffic signals, control centres
- Modern ITS: Enabled by technology, i.e., GPS, mobile applications, vehicle to infrastructure communication
- HC-ITS: Aware of humans and adapts to their behaviors, i.e., emotion-aware systems, real-time human feedback loops (see Figure 9.2).

Essentially, this represents a wider societal trend toward inclusivity, personalization, and ethical design.

3.2 Human Factors in Transportation Systems

Human factors encompass the psychological, physiological, and behavioral characteristics of people actively involved in transportation systems. The consideration of human factors is necessary for system designs that ensure absolute safety and comfort for users while reducing all opportunities for human error.

Areas of human factors include the following:

- Cognitive Load and Attention-to-Interface: The interface should minimize the opportunities for distraction and overload.
- Time and Reaction and Predictability Systems: The systems should be operable for the use of a wide range of user reactions.

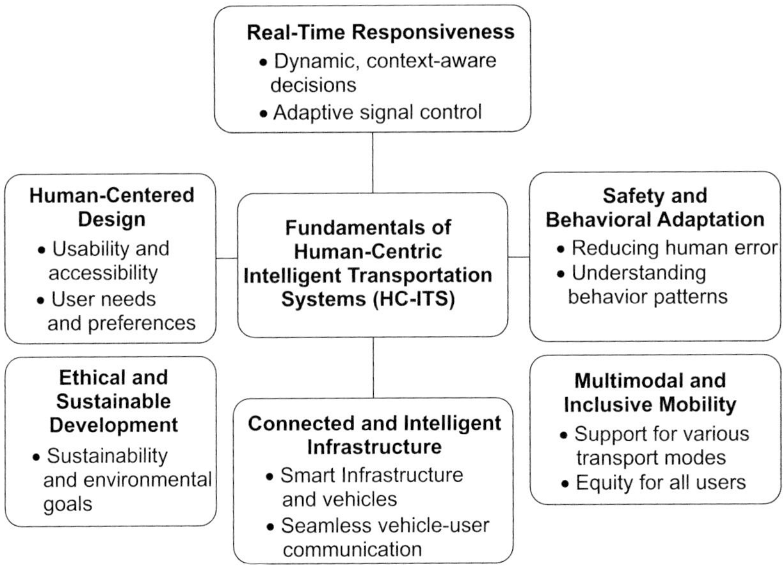

Fig. 9.2 Evolution of HC-ITS.

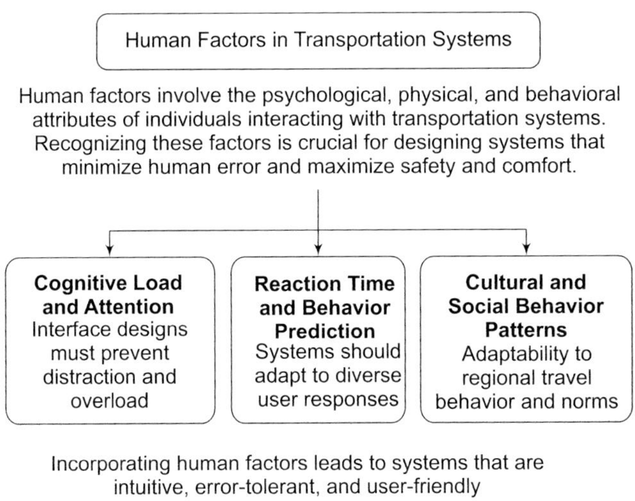

Fig. 9.3 Human factors in transportation systems.

- Inclusive accessibility: Integrated inclusivity should be considered for those who are elderly, differently abled, or not tech savvy.
- Cultural and Social Behavior Patterns: Systems should accommodate local travel patterns, norms, and behaviors across regions.

Human factors integration leads to making systems that are intuitive, resilient to user error, and user friendly (see Figure 9.3).

3.3 Role of AI, IoT, and Smart Infrastructure

AI, the IoT, and smart infrastructure form the technological basis of HC-ITS.

AI can provide sufficient real-time decision-making and communication, traffic predictions, and tailored recommendations. AI can utilize large data streams for safety monitoring, autonomous driving, and adaptive routing purpose.

IoT connects vehicles, traffic signals, road sensors, and user devices to exchange insight data in real time. The results the information exchange can lead to occurrences, traffic congestion, and has the ability of being manipulated to deal with ever-changing environments.

Smart infrastructure incorporates technologies such as intelligent traffic signals, smart road signs, and roads equipped with sensors, working together with vehicles and control systems to optimize benefits and mitigate harm (see Figure 9.4).

The effects of these technologies individually and collectively produce solutions that allow transportation systems to be optimally proactive and considerate of what is relevant to people.

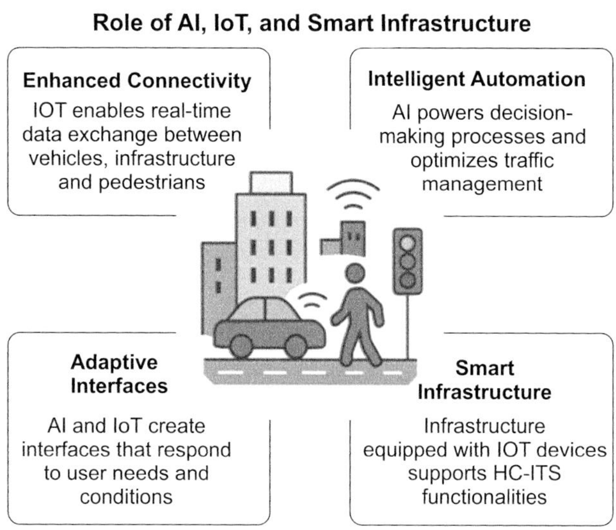

Fig. 9.4 Role of AI, IoT, and smart infrastructure.

3.4 Need for Human-Centered Design Principles

Human-centered design (HCD) is important to make sure that transportation technologies are not only usable but also user friendly, ethical, and inclusive. HCD incorporates the following aspects:

- Empathy and user participation: Involving users throughout the development process.
- Usability and simplicity: Avoiding complex interfaces and decision-making.
- Safety and emotional comfort: Designing systems that engender trust and reduce anxiety.
- Adaptability: Allowing for diverse preferences, needs, and abilities.

If you do not utilize HCD principles, you risk any transportation technology, regardless of intelligence, failing their intended user. But an HC-ITS is both intelligent and humane (see Figure 9.5).

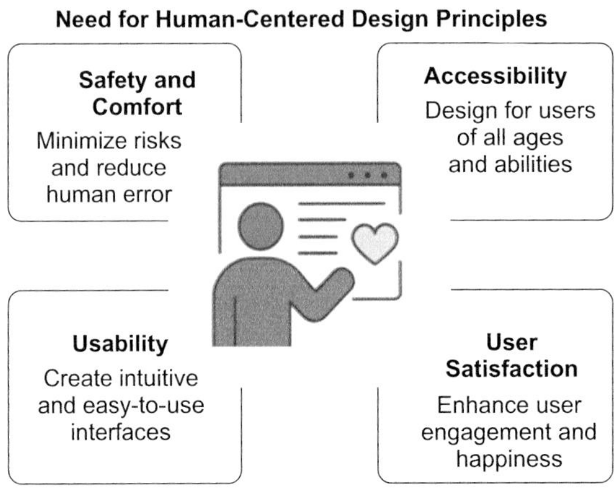

Fig. 9.5 Need for human-centered design principles

4. Key Performance Metrics for HC-ITS

To assess the performance of HC-ITS, various quantitative and qualitative performance metrics are used that reflect not only the efficiency of operation of the system but also how well it meets human needs and societal objectives, and how sustainable it is in the long term (see Figure 9.6).

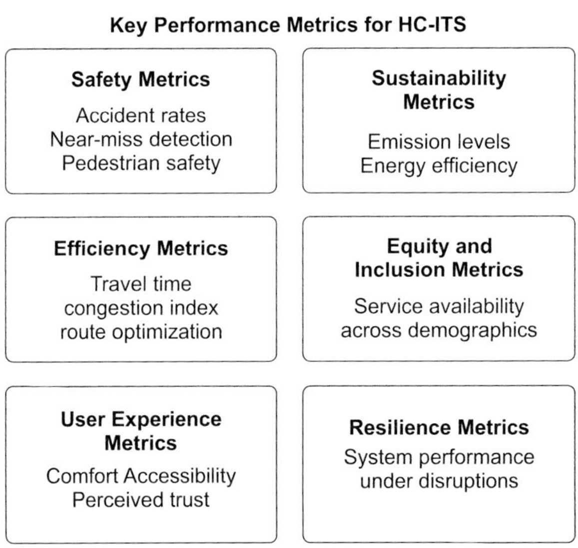

Fig. 9.6 Key performance metrics for HC-ITS.

Table 9.3 reflects on the key performance metrics for HC-ITS.

Table 9.3 Key performance metrics for HC-ITS.

Metric Category	*Description*	*Examples of Metrics*	*Impact Area*
Safety Metrics	Check how the system cuts risks and keeps all road users safe, including people walking and biking.	• Crash rates • Times of close calls • Score for pedestrian injuries	Health of public, trust of users
Efficiency Metrics	Look at how well the system makes the best use of time, space, and resources.	• Time to travel on average • How bad traffic gets • How well routes are planned	Handling traffic, city output
User Experience Metrics	Look at both feelings and facts about how people use the system.	• How comfy it is • How easy it is to use • How much users trust it	Including every-one, making peo-ple happy, getting people to use it
Sustainability Metrics	Keep an eye on how the transport system affects the environment and uses energy.	• CO_2 output per km • How much fuel or power it uses • How loud it is	Rules for the environment, travel that is good for Earth
Equity & Inclusion Metrics	Make sure services and structures are spread across all groups and areas.	• How well it serves forgotten areas • How easy it is for old or disabled folks to use • Difference in use between men and women	Fairness in society, including everyone in tech
Resilience Metrics	See how tough the system is when things go wrong, like crashes, disasters, or cyberattacks.	• How often it stops working • How it functions in emergencies • How fast it gets back to normal	Dealing with disasters, keeping things running

4.1 Safety Metrics

These metrics assess how well the transportation network keeps accidents at bay and safeguards its users. They pay special attention to those most at risk on the roads, like people on foot or bikes.

- Accident Rates: Measures the frequency and severity of collisions or crashes.
- Detecting Near Miss: Identification of potential harmful events that did not lead to a crash but indicated a potential for risk.
- Pedestrian Safety: Observing behavior at crosswalks, vehicle–pedestrian encounters, and pedestrian priority systems (see Figure 9.7).

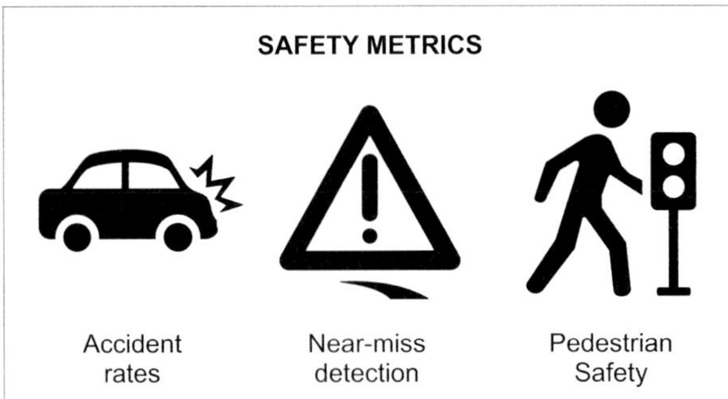

Fig. 9.7 Safety metrics.

4.2 Efficiency Metrics

Efficiency metrics provide an indication of how well transportation systems perform in terms of reducing overall travel time, avoiding congestion, and directing trips along better routes.

- Travel Time: Average length of trips during regular and peak periods.
- Congestion Index: Measurement of the density of traffic and delays in corridors (see Figure 9.8).
- Route Optimization: Ability to reroute vehicles to follow the quickest or the most fuel-efficient path.

Fig. 9.8 Efficiency metrics.

4.3 User Experience Metrics

These metrics capture customer satisfaction, passenger comfort, and user perception of trust.

- Comfort: Level of vibration or ride smoothness in the transportation or comfort associated with climate control.
- Accessibility: Use of the transportation system by differently abled populations, seniors, etc.
- Perceived Trust: Users' trust of the system reliability, safety, and perceived equity (see Figure 9.9).

Fig. 9.9 User experience metrics.

4.4 Sustainability Metrics

Sustainable performance is addressed by measuring environmental performance indicators and energy use indicators of the system over time.

- Emission levels: Measured levels of CO_2 and other vehicle pollutants (e.g., NOx, volatile organic compounds [VOCs]) recorded.
- Energy Efficiency: As determined by fuel use per seat mile or whether electrically operated or "green" energy (see Figure 9.10).

4.5 Equity and Inclusion Metrics

These indicators verify whether or not the award of transportation services occurs fairly, leading to service delivery for all members of our society.

- Service Availability: Distribution, coverage of low-income, rural, or underserved urban population segments.
- Demographic Inclusion: Access to services across demographic groups irrespective of their age, gender, ability levels, and socio-economic status (see Figure 9.11).

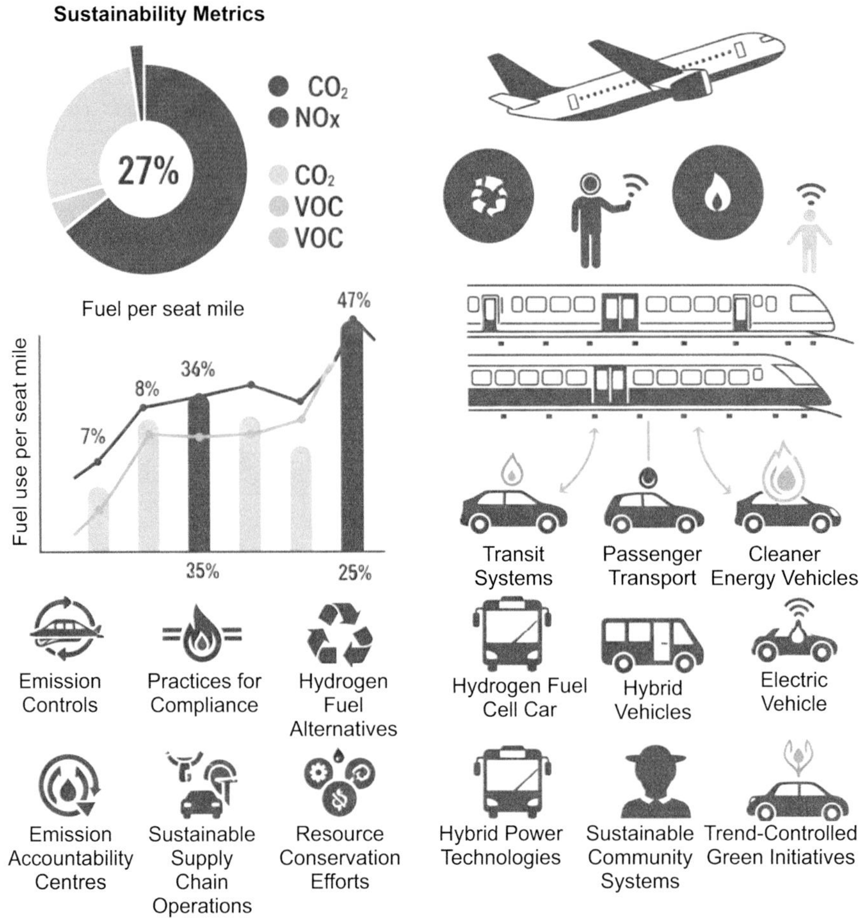

Fig. 9.10 Sustainability metrics.

4.6 Resilience Metrics

These metrics define the resilience of the system including its ability to function at all when including and considering the stress of an emergency situation like a natural disaster.

- System Downtime: The number and duration of service interruptions.
- Performance during Disruption: The ability to continue functioning when disabling events are occurring, such as an accident, a cyberattack, or severe weather (see Figure 9.12).

Performance evaluations of HC-ITS should utilize description metrics focused on transportation, but they also need to measure the human-centered principles of equity, trust, and environmental stewardship. These metrics will ensure that intelligent transportation systems benefit not only individuals but also society as a whole.

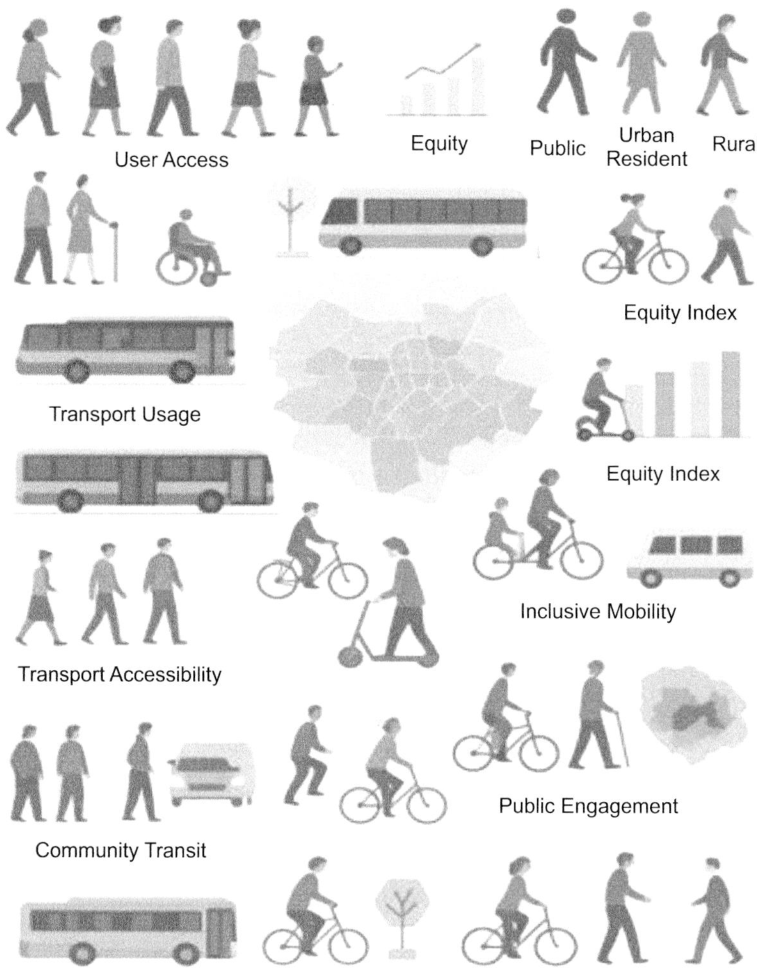

Fig. 9.11 Equity and inclusion metrics.

5. Assessment Frameworks and Methodologies

Rigorous and diverse methods of evaluation are required to assess performance and effectiveness of Human-Centric Intelligent Transportation Systems (HC-ITS). These frameworks and methodologies allow stakeholders to assess the extent which human-centered needs, such as safety, comfort, fairness, and trust, and technical goals are being achieved.

Table 9.4 provides a comprehensive overview for assessment frameworks and methodologies for HC-ITS designed for inclusion.

5.1 Comparative Evaluation Techniques

In this method, multiple solutions or models, for example, HC-ITS, are compared based on predetermined metrics (travel time, safety, energy use). (see Figure 9.13).

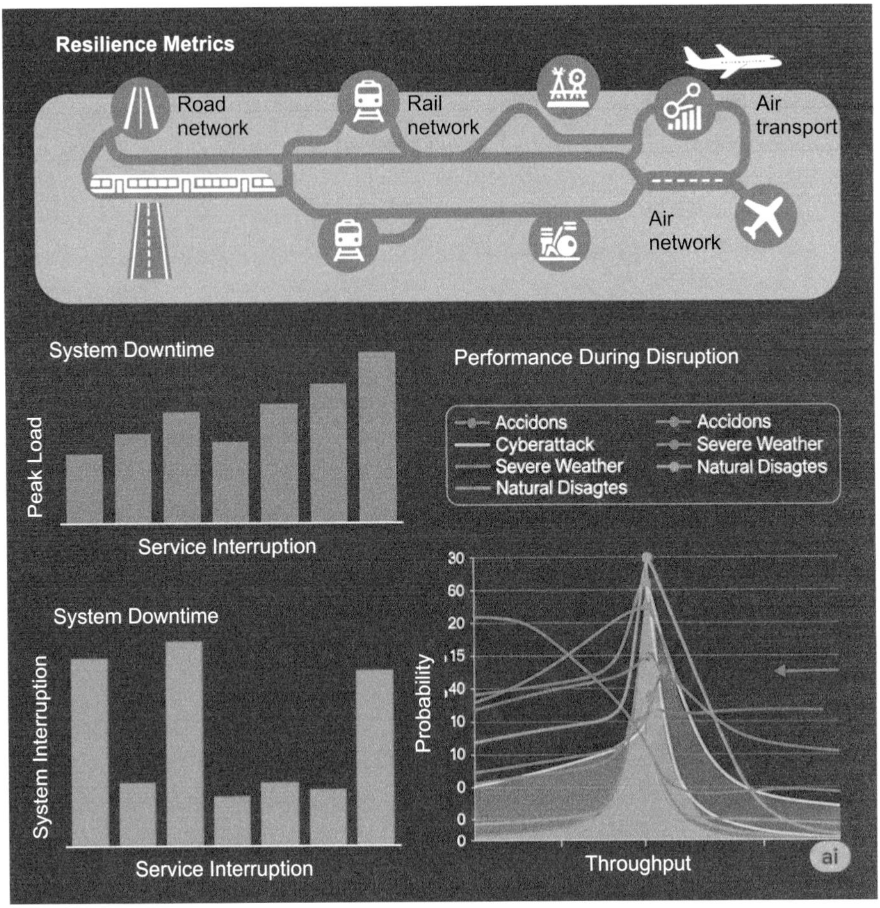

Fig. 9.12 Resilience metrics.

Table 9.4 Evaluation frameworks and methodologies for HC-ITS.

Evaluation Method	*Description*	*Use Cases*	*Strengths*	*Limitations*
Comparative Evaluation Techniques	Side-by-side performance analysis of multiple HC-ITS systems, tools, or strategies	Benchmarking routing algorithms, comparing smart signal designs	Enables decision-making, clear performance trade-offs	May oversimplify real-world complexities
Simulation vs. Real-World Testing	Uses both simulated environments and field deployments to assess performance under various conditions	Urban traffic flow modeling vs. pilot smart intersections	Cost-effective, scalable (simulation), realistic data (real-world)	Simulation-to-reality gap, field tests are expensive and slower

Contd.

Table 9.4 *Contd.*

Evaluation Method	*Description*	*Use Cases*	*Strengths*	*Limitations*
Human-in-the-Loop (HITL) Evaluation	Includes users in system testing to assess interaction quality, trust, and adaptability	AV override tests, human feedback on interface design	Captures real behavior, trust dynamics	Hard to scale, requires privacy safeguards
Multi-Criteria Decision Analysis (MCDA)	Structured approach to evaluate alternatives based on multiple quantitative and qualitative metrics	Selecting optimal transport solutions based on cost, equity, and sustainability	Holistic, supports complex decisions	Subjective weight assignment, requires domain expertise
Big Data & Real-Time Analytics	Leverages real-time sensor and usage data to evaluate and adapt HC-ITS performance	Congestion detection, travel time estimation, user behavior prediction	Scalable, responsive, enables predictive insights	Data privacy concerns, high infrastructure and storage demand

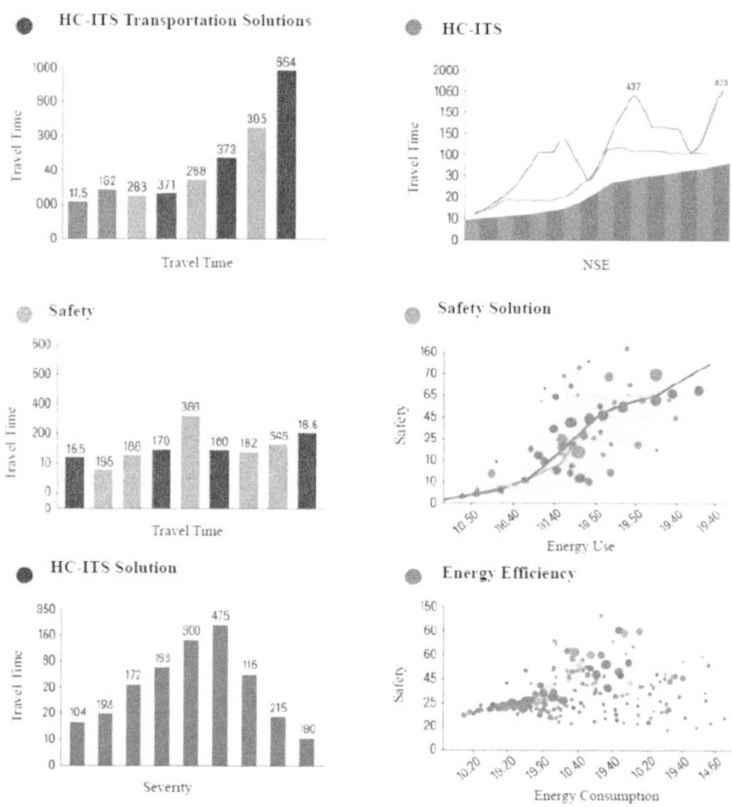

Fig. 9.13 Comparative evaluation techniques.

This can help decision-makers to:

- Reference competing systems or technologies.
- Identify the best solution, given real-world constraints.
- Identify trade-offs (speed vs. safety, or cost vs. sustainability).

5.2 Simulation vs. Real-World Testing

Simulation Evaluation: Evaluate system behavior in a virtual world under safe, scalable, and controlled conditions. This is appropriate for early-stage design and can be good for stress testing.

- Examples include SUMO, VISSIM, CARLA.
- Benefits include safe, repeatable, and scenarios that allow for even more exploration.

Real-World Testing: Evaluate when deploying HC-ITS components in an operational environment, and real users and vehicles can be measured against for actual performance (see Figure 9.14).

- Examples include the operation of smart intersections or autonomous shuttle pilots.
- Benefits include high fidelity to real-world operation and testing out human interaction in unexpected conditions.
- The best practice is a combination of the two.

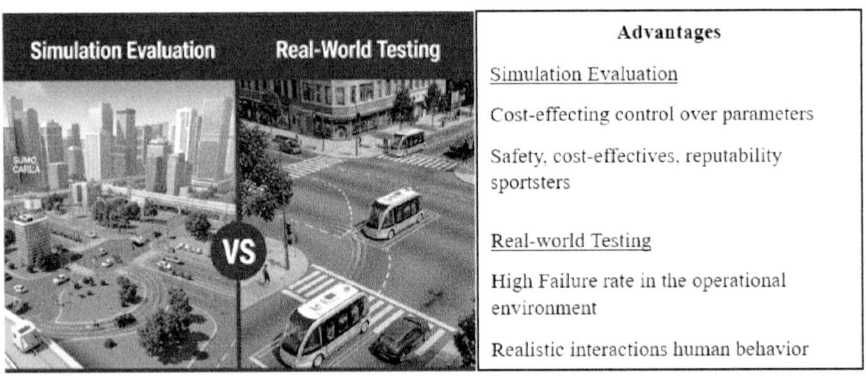

Fig. 9.14 Simulation vs. real-world testing.

5.3 Human-in-the-Loop Evaluation

In a human-in-the-loop evaluation, human users are engaged in the testing process to help evaluate the quality of interactions, trust, usability and emotional response (see Figure 9.15).

- Important considerations for validating include:
  - User interfaces
  - Driver-assistance systems
  - Pedestrian-aware smart signals
- Benefits include realistic behaviours, preferences, and acceptance levels.

Fig. 915 Human-in-the-loop evaluation.

5.4 Multi-Criteria Decision Analysis (MCDA)

MCDA is applied to evaluate a variety of criteria, often conflicting, related to HC-ITS performance, such as safety, cost, accessibility, and environmental implications (see Figure 9.16).

Methods: Analytic hierarchy process (AHP), OPSIS–Technique for Order Preference by Similarity to Ideal Solution, PROMETHEE

Advantages:

- Structured decision-making
- Recognition of the subjective nature of judgments
- Prioritization of alternatives when no single metric prevails

5.5 Big Data and Digital Analytics for Evaluation

In the real world, HC-ITS have enormous volumes of data from vehicle, user, and sensor data. Big data can be used for evaluation purposes (see Figure 9.17), including the following:

- Real-time performance evaluation to monitor issues such as congestion or system health
- Predictive analytics for forecasting potential failures or safety concerns
- Insights for users that personalize their experience and improve adaptive traffic control capabilities

Platforms include stream processing platforms, such as Apache Kafka and Spark, and machine learning techniques for both behavioral prediction and pattern recognition for human travel behavior and mobility trends.

Table 9.5 *Contd.*

Application Area	*Description*	*Human-Centric Benefits*	*Key Technologies Involved*
Emergency and Incident Management	Rapid detection and response to traffic incidents, natural disasters, or vehicle failures	Faster medical response, alerting systems for nearby users	IoT sensors, predictive analytics, dispatch systems
Assistive Systems for Vulnerable Users	Tailored mobility solutions for elderly, disabled, and non-tech-savvy users	Equity, autonomy, and improved accessibility	Voice interfaces, assistive navigation, inclusive UI

6.1 Smart Urban Mobility Systems

HC-ITS help create integrated, effective, and user-centred changes in urban mobility (see Figure 9.18), such as the following.

- Real-time travel information on smartphones and apps.
- Shared mobility (e.g., ride-sharing, e-scooters).
- Dynamic pricing (for congestion impacts and equity).

Goal: Enhance travel experiences, reduce congestion, and provide flexible mobility to meet individuals' needs.

Fig. 9.18 Smart urban mobility systems.

6.2 Adaptive Traffic Signal Control

Adaptive traffic signal control refers to regulated, real-time adjustment of traffic light settings based on AI and sensor data related to the following:

- Volume of traffic, including bicycles.
- Pedestrian crossing activity.
- Presence of emergency vehicles.

It results in less time delay, reduced capacity loss, and improved pedestrian safety at crossings, especially for children, seniors, and other vulnerable groups (see Figure 9.19).

Fig. 9.19 Adaptive Traffic Signal Control.

6.3 Connected and Autonomous Vehicles (CAVs)

With HC-ITS, CAVs can do the following:

- Use the HC-ITS system for vehicle-to-vehicle (V2V), vehicle-to-infrastructure (V2I), and vehicle-to-human communication.
- Learn human discomfort to optimize routes.
- Use real-time hazard detection and level of pedestrian priority to classify abstracted future hazards.
- Learn patterns of human driving behaviors to manage their coexistence with human drivers and increase their own safety by analyzing and predicting human driving behaviors (see Figure 9.20).

6.4 Multimodal Transportation Planning

HC-ITS will integrate a variety of transport modes in multimodal transport planning, including buses, trains, bicycles, footpaths, and integrated multimodal ticketing systems. It will provide suggestions for real-time multimodal trip routes.

Fig. 9.20 Connected and autonomous vehicles (CAVs).

These will enable users to switch modes more often with less annoyance, thus reducing their reliance on personal vehicles and increasing social, environmental, and economic sustainability (see Figure 9.20).

Fig. 9.21 Multimodal transportation planning HC-ITS.

6.5 Emergency and Incident Management

In the case of an incident or any natural disaster, the HC-ITS has the potential to:

1. Detect and report accidents in real-time.
2. Know the highly preferred routes of emergency responders.
3. Warn the other surrounding traffic to reduce secondary crashes.

This capability increases awareness of a situation and gets aid to where it is needed much faster, especially if human life is in jeopardy (see Figure 9.22).

6.6 Assistive Systems for Elders and Disabled

HC-ITS helps ensure equity of mobility by providing the following:

1. Interfaces available for its transport for wheelchair riders, when possible.

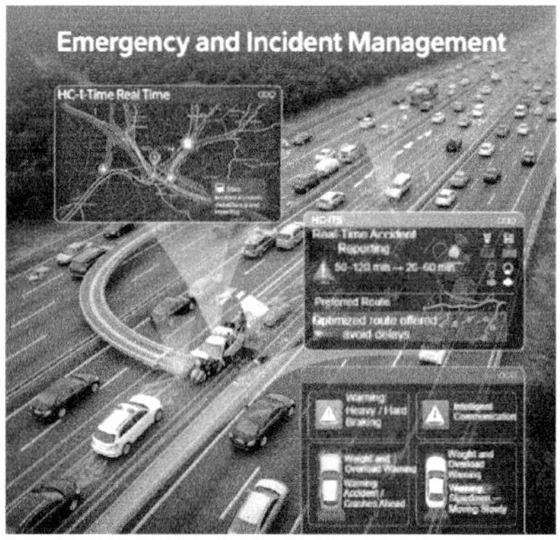

Fig. 9.22 Emergency and incident management.

2. Apps capable of performing command through voice.
3. Navigation tools to help visually impaired users.

Assistive systems as such will provide mobility inclusion, and subsequently, mobility breakthrough for vulnerable groups' transport accessibility (see Figure 9.23).

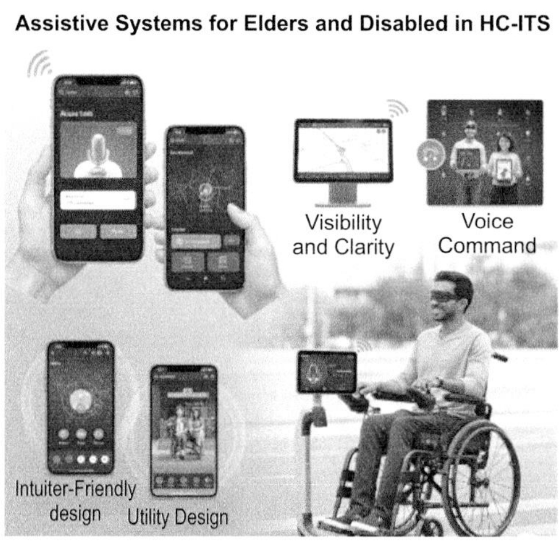

Fig. 9.23 Assistive systems for elders and disabled.

HC-ITS, as a system, can facilitate public safety, facilitate accessibility, and produce a mobility system responsive to humans.

7. Analytical Models for Assessing Performance

In order to assess the performance of HC-ITS, as well as maximize that performance, you will likely rely on various analytical models. These models help with data-driven decision-making, capturing system behavior and anticipating outcomes by forecasting and arranging a human-responsiveness mode according to human and environmental variables (see Table 9.6).

Table 9.6 Analytical models for performance evaluation in HC-ITS.

Model Type	*Description*	*Typical Applications*	*Strengths*	*Limitations*
Statistical & Probabilistic Models	Use historical and real-time data to uncover trends, predict probabilities, and assess uncertainties	Crash prediction, travel time estimation, demand modeling	Simple, interpretable, effective for trend analysis	Limited adaptability to real-time dynamics
Machine Learning-Based Models	Data-driven models that learn from past patterns to predict or classify system behavior dynamically	Congestion prediction, signal timing, user behavior modeling	High accuracy, adaptive, scalable to large datasets	Require large, high-quality data, low interpretability
Fuzzy Logic & Decision Trees	Handle vagueness and complexity by simulating human-like reasoning through rule-based systems	Adaptive signal control, comfort evaluation, qualitative assessments	Interpretability, effective in uncertain conditions	May oversimplify complex interactions
Game Theory & Behavioral Models	Model multi-agent interactions and strategic decision-making with human-centric behavioral assumptions	Autonomous vehicle interactions, shared mobility pricing	Captures strategic and social behavior dynamics	Requires accurate behavior modeling, high computational cost
Graph-Theoretical & Network Models	Represent transportation systems as graphs to analyze structure, flow, and connectivity	Route optimization, bottleneck identification, multimodal planning	Visual, mathematically rich, efficient for large-scale systems	Static unless integrated with real-time data

7.1 Statistical and Probabilistic Models

Statistical models understand trends and make inferences based on historical and real-time data.

- Statistical Models: Look at a relationship of two or more variables at a time (e.g., solve a travel time equation while including weather as one variable in the regression analysis).
- Probabilistic Models: Look at uncertainty and risk (e.g. probability of crash, delay likelihood) (see Figure 9.24).

Applications:

- Demand forecasting
- Risk and safety analysis
- Travel-time estimates based on various conditions

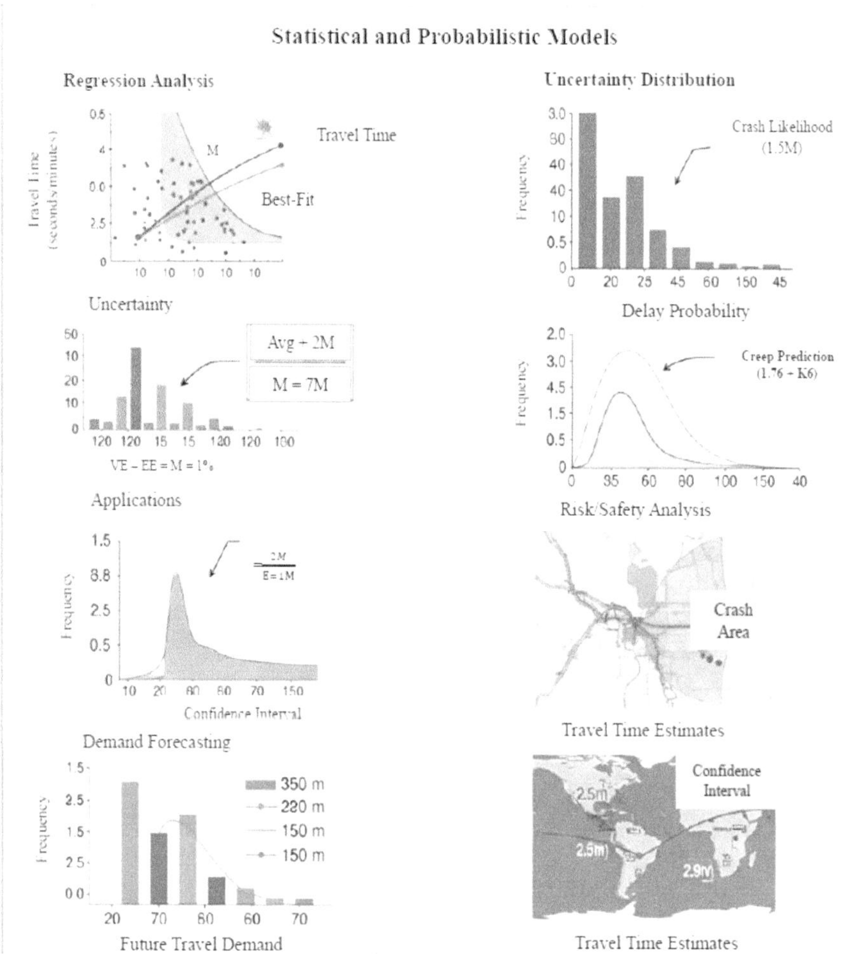

Fig. 9.24 Statistical and probabilistic models.

7.2 Machine Learning-Based Predictive Models

Real-time prediction and adaptive control are enabled by using machine learning (ML) to identify patterns in large sets of data (see Figure 9.25).

- Supervised Learning: Predicting congestion or travel demand.
- Unsupervised Learning: Spotting anomalies such as unexpected behaviors of traffic.
- Reinforcement Learning: Optimizing route planning or signal timing.

Benefits:

- Self-learning capabilities
- Personalized user experience
- Scalability to any city/region

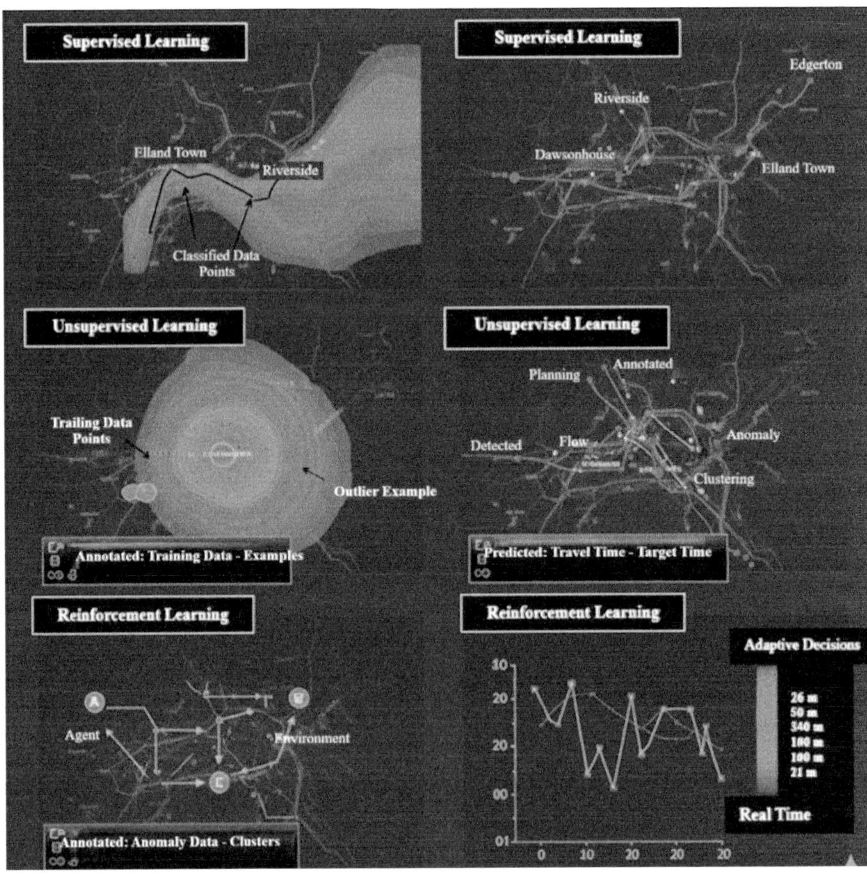

Fig. 9.25 Machine learning-based predictive models.

7.3 Fuzzy Logic and Decision Trees

Fuzzy logic and decision trees provide great handling of uncertainty, vagueness, and large decision spaces which are non-linear (see Figure 9.26).

- Fuzzy Logic: Gives flexibility as it allows us to represent imprecise inputs such as very heavy traffic or moderate delay, which assists us in determining the best control option.
- Decision Trees: Provide explanation, in the form of rules, to predict future events (e.g. if-then rules associated with adaptive signals).

Use cases:

- Adaptive traffic signal control
- Comfort level estimation for passengers
- Route recommendation systems

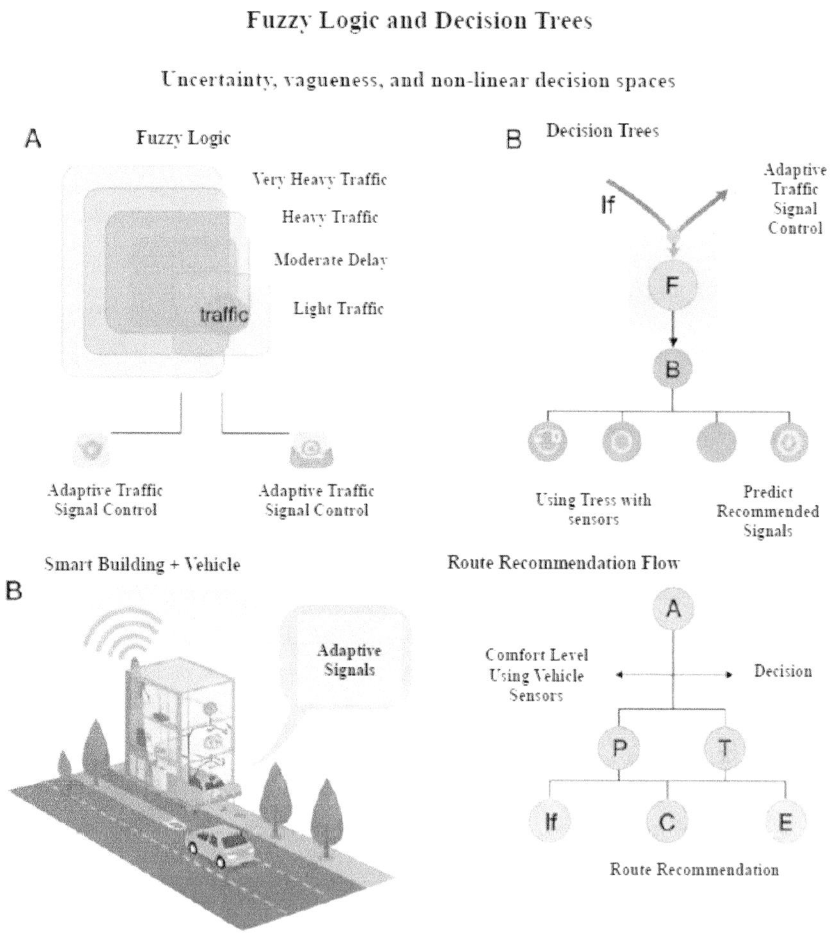

Fig. 9.26 Fuzzy logic and decision trees.

7.4 Game Theory and Behavioral Modeling

Game theory can model and analyze the interactions and behaviors of two or more agents, for example, vehicles, pedestrians, or service agencies, especially if their agents' objectives may conflict with those of other agents (see Figure 9.27).

Behavioral modeling considers psychology and how humans think or do things (e.g., how humans respond to congestion pricing or lane changes).

Nash equilibrium and modeling strategies are used for scheduling of autonomous vehicles and simulations with multiple agents.

Some examples:

- Smart parking, toll pricing
- Changed behavior with shared mobility, and coordination of vehicles fleets
- Conflict resolution at intersections

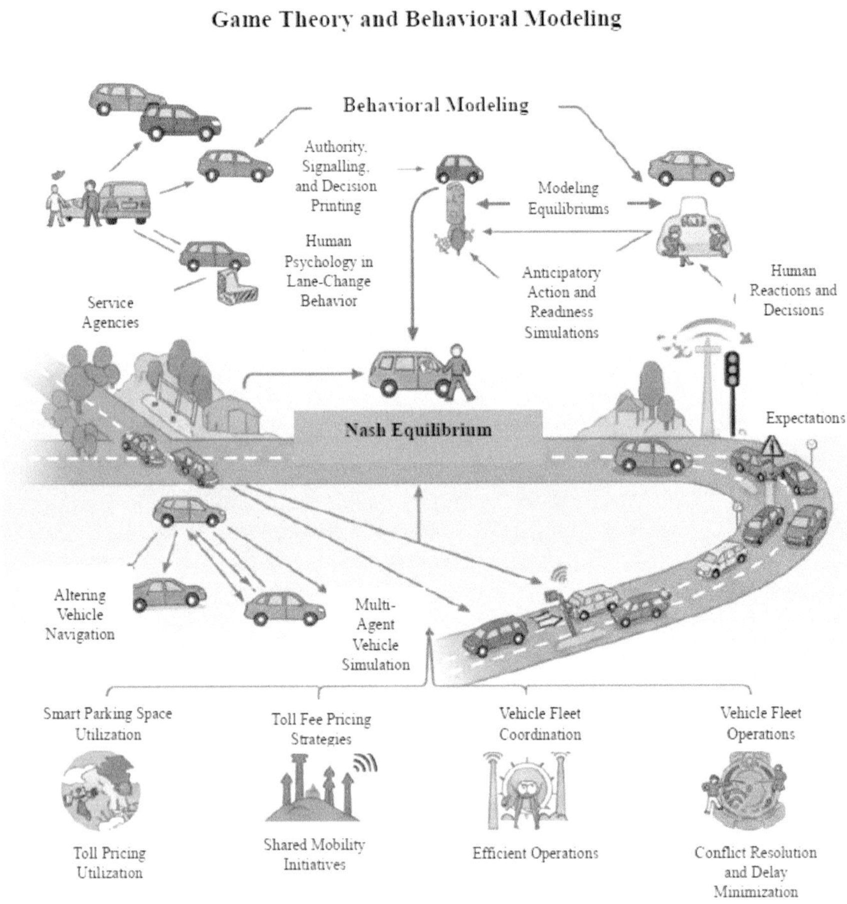

Fig. 9.27 Game theory and behavioral modeling.

7.5 Graph-Theoretical and Network Models

Healthier and safer transportation networks can be constructed using graph theory, with intersections as nodes and roads as edges (see Figure 9.28).

- Shortest path algorithms: Used for route recommendations to optimize paths.
- Centrality measures: To identify central intersections or slim vulnerable locations in a region.

- Analysis of a dynamic network: To track the system for maintenance or even disasters.

The benefits of using graph-theoretical and network models:

- Network-wide optimization
- Identification of bottlenecks
- Improved planning processes for multimodal transport systems

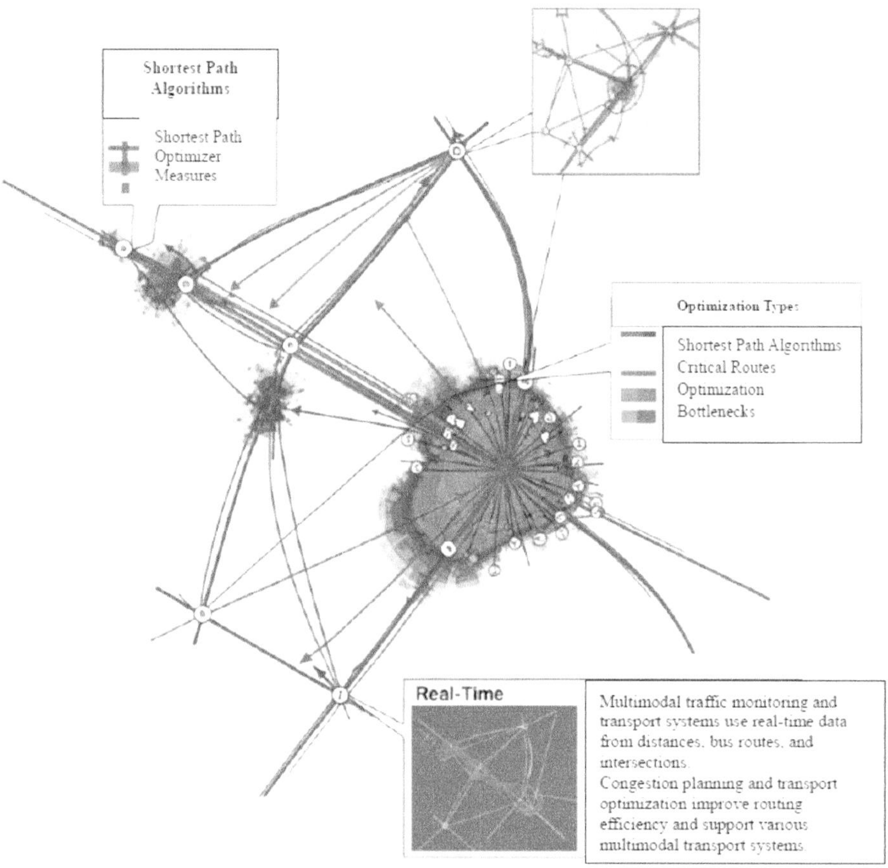

Fig. 9.28 Graph-theoretical and network models.

Each model analysis provides a different approach to understanding, evaluating, and improving HC-ITS, and each modeling domain has advantages. Overall, the most effective evaluative approach would include a hybrid model that integrates statistical approaches with machine intelligence and behavioral modeling that provides flexibility in addressing HC-ITS.

Table 9.7 Challenges in evaluating HC-ITS.

Challenge Area	*Description*	*Implications*	*Example Scenarios*
Dynamic Human Behavior	Human actions vary by context, emotion, culture, and experience	Difficulty in building consistent, adaptive evaluation models	Variability in pedestrian responses or driver attention
Privacy & Data Governance	Evaluation relies on sensitive data (location, biometrics, preferences)	Legal and ethical barriers to real-time monitoring and data sharing	GDPR compliance, anonymization challenges in surveys
Lack of Standardized Metrics	No universally agreed-upon human-centric evaluation criteria	Inconsistencies across studies and cities, hindering benchmarking	Differing comfort or trust metrics in AV evaluations
Technological & Infrastructure Gaps	Inadequate sensors, outdated systems, or poor connectivity in some regions	Limits real-world deployment and testing of advanced HC-ITS	Rural areas lacking IoT or cloud support
Integration with Legacy Systems	HC-ITS must coexist with traditional traffic systems and manual controls	Causes data incompatibility, reduced efficiency of smart components	AVs navigating non-smart traffic lights
Ethical & Policy Constraints	Absence of clear laws and social consensus around autonomy, fairness, and algorithmic bias	Risk of misuse, resistance to adoption, or unintended discrimination	Biased routing decisions disadvantaging low-income areas

8. Problems Evaluating HC-ITS

Assessing HC-ITS is complex because of the convergence of continuously evolving technologies, varied human behavior, and societal norms. Even though HC-ITS has the potential to revolutionize transportation, a number of fundamental problems exist that make consistent evaluative assessment difficult (see Table 9.7).

8.1 Dynamic Human Behavior and Variability

Human behavior is intrinsically dynamic and contextually driven, making evaluation problematic.

- Responses to alerts, detours, or automated assistance all vary.
- Cultural, social, psychological, and situational factors impact human interaction with and responses to systems (see Figure 9.29).
- Individual variances in user preferences and adaption to learning curves affect acceptance.

Implication: Models and metrics must continually adapt to the agrarian dynamics of human behavior, and this is a significant methodological and technical challenge.

Fig. 9.29 Dynamic human behavior and variability

8.2 Privacy and Data Governance Issues

HC-ITS relies heavily on user data from sensors, mobile devices, and vehicles. This raises the following privacy and data governance issues:

- Misuse of data/risk of data breaches
- Informed consent from users
- Adherence to a relevant legal framework (e.g., GDPR, data localization laws)

Implication: Evaluation systems must be built around strong principles of data privacy and ethical usage protocols. Providing "grey area" data at a granular level may not be permissible (see Figure 9.30).

Fig. 9.30 Privacy and data governance issues.

8.3 Lack of Standardized Metrics

There is currently no agreement on common standards by which to measure human-centred aspects of ITS (see Figure 9.31).

- Trust, comfort, and inclusivity metrics tend to be subjective or poorly defined.
- With no benchmarking, values on a per region or system basis are meaningless in scope.

Implication: Evaluation outcomes may not be generalizable or replicable across different contexts.

Fig. 9.31 Lack of standardized metrics.

8.4 Gaps in Infrastructural and Technological Foundations

In many regions at LU–Land Use, the critical underlying infrastructures to support HC-ITS are insufficient, including the following:

- Sensor coverage
- Internet connectivity
- Interface with smart systems in various physical or virtual locations

Implication: These foundational issues will inhibit the ability to conduct real-world (pilots) and/or the number of evaluation occurrences across various geographical/ elevational contexts (in particular underdeveloped from a rural perspective) (see Figure 9.32).

Fig. 9.32 Gaps in infrastructural and technological foundations.

8.5 Integration with Legacy Systems

Many transportation systems are largely dependent on legacy systems.

- They are difficult to redevelop and retrofit to include contemporary AI, IoT, or cloud-enabling components.
- They are in a graceful period with some older systems but surviving on old evaluative platforms.

Implication: Poorly designed evaluation findings may be skewed or influenced by legacy system functionality or functionality of only existing outcomes (see Figure 9.33).

Fig. 9.33 Integration with legacy systems.

8.6 Ethical and Policy Constraints

There are important ethical considerations related to evaluating HC-ITS (see Figure 9.34), such as the following:

- Equity in AI-based decision-making (e.g., who has priority in routing)
- Algorithm-based biases that affect people from marginalized groups
- Policy vacuums concerning autonomous vehicle trials and social monitoring

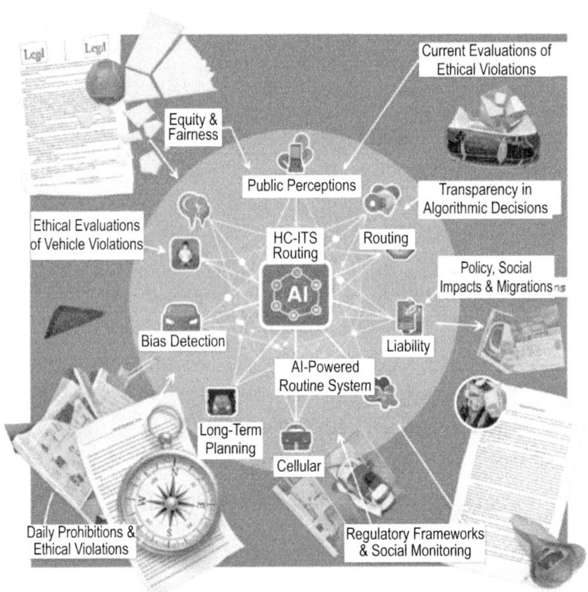

Fig. 9.34 Ethical and policy constraints.

Implication: Currently without clear regulatory frameworks to support their evaluation, it is possible for such evaluations to elide ethical violations or social inequities. Evaluating HC-ITS is more than a technical problem; it is a socio-technical problem that balances accuracy, fairness, privacy, and responsiveness. To address this problem will require a multidisciplinary approach, strong policy supports, and a commitment to evolving practices in evaluation science.

9. Future Directions and Research Opportunities

We can continue to improve HC-ITS evaluation frameworks as these systems evolve in order to maintain relevance, responsibility, and usefulness. The future of HC-ITS evaluation will be about building systems that are accountable, engaging, participatory, and systematic—centering human experience in all design and measurement practices (see Table 9.8).

Table 9.8 Future directions and research opportunities in HC-ITS.

Future Focus Area	*Description*	*Research Opportunities*	*Potential Impact*
Explainable and Transparent Evaluation	Developing interpretable AI and evaluation systems to ensure user trust and regulatory acceptance	Design of human-readable dashboards, rule-based scoring systems	Enhanced public trust, regulatory compliance
Integration of Digital Twins & VR	Using real-time digital replicas and immersive simulations for system design and testing	Scenario simulation for emergency response, user behavior testing in VR	Safer deployment, cost-effective prototyping
Emotion-Aware and Adaptive Systems	Embedding emotional intelligence and cognitive load detection in transport evaluation	Algorithms that adapt based on driver fatigue, stress, or confusion	Reduced human error, personalized mobility experience
Crowdsourced & Participatory Sensing	Leveraging user feedback and data from personal devices for real-time system improvement	App-based hazard reporting, ride feedback integration in analytics	Increased user engagement, richer evaluation data
Standardization of Metrics & Protocols	Creating universal guidelines and key performance indicators (KPIs) for HC-ITS across cities, platforms, and vendors	Development of global human-centric performance frameworks	Comparable assessments, accelerated adoption

Contd.

Table 9.8 *Contd.*

Future Focus Area	*Description*	*Research Opportunities*	*Potential Impact*
Ethical AI and Responsible Innovation	Embedding ethics, fairness, and social responsibility into system design and evaluation	Bias mitigation, value-sensitive design approaches	Inclusive, accountable intelligent transport systems
Interdisciplinary Collaboration	Bridging engineering, psychology, urban planning, and public policy domains	Joint policy-tool development, co-creation of evaluation criteria	Holistic, human-aligned system development

9.1 Beginning to Develop Explainable and Transparent Evaluation Systems to Support HC-ITS

Given the increasing presence of AI in HC-ITS, there is a need for evaluative frameworks that can be explained and are also accountable.

- AI models that inform decision-making need to be explainable to system developers, users, and policymakers in deciding when and how to take specific actions.
- The evaluation systems must be transparent as to the basis for performance scores and the resulting outcome(s).

Research Opportunity: Designing interpretable AI models and human readable evaluation dashboards that increase trust and accountability.

9.2 Digital Twins and Virtual Reality as Support for Simulation

Digital twins are real-time, virtual copies of a physical transport system that facilitate testing of transport systems at scale and under safe conditions.

- Virtual reality (VR) and augmented reality (AR) are useful for simulating user behaviour within an immersive environment
- VR and AR can be used by researchers to test the performance of HC-ITS under circumstances or scenarios that cannot occur without actual deployment of a system.

Research Opportunity: Use of simulation environments for testing of behavior, evaluation of experience of use, and predicting when challenges might occur before actual deployment.

9.3 Human-Centered AI as Emotion-Aware Systems

Next-generation HC-ITS should integrate emotion-aware computing into HC-ITS and adaptive systems to support user autonomy needs. For instance:

- AI that recognizes driver stress, fatigue, or frustration
- Interfaces that adapt to the emotional or cognitive state of users

Research Opportunity: Design systems that consider not just the attributes of technical performance but also the emotional well-being and cognitive comfort of users.

9.4 Crowdsourced and Participatory Sensing Models

There are ways to engage the user community in data collection and evaluation that offers useful real-time feedback, for instance:

- Crowdsourced apps to report hazards, accessibility challenges, or service satisfaction
- Participatory sensing from mobile devices and wearable technology

Research Opportunity: Implementing crowd intelligence in frameworks that assess, monitor, and improve HC-ITS with negligible overhead.

9.5 Standardization of Evaluation Protocols

At this moment in time a significant barrier to progress in HC-ITS is that there are no global standards for evaluating HC-ITS.

- There is a need for common metrics and benchmarks for evaluation
- There should be interoperable protocols for cities, countries, and vendors to work together and to share best practices.

Research Opportunity: Establishing internationally accepted HC-ITS evaluation standards, data formats, and ethical guidelines.

The future of HC-ITS evaluation is the convergence of human-centric and intelligent technology, creating ethically sound, explainable, emotion aware, and participatory systems. Interdisciplinary collaborative work across engineers, designers, psychologists, public policy people, and urban planners will be vital in implementing all of these innovations.

10. Conclusion

The implementation of HC-ITS is an exciting opportunity to further align mobility with human values, personal safety and inclusion as transportation systems continue to evolve to meet the complex demands of societies today. This chapter has explored the fundamentals, key performance indicators, evaluation approaches, applications, analytical models, challenges, and future research directions that are critical for designing and assessing next generation systems.

The core tenets of HC-ITS utilize understanding human behaviors, minimizing risk, and improving the user experience through embedding AI, IoT, and smart infrastructure. A set of performance indicators that account for safety, efficiency, sustainability, equity, and resilience ensure that the systems are not only operable but also socially responsible. The emphasis on multiple evaluation approaches to our understanding of impact, including human-in-the-loop approaches, rooftops, real-world trials, and data-led practices, demonstrates the need to understand what outcome measures in truly human-centered systems.

The evaluation of HC-ITS requires substantive understanding of challenges in terms of human behavior, privacy, infrastructures, and lack of standardization. These challenges make it difficult to seamlessly implement and assess HC-ITS without some compromise on any of the aforementioned aspects. Dealing with these challenges will require an immense collaborative effort between disciplines, encapsulating the precision of engineering, the insight of the social sciences, and ethical considerations.

In the future of HC-ITS evaluation, we can expect evaluations to be transparent, adaptive, and participatory. Technologies such as emotion-aware AI, digital twins, explainable models, and participatory sensing will transform how we measure and improve performance. Furthermore, collaboration across borders will define how organizations develop common evaluation standards and tools and appropriate policy frameworks for international scaling and interoperability.

In summary, evaluating HC-ITS is more than a technical measurement—it is a human-centered aspirational project. It requires us to take seriously how we measure not only how a system moves people but also how systems improve lives. The move from transportation networks to mobility systems requires the integration of human values in core design processes to create safe, just, intelligent, and future-ready systems.

References

Al-Khateeb, K. and J.A. Johari. (2023). Human-centric intelligent transportation systems: a review of enabling technologies and future research directions, *IEEE Access*, *11*, 76420–76438. https://doi.org/10.1109/ACCESS.2023.3278912.

Amini, A., Gilitschenski, I., Phillips, J., Moseyko, J., Banerjee, R., Karaman, S., & Rus, D. (2022). Deep learning for autonomous vehicle control: Algorithms and performance metrics, arXiv preprint arXiv:2202.05997. https://arxiv.org/abs/2202.05997.

Chen, Y., S. Wang and H. Zhang. (2022). A human-in-the-loop framework for evaluating autonomous driving systems, *IEEE Transactions on Intelligent Transportation Systems*, *23*(10), 17204–17218. https://doi.org/10.1109/TITS.2022.3141230.

García, P., et al. (2023). Multicriteria decision analysis for evaluating sustainable transportation solutions: a case study, *Journal of Cleaner Production*, *391*, 136238. https://doi.org/10.1016/j.jclepro.2023.136238.

National Institute for Standards and Technology (NIST). (2022). Framework for evaluating the usability of intelligent transportation interfaces. https://www.nist.gov/publications.

Papadimitratos, P., et al. (2021). Human-centric vehicular networking and mobility management, *IEEE Communications Magazine*, *59*(11), 88–94. https://doi.org/10.1109/MCOM.001.2000314.

Rasouli, A. and J. K. Tsotsos. (2020). Autonomous vehicles that interact with pedestrians: a survey of theory and practice, *IEEE Transactions on Intelligent Transportation Systems*, *21*(3), 900–918. https://doi.org/10.1109/TITS.2019.2901817.

Raza, A., et al. (2022). Ethical and privacy issues in smart transportation systems, *Computer Communications*, *180*, 255–266. https://doi.org/10.1016/j.comcom.2021.08.005.

Shladover, S. E. (2020). Connected and automated vehicle systems: introduction and overview, *Journal of Intelligent Transportation Systems*, *24*(3), 190–200. https://doi.org/10.1080/15472450.2019.1581861.

Sliwa, B., et al. (2021). A review on simulation frameworks for transportation systems with human-centric mobility, *Simulation Modelling Practice and Theory*, *106*, 102180. https://doi.org/10.1016/j.simpat.2020.102180.

Törngren, M. and A. Selberg. (2022). Digital twins for human-centric cyber-physical systems, *Springer Lecture Notes in Computer Science*, 13071, 3–19. https://doi.org/10.1007/978-3-030-98542-6_1.

UNECE (United Nations Economic Commission for Europe). (2023). Policy framework for inclusive, sustainable, and human-centric transportation. https://unece.org/transport.

World Economic Forum. (2021). *Guidelines for Safe and Ethical Use of AI in Transport*. https://www.weforum.org/reports.

Zhou, X. and Y. Xie. (2020). Integrating machine learning in intelligent transportation systems: a review, *Transportation Research Part C: Emerging Technologies*, *117*, 102667. https://doi.org/10.1016/j.trc.2020.102667.

CHAPTER 10

Energy Efficient Wireless Body Area Network Technology for Prevention and Detection of Hasty Cardiac Arrest using IoT

Gousia Thahniyath[1], *Rajesh T. M.*[2], *Girisha G. S.*[3], *Shaila S. G.*[4] and *Ambeshwar Kumar*[5]*

This chapter focuses on providing a remote, real-time heart-health monitoring system for a patient using a secure underlying wireless body area network (WBAN). WBAN works on the principle of radio frequency (RF) communication among sensor nodes placed on the patient's body that continuously monitor the patient's physiological measures, like pulse rate, heartbeat rate, blood pressure, and oxygen level in the blood, and transmit the patient's heart parameters statistics to the gateway for real-time cardiac-event detection. However, WBAN is susceptible to various cyberattacks due to its wireless nature of communication. Hence, it also becomes crucial to have a secure underlying WBAN to prevent any tampering with the patient's health data. Hence, the chapter addresses the reliability and latency concerns commonly faced by WBANs, thereby effectively transmitting the patient's health information to the gateway. The gateway uses the deep learning methodology that analyzes and processes the patient's data to detect any warning signs of a possible heart attack in a patient, sending an alert to the health's professional dashboard for a prompt action that will save the patient's life by preventing and detecting sudden cardiac arrest (SCA) and sudden cardiac death (SCD).

[1-4] Department of Computer Science and Engineering, Dayananda Sagar University, Bangalore, India.
[5] Department of Computer Science and Engineering, GITAM University, Visakhapatnam, India.
https://orcid.org/0000-0001-5688-7366[1]; https://orcid.org/0000-0001-9258-7870[2]; https://orcid.org/0009-0008-1313-3239[3]; https://orcid.org/0000-0002-0810-8767[4]; orcid.org/0000-0001-5393-6361[5]
* Corresponding author: ambeshwar.kumar@gmail.com

DOI: 10.1201/9781003659822-10

1. Introduction

This chapter proposes a low latency and secure wireless body area network (WBAN) system where sensor data from WBAN is sent to the gateway for analysis of the patient's vital health parameters and then transferred to a cloud-based system for deeper examination and storage.

WBAN uses wireless radio frequency (RF) technology that communicates among the motes placed on the patient's body. Biosensors (e.g., pulsed oxygen meter, electromyogram (EMG), electrocardiogram (ECG), and SpO2 sensor board Mica2 motes are placed on the patient's body (Li et al., 2017).

The WBAN IEEE 802.15.6 standard (Salehi et al., 2016) provides short-range, least-power communication among the sensors. In order to capture the environmental conditions, like temperature, dust, motion, and light of the patient's room, environment sensors (i.e., a MicaZ board) are placed on the walls of the patient's rooms. The biosensors and the environment sensors form an internet of things (IoT) network. The environment data and the data from the biosensors are reliably transferred with low latency to the gateway using the shortest path routing and secure transmission methods that are employed in the proposed methodology. The data from environment sensors and biosensors will help to analyze the vital health parameters of the patient along with the environmental condition of the patient's room. At the gateway, deep learning techniques analyze the heart's vital signs as sent by sensors to detect any level of risks involved. The analyzed data is transferred to the IoT cloud for further analysis through mobile data technology like General Packet Radio Service (GPRS) or long-term evolution (LTE). A web-based protocol like Hyper Text Transfer Protocol (HTTP) and a lightweight protocol for IoT applications like MQTT Message Queuing Telemetry Transport (MQTT) are used in the IoT cloud to provide medical professionals with timely and visual ECG data for additional diagnostics.

WBAN's main aim is to bring convenience to the patient, but WBAN, being wireless in nature, faces reliability concerns, and sensors placed on the body of the patient are very tiny and thus have a very low battery capacity. Hence, it is crucial that the sensors are kept alive for a longer time for efficient data transmission. There is a need to combine strategies for addressing energy efficiency, latency, and reliability concerns in WBAN. In this chapter, we propose a low-latency, reliable, and energy-efficient routing protocol for the underlying WBAN to make the purpose of remote health monitoring effective.

The rest of the chapter is organized as follows: Section 2 discusses the WBAN network, focusing on the overview of WBAN, medium access control (MAC) layer, challenges, and MAC protocols in WBAN. Section 3 is a literature survey on heart attack detection using wearable sensors. Section 4 is the proposed methodology. Section 5 is the Conclusion.

2. Wireless Body Area Network (WBAN): An Overview

WBAN is a technology where the biosensors communicate wirelessly using RF technologies. Biosensors (e.g., pulse oximeter, EMG, EKG) form the IoT network, and they are used to measure the patient's vital signs. The WBAN IEEE 802.15.6

standard provides short-range and low-power communication solutions (Salehi et al., 2016). BAN specifications according to IEEE-BAN-Summary (Gavilovska, 2010) are shown in Table 10.1.

Table 10.1 BAN specification.

BAN Parameters	*Specifications*
Date range	10 KB/s–10 Mb/s
Distance	2–5 m
Density	2–4 nets/m^2
Set-up-time	1 sec
Latency	10 ms
Power usage	approx. 1 mW/Mbps
Network size	Up to 100 devices

Source: https://www.cse.wustl.edu/~jain/cse574-08/ftp/ban/index.html.

The various topologies that are supported in WBAN are tree, bus, point-to-point, star, mesh topology. The star topology is the most suitable topology used in WBAN as it is best suited to reduce latency and is used in high bandwidth. Here, the network is controlled by a master node, which has more power than other nodes. It is usually small and uses a single hop for communication. Failure of the master node leads to failure of the network which then leads to low reliability. In point-to-point topology, communication is done directly among the peers concerned, who are connected to each other. Bus topology has a single bus shared among all the nodes. The bus should use collision avoidance when motes try sending the packets across the link at the same time. Failure of the link leads to the disconnection of the network. In linear topology, the nodes are connected to each other in a linear way. The communication is done two-way Braem, B. et al. (2011). The terminal nodes connect to only one next neighbor mote. The moves are dependent on each other to forward the message. In ring topology, the nodes are connected in a ring-like structure. Failure of one node disconnects the networks. Tree topology has nodes arranged in different levels starting from the root node. The levels closer to the root nodes have more power and processing power than other lower-level nodes. Because of the hierarchical nature of the nodes, it is easy to identify the faults. The tree topology is also scalable, but it becomes complicated as we add more nodes. Mesh topology connects the nodes to each other in a manner that all motes are linked to each other, thereby having multiple routes to the destination. Whenever reliability is a concern, mesh topology is the best option. A high number of connections can lead to a complicated network.

WBAN MAC Layer: The MAC protocol provides sharing of the communication channel among all users efficiently with fair and equal access to all the users, thereby providing an efficient usage of resources. The MAC layer maintains the balance between packet delivery and energy efficiency. A good MAC layer protocol reduces energy consumption in WBAN because of various reasons like collision of packets, over hearing of nodes, idle listening, transferring control packets, and transition between sleep and active state. In the carrier sense multiple access with collision avoidance (CSMA/CA), prior to the transfer of the data, node priorities are set. This approach is easily adaptable as there are no time synchronization limits. The other approach is the time division multiple access (TDMA) approach, where time slots are given to each user. Its best use is for the energy-efficient MAC protocol.

***Challenges in WBAN*:** The major challenges faced in establishing a secure, reliable WBAN are as follows:

- **Size Constraint:** WBAN is miniature in size hence it has low bandwidth and low computation capability.
- **Security Challenges:** Security is a very important factor in body area network, as the health information of a patient must reach the health professional in a reliable manner.
- **Data Confidentiality:** The biosensor's data is critical for safeguarding the health information of the patient and should not be accessible to unauthorized users via various attacks like eavesdropping. The attackers can access the data by correlating the data streams.
- **Data Integrity:** The data is subjected to malicious behavior during transmission, leading to alteration of the data.
- **Authentication:** The sensor node can be impersonated by a malicious node.
- **Limited Resources:** The sensors comprising BAN are very small in size with limited processing power and storage. Hence, complicated security algorithms that require more processing power and storage cannot be used in WBAN.

Quality of service should address these challenges to provide a reliable, energy-efficient network with minimum delay and minimum electromagnetic interference. As a result, a dependable connection requires the development of an efficient routing protocol that addresses the concerns.

***Exploring MAC Protocols for WBANs*:** WBAN previously used the IEEE 802.15.4 protocol as it achieves low-power communication and low power rate. Many studies, including Timmons et al. (2004), have analyzed the performance of IEEE 802.15.4 non-beacon systems. They found that the protocol worked better only for a low data rate, which may not be always practical as WBAN data rate fluctuates between 10 Kbps to 10 Mbps. Li, C. et al. (2017) concluded that IEEE 802.15.4 is better when throughput and latency are the main criteria, but it leads to a higher cost of power consumption. IEEE 802.15.4 runs in the unlicensed 2.4 GHz range, and there is a risk of involvement from IEEE 802.11 devices and microwave ovens.

Li and Tan (2009) used information on the heartbeat's rhythm to synchronize the nodes and execute TDMA in their work. Fixed slots are given to each of the participating nodes for transaction. This reduces the energy consumption. The time slot in TDMA is created by the heartbeat of 60–200 beats per minute, which causes the peak interval to be in the range of 300–100 ms. The resynchronization scheme is implemented. But when traffic is low, there is a loss of bandwidth. H-MAC is not adapted to the changes in the traffic.

In order to accommodate dynamic traffic, BodyMAC introduced a downlink frame and another uplink frame dedicated to normal traffic (Fang et al., 2009). The uplink frame has the contention access period (CAP) and the contention free period (CFP). The CAP allows smaller packets to be sent and employs CSMA, which could lead to an increase in energy use. However, in CFP guaranteed slots are given by the coordinator for on-demand traffic in order to avoid unnecessary wake-up periods. Wake-up radio is accommodated, which wakes up the node only when required so that on-demand traffic can be accommodated. For regular WBAN traffic, a reservation-based dynamic TDMA (DTDMA) protocol (Li et al., 2009) is proposed, in which slots are allotted to nodes that have packets, and those packets that are buffered are forwarded to others when the data transmission/reception is complete. In the DTDMA protocol, the novelty is that the super frame structure has CFP to send normal packets, the period of which can be adjusted. Then CAP, which is fixed, is used to send short command messages and the inactive period is used to save energy. The inactive period depends on CFP and can be adjusted accordingly. This structure reduces energy consumption and data loss rate. Emergency packets are not considered.

3. Literature Survey on Heart Attack Detection System Using Wearable Sensors

Basheer et al. (2021) have proposed a hybrid fuzzy-related decision tree approach that provides a better accuracy rate and prediction rate in the starting phase. However, it consumes a lot of time and is sensitive to noise.

Sarmah (2020) has proposed a methodology for cardiac disease analysis. The data from the biosensors is securely communicated to the cloud using Advanced Encryption Standard algorithm and is effectively analyzed using the deep learning modified neural network (DLMNN) classifier. The model proposed does not do real-time analysis of the data.

Leijdekkers et al. [2] (2018) highlighted the delay that occurs from the onset of a heart attack to the notification of a specialist. An ECG sensor is worn by the client, and a mobile is used to convey to the specialist or caretaker when the client might be experiencing a heart attack. Two ECG chronicles on the mobile verify the symptoms. This helps the client to get correct assistance through automated calls. The ECG is documented and dichotomized steadily at the mobile tool, making use of a 2-terminal and 1-lead coronary heart monitor. The control applied is known as coronary heart rate abnormalities.

Gurjar et al. (2018) have found a technique to note the occurrence of a heart attack through a sensor that monitors heartbeat. This can be used in all places without any restriction. The sensor is connected to a microcontroller. You can learn the pulse and send it over the internet. The client can set a heartbeat limit, and a trace will be performed when the heartbeat exceeds the margin. An affected person uses a transmitting circuit, and specialists, doctors, and caretakers use alternative circuits. Heart rate sensors are used to recognize a patient's current pulse rate and display it on a liquid crystal display (LCD) screen.

Patel et al. (2018) have proposed a different method to detect a heart attack by observing pulses on the IoT. In this method a heart-beat sensor, Arduino board, and a Wi-Fi module are used. The sensor detects pulse reading and displays the heart on the LCD screen. The Wi-Fi will transfer the data over the internet. The framework lets in a hard and fast factor which could assist in determining if a patient is sound or now no longer, with the aid of checking pulse and setting factors. By checking the heartbeat of the affected person and whether the heartbeat is going above or below the set value, an alarm message is sent. An android software display is intended to observe the heartbeat of an affected person and display it effectively.

Santhanakrishnan et al. (2019) proposed a growing framework as a good way to decrease the death rate due to coronary heart attack through the early stage of coronary heart attack. In this framework, a pulse-sensor, GSM, and GPS are used. The pulse sensor shows the pulse regularly. The GSM and GPS provide the history of records. The threshold is fixed appropriately. When the threshold is exceeded or falls below the set limit, the microcontroller logs the data and connects to the nearest healthcare centre.

Ali and Ghazal (2017) have suggested an approach that checks the effect on a vehicle when a driver experiences a heart attack. This works with IoT technology. The researchers proposed a voice-controlled program and a motion control program to detect cellular heart attacks. Both accomplished fusion sensors, which has improved their customer reputation and accessibility. The mobile heart-monitoring device primarily considers how the user can operate it inside a vehicle, while the second mode supports use outside the vehicular environment.

Selvathi, D et al. (2017) believed that heart failure could be predicted only after a heart attack and that there was no specific system. Heart failure is detected by measuring a patient's heart rate, which is the heart rate per minute. Critical fluctuation of the heart rate (60–90) is a sign of a heart attack. The microcontroller records these results and checks the heartbeat. In the event of a stronger impact or below a certain size, the microcontroller activates the cellular module and sends an alarm signal to special contacts on the microcontroller.

Song et al. (2014) have proposed a technology in which sensors are transportable and may be placed, in addition to gadgets such as mobiles. The sensors monitor the continuous state of the frame, and save or send the result to the expert. Under these pressures, you can help people provide more care, such as signaling dangerous diseases or helping people prepare for disaster-related issues. In order to persuade a progressive win-win situation with challenges and difficulties, the researchers provide a high-quality summary of the great improvements made to the phone frame and touch texture. In addition, he conducted a more comprehensive investigation to check the relevant detection and calculation systems to find the development pressure of each sub-region.

Studies have focused on the possibility of a heart attack in a patient. The underlying WBAN network is exposed to vulnerabilities that lead to low latency, security issues, network getting collapsed, and false diagnosis of patients' health. There is very little research that covers most of the major difficulties that WBAN encounters when it comes to improving its efficiency. Hence, in our work, we present a lightweight, energy-saving, and reliable routing protocol that effectively reduces the latency incurred in the communication and increases the lifetime of the IoT system.

$$E_{tranx} = E_{transelec} + E_{amp} * dist^2 \quad ...(4)$$

$$E_{recvd} = E_{recvdelec} \quad ...(5)$$

where $E_{transelec}$ is the energy for radio transmission. $E_{recvdelec}$ is the energy needed for receiving. E_{amp} is the energy for amplifying the signal. $dist^2$ is the distance between the nodes.

The cost function (CF) parameter for each energy category is computed as follows:

$$CF = \frac{AvGHopSize}{N * Eresidual} \quad ...(6)$$

where N is the count of sensors, *AvGHopSize* is the average hop size to the base station, *and Eresidual* is the leftover energy of the sensor node in that cluster.

Sensor having minimum CF will be chosen for cluster head (CH) based on the amount of energy remaining.

Enhancement of Cluster-Head-Selection Scheme: The cluster-head-selection scheme that has been proposed enhances the importance of the ECG sensor that is attached to the sensor. If $S(n).value$ is the importance level of each sensor, then the updated cost function is as follows:

$$CF = \frac{AvGHopSize}{N * Eres} * \frac{1}{S(n).value} \quad ...(7)$$

- **Cluster Formation**

The CH broadcasts the data to non-cluster-head devices. The CH generates a TDMA plan for every client node and transmits it to all nodes. During this schedule the client node is supposed to interact with the cluster head. Non-cluster-head-mote elects itself as CH. Once the CH gets its client, it is able to form a cluster.

- **Reliable Routing (Link Reliability and Path Reliability)**

Link-Cost Function: To find the next hop node, each node utilizes the link-cost function. This function includes leftover energy of the node, a leftover buffer size of the node's queue, and lin-reliability between nodes *i* and *j*, respectively. The link-cost function $Cost_{ij}$ is defined in eq. (8). For the next hop, the node with the highest maximum link-cost function is chosen. In order to reach the sink, each sensor node considers the link cost function by reaching out to its neighbors.

$$Cost_{ij} = \frac{Eresidual_j}{Einitial_j} * coef\,f_a + \frac{Buf\,fercrn_j}{TotalBuf\,fer_j} * coef\,f_b + Linkreliability_{ij} \quad ...(8)$$

A dynamic method of calculating link reliability is calculated in eq. (9) in which the balance between the effects of the baseline factor and the observed success rate of packet transfers is controlled by the parameter γ

$$Linkreliability_{ij} = (\gamma - 1)\, Lnkreliability_{ij} + \quad ...(9)$$

$$\left(\frac{number\ of\ succesfulpackettransmission}{totalnumberoftranmission + retransmission}\right)$$

Link Reliability: This is achieved by using a watchdog mechanism to find the number of successful packet transmission ratios. To have reliable routing, it is very important to establish trust among nodes. The target node is monitored using a watchdog mechanism for any suspicious behavior. Watchdog mechanism is considered where the node monitors the target node. It monitors whether the target node forwards the data to the next hop. If it does not forward, then the node is malicious. A sequence number is used for each transfer packet. The correspondent node sends the acknowledgment of the received packet within a scheduled time. If the sender node does not get acknowledgment, retransmission of the packet is done.

Path Reliability (Stable Path Selection): The reliability of data in WBAN is very critical (Gousia et al., 2022). The path selected for the data transmission should be stable. Hence for each selected path, the path reliability is obtained depending on the overall node distribution. The path reliability selects a stable path, hence obtaining end-to-end reliability.

$$AvgPathreliabilty = 1/n \sum_{i=1}^{n} Linkreliability_{ij} \qquad ...(10)$$

Eq. (10) calculates the average reliability of all the links along the path.

$$\text{standard deviation} = \sigma LR = \sqrt{\frac{\sum_{i=1}^{n}(Linkreliability_{ij} - AvgPathreliabilty)^2}{n}} \qquad ...(11)$$

This standard deviation provides an indication of the degree to which the mean reliability of all the links along the path differs from the reliability of individual links. Greater variability in connection dependability throughout the way is indicated by a higher standard deviation, which may influence the overall reliability of data transmission along that path.

- **Increased Network Lifetime**

Increased network lifetime is achieved by maintaining balance of energy among nodes. IoT devices in WBAN are limited in power, so placing these devices for a longer duration is the main challenge in resource management. In order to avoid overloading a particular node, $\frac{Eresidual}{Einitial}$ gives the node energy ratio. Those nodes with high node energy ratio are selected for the next hop in the route section procedure. The residual energy *Eresidual* balances the energy consumption among nodes. The current buffer size (Buffercrn) used minimizes the latency.

The TDMA schedule that is used in CF reduces the energy wastage across BAN and increases the network lifetime.

4.2 Analysis Model of Physiological Signals Collected at the Gateway

- **Authentication of the User:** Since there is a possibility that the malicious modes forget the ethernet address, eavesdrop on the communication channel, change, replay, insert bogus data, and send and receive packets at various power levels, it is very important to authenticate the user. Since the sensors in WBAN exhibit size constraint, it is limited in processing power. It is highly recommended that lightweight schemes be incorporated, which will not drain off the sensor node battery. A non-cryptographic node authentication scheme is suggested at the control unit to filter out signals received from malicious nodes.

A Non-cryptographic Node Authentication Scheme by the Control Unit

Measuring the received signal strength (RSS) variations: The control unit (gateway), sends a control packet to the sensors. The sensors are required to send a reply within a designated time. The control unit calculates the average RSS (*AvgRSS*) for each of the sensors comprising BAN as follows:

$$\forall_{S(j=1)}^{S(j=N)} RSS1, RSS2, RSS3.....RSSSN \qquad ...(12)$$

Calculate the Average Received Signal Strength (*AvgRSS*) as

$$AvgRSS = \sum_{I}^{N} \frac{RSSI}{Nt} \quad \text{for} \quad (I = 1 \text{ to } N) \qquad ...(13)$$

Nt is the discrete time point which is predefined. The sensor node whose RSSI varies too much from the average RSS calculated in eq. (13) is not considered by the control unit.

The data that is gathered from the physiological sensors is transferred to the gateway. The gateway (a personal server) processes the data collected from various sensors comprising WBAN to form a better insight of the data.

- **Deep Learning Model at the Gateway:** An analysis model at the gateway uses deep learning techniques to classify the abnormalities in the received vital signs of the patient as illustrated in the analysis model depicted in Figure 10.1.

The automatic output of classified ECG gives us the result which shows the type of heart attack. We can also predict the other two types of heart attacks by various methods. We currently have three methods.

- **Measurement of Height and Width of the ECG Graph:** Vertically, the ECG graph measures the height (amplitude), i.e., the voltage of a given wave or deflection. Horizontally, the ECG graph measures the width, i.e., it records the time of the given wave. Calculating the height and width of a heart disease patient and then comparing it with a normal ECG report predicts whether it is a mild attack or a severe heart attack.
- **Hierarchical Condition:** A risk adjustment technique that produces risk ratings for Medicare beneficiaries who are elderly or disabled is known as hierarchical condition. Basically, this method helps us check for risk factors.
- **Length:** Between two QRS complexes, count the 1 mm squares. The ECG paper travels at a speed of 25 mm/sec through the ECG printer, resulting in:

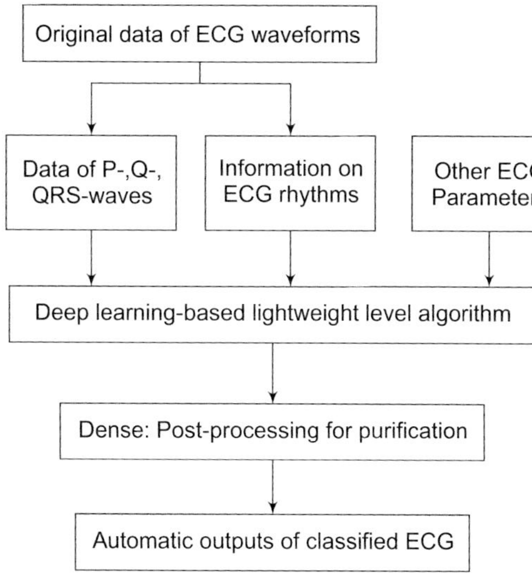

Fig. 10.1 Analysis model using deep learning.

HR (b/m) = (25 mm/sec * 60 min/sec)/number of squares.

[HR is heart rate]

This method is also like the height and width method, but here the heart rate is calculated for a heart patient's ECG report and then compared with a normal ECG report as shown in Figure 10.2. Therefore, we can also predict the grading of heart attacks with this method.

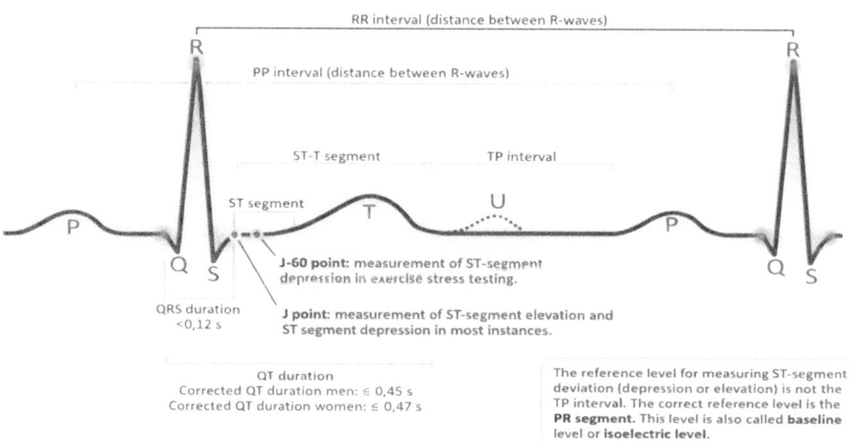

Fig. 10.2 ECG report.

- **Grading of the ECG Report:** The purpose of this process is to classify the type of heart attack, i.e., whether it is a mild or severe attack. This can be graded by using the ECG reports of people/a person. ECG interpretations usually start with an assessment of P waves. "ST segment", which appears on the ECG, is the display of a normal heartbeat pattern. There is also another type of heart attack known as silent heart attack, which can be easily predicted by imaging and blood test results. The test is done taking the abnormal (heart attack) ECG report.

- A normal ECG image as input is converted to gray scale as shown in Figure 10.3.

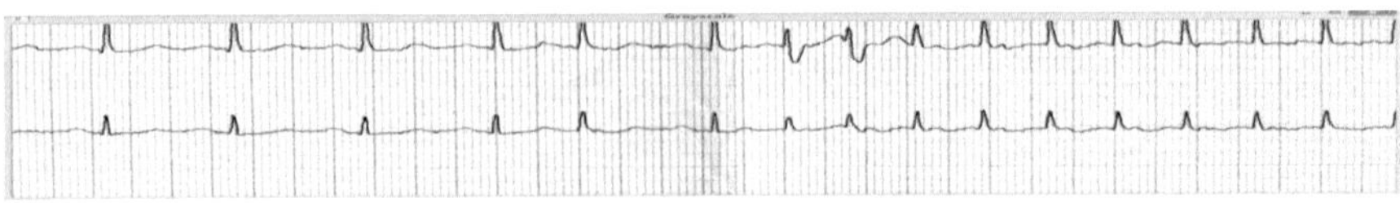

Fig. 10.3 Grayscale ECG.

- It is then plotted in the form of graph as shown in Figure 10.4.

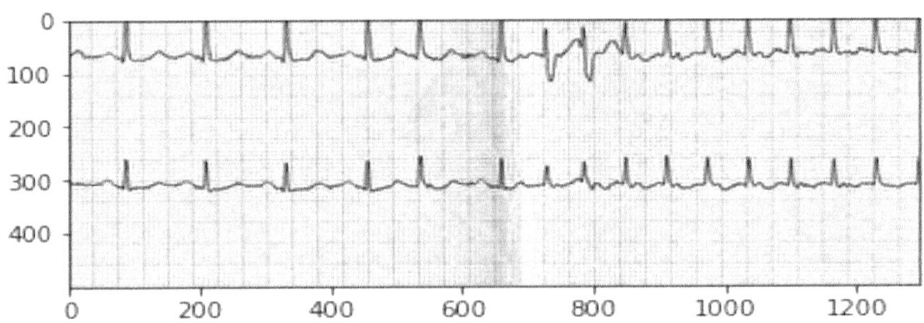

Fig. 10.4 Graph Plot of ECG.

- Binarization process is done to remove the grid in the image as shown in Figure 10.5.

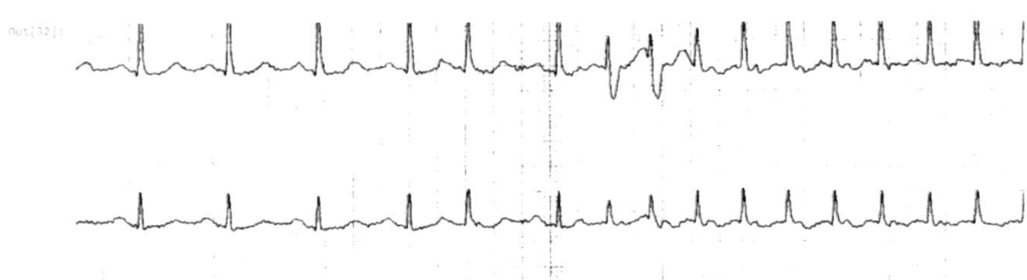

Fig. 10.5 Binarization.

- Crop the region of interest, which is shown in Figure 10.6.

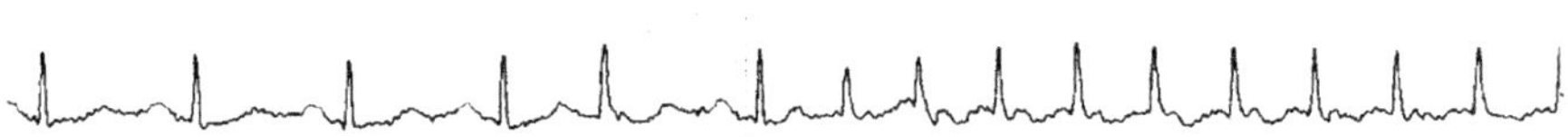

Fig. 10.6 Extracted region of interest.

- Figure 10.6 is plotted using matplotlib as seen in Figure 10.7.

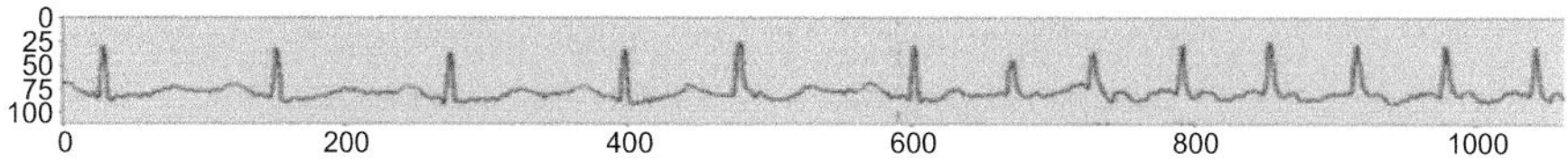

Fig. 10.7 Plotting region of interest.

After changing the axis value, we are done with the complete preprocessing stage.

- Image information is shown in Figure 10.8.

```
PNG
L
(344, 108)
```

Fig. 10.8 Image formation.

- Normalization: Making the input range between 0–1 (see Figure 10.9).

```
Mean: 245.061, Standard Deviation: 49.352
Mean: 0.577, Standard Deviation: 0.116
Min: 0.000, Max: 0.601
```

Fig. 10.9 Normalization.

- Mean filtering: Applying the linear filter to the image (see Figure 10.10).

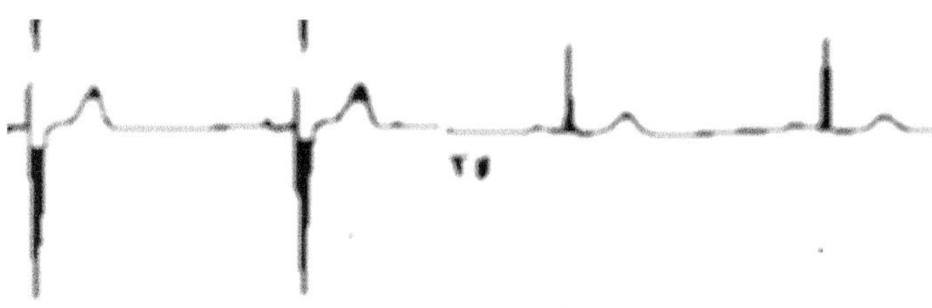

Fig. 10.10 Mean filtering.

- Laplacian filter (for edge detection): This is done for edge detection; this filter enhances the edges in the image (see Figure 10.11).

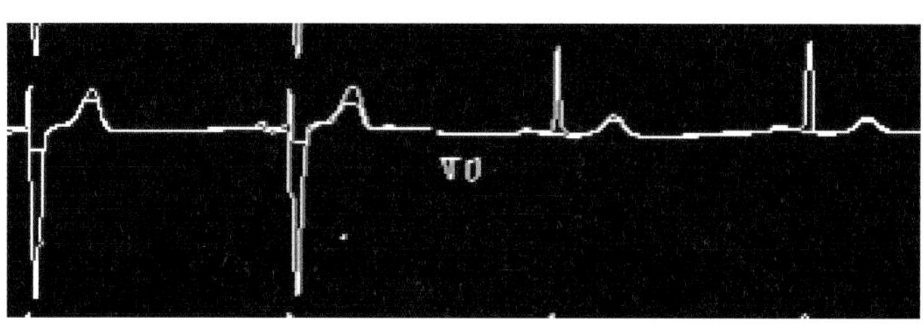

Fig. 10.11 Laplacian filter.

- Soble Filter (Edge Detection): This filter cancels the line or edge present in *x* axis, i.e., it enhances the peak in the image (see Figure 10.12).

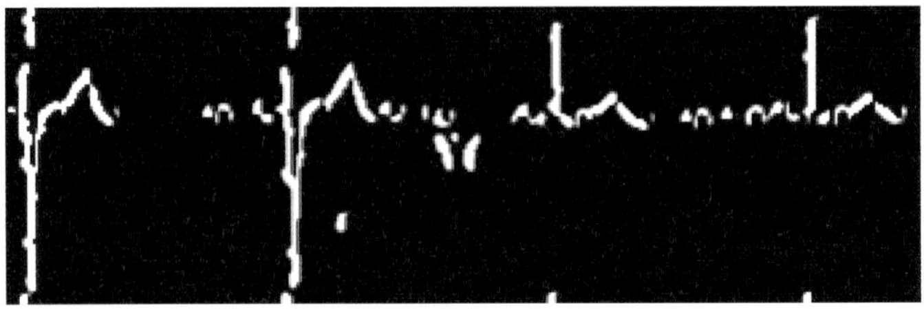

Fig. 10.12 Soble filter.

- Global Standardization: Making the input range between 0–1 (see Figure 10.13).

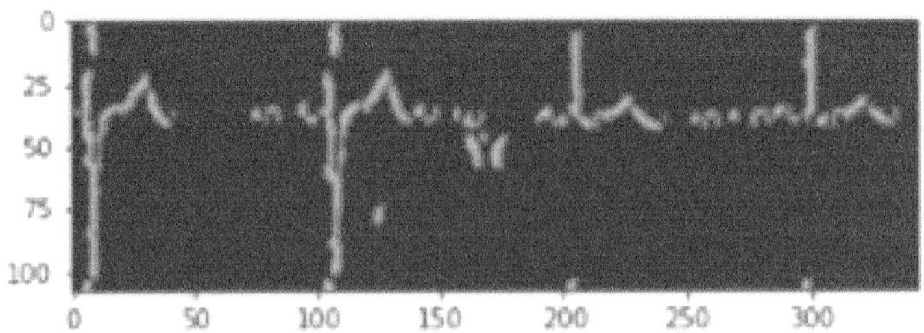

Fig. 10.13 Global standardization.

- Edge Detection: This provides the point from where the peaks start (see Figure 10.14).

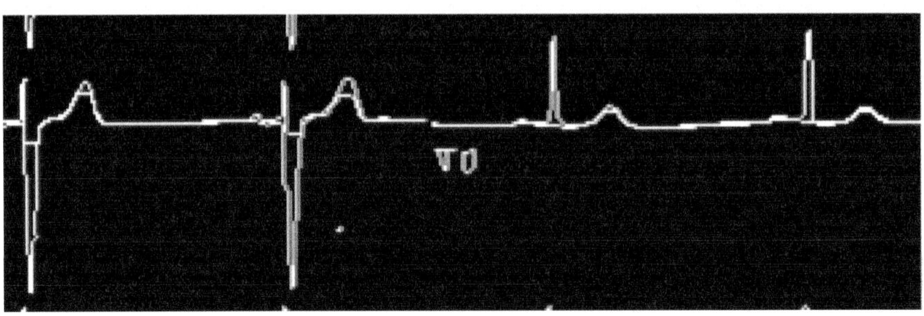

Fig. 10.14 Edge detection.

- RR Interval: Here, we are manually choosing the respective peaks and calculating their intervals (see Figure 10.15). It gives us the heart rate. It should be between 60–100 for a normal heart.

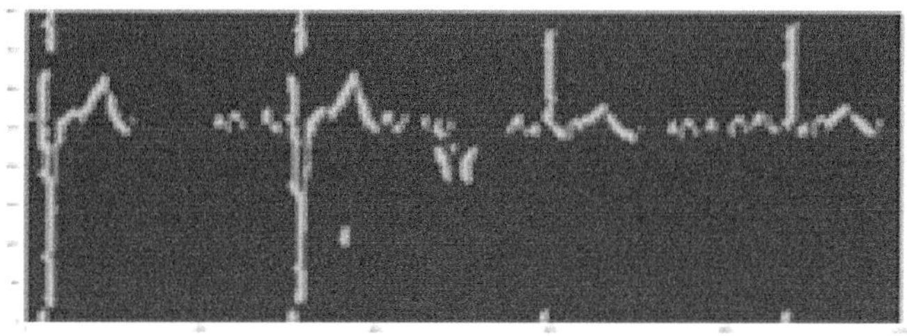

Fig. 10.15 RR interval.

```
Please click 3 points
you clicked: [(599.1120767014593, 374.5831955792755), (873.0458602937701, 383.08458886317476)]
(599.1120767014593, 873.0458602937701) (374.5831955792755, 383.08458886317476)
the RR Interval is: 43.80620689655174
```

- The calculations of the interval:

```
RR= 43.80620689655174

PR= 0.264487791054645

QRs 0.2597647947858121
```

The normal range of the parameters are as follows:

RR should be between 60–100

PR=0.12–0.20

QRS<=0.12

- **The results obtained suggest possible abnormalities in the electrical conduction system of the heart.**

4.3 *Security during Long-Range Communication*

The data collected by the gateway (e.g., mobile phone) is transferred to the IoT cloud via Wi-Fi for advanced analysis. Security is needed when the data is transferred via a public network like the internet. The cryptography method is employed so that data packets are reached without being tampered with. Various cryptography methods that can be used for safe transactions are Triple-DES, AES, RC5, RC6, and Blowfish. The data is stored in the IoT-cloud using non-relational databases (e.g., Redis) that enhance the flexibility and I/O speed of the database.

4.4 *Security in IoT Cloud*

Security here plays a vital task. The patient's stored data should not be leaked or changed. As resources (servers) are shared among multiple tenants, each tenant's data has to be secured from other tenants and attacks.

Louk M et al.(2015) encryption Approach for the security of data in IoT -Cloud: The homomorphism suggested by Rivest et al. (1978a) and the multiplicative homomorphism given by Rivest et al. (1978b) can be used. Homomorphic encryption is shown in Figure 10.16.

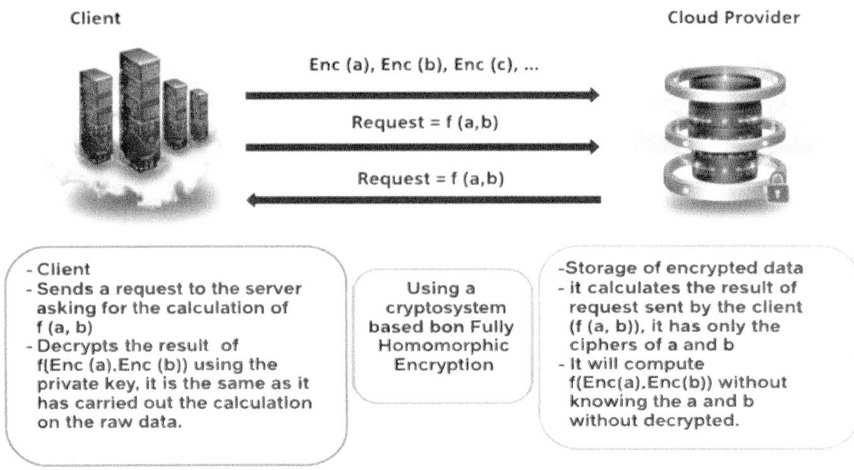

Fig. 10.16 Homomorphic encryption

4.5 Data Transmission from Cloud to the Concerned Health Person via Hypertext Transfer Protocol (HTTP) and Message Queuing Telemetry Transport (MQTT)

In our proposed system vital parameters of the patient like pulse rate, body temperature, and other environmental parameters such as humidity, temperature, and level of various gases like CO and CO2, measured through the various sensors, are uploaded to the MQTT server. Because it uses cryptographic protocol, the MQTT protocol is a safer, lighter, and longer latency protocol than the HTTP protocol. The data is saved in the MQTT server on a constant basis. The benefit of combining the MQTT and encryption protocols is that along with the necessary data, no confusing data is stored.

Karthiga, I. et al. (2015) the processing layer publishes the ECG sensor's heart rate readings using the MQTT publish technique and sends the same to the application layer. MQTT is an open-source protocol which is used to communicate data between two devices over Transmission Control Protocol (TCP) and does not require much bandwidth or processing time. It can control the number of times the messages are sent thereby improving the quality of service. Client and server are the two important concepts in MQTT. The messages are sent by using either topics, publisher, subscriber, or broker. Topics categorize the different messages. In the publisher, the device is configured to send messages to data. Subscribers are configured devices to receive messages on specific topics.

- **Methodology in MQTT**

Client and server are the two important concepts in MQTT. These can exchange data using topics, publisher, subscriber, and broker as shown in Figure 10.17.

- Topics: Categorizes the type of messages that are probably sent.

For example, when a sensor measures temperature, the subject is probably described as "TEMP," and the sensor sends messages categorized as "TEMP."

- Publishers: You can configure your device to send messages that contain data. A sensor that measures temperature publishes this data.
- Subscribers: Systems are configured to acquire the simplest messages of a particular topic. They also can enroll in a couple of topics.
- Broker: This is the server. It sends published messages to servers or clients that subscribe to a specific topic.

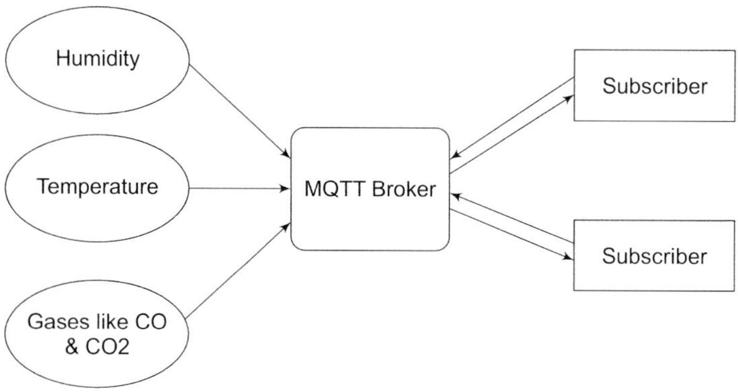

Fig. 10.17 Architecture of MQTT.

- **Setup Overview**

In this section, we see a technique for sending the response time and packet size over similar payloads through MQTT and HTTP as shown in Figure 10.18. To do this, it uses a registry on one of the cloud IoT cores that accepts both HTTP and MQTT connections. Here, the cloud function logs the payload. The quiescing tool is simulated on a notebook running each MQTT sponsor and HTTP sponsor. After that, it measures the reaction time and traces the packets sent over the network.

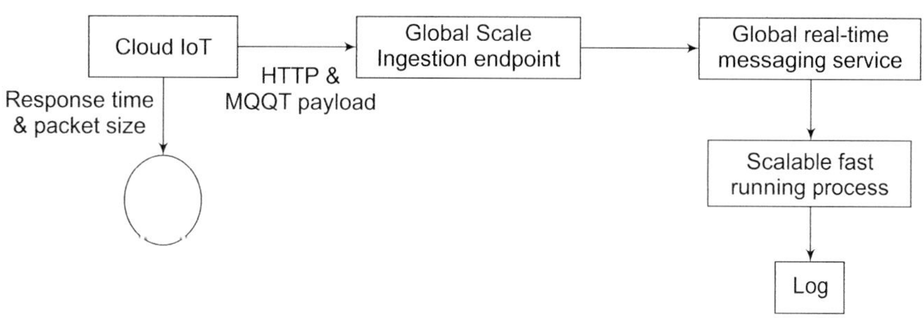

Fig. 10.18 Setup overview.

5. Conclusion

This chapter proposes a novel remote heart-monitoring system that is secure, reliable, and quick. The proposed method implements a secure WBAN for reliable and accurate heartbeat monitoring of the remote patient. Minimum cost function, i.e. maximum energy left in the sensor nodes and central-based approach for the selection of cluster head, prevents the draining of the battery of a particular sensor node and reduces the uneven distribution of the cluster head, respectively. The reliable and stable route in WBAN using the watchdog mechanism and overall node distribution, respectively, brings novelty to the algorithm. Buffers used in the network impose order on the traffic and compensate for the propagation delay that is inherent in the wireless network with a feedback system. This approach is used for quick delivery of patient health statistics. A non-cryptographic node authentication scheme is proposed for WBAN as it incurs low computational overhead. The WBAN model is integrated with a novel deep learning analysis model at the control unit for the identification of abnormalities in the heart rate. The analyzed data is uploaded in the cloud. A homomorphic encryption approach for data security is proposed for cloud computing. Data transmission from the cloud to the patient concerned is done via HTTP and MQTT. The proposed approach can increase reliability and network lifetime. Thus, the patient can be kept continuously under surveillance for the maximum period. The health professional can detect any abnormalities in the patient's heart reading well in advance and be able to take prompt and appropriate action to save the patient's life.

References

Ali, S. and M. Ghazal. (2017). Real-time heart attack mobile detection service (RHAMDS): An IoT use case for software defined networks, *2017 IEEE 30th Canadian Conference on Electrical and Computer Engineering*, 1–6.

Basheer, S., A.S. Alluhaidan and M.A. Bivi. (2021). Real-time monitoring system for early prediction of heart disease using internet of things, *Soft Computing*, *25*, 12145–12158. https://doi.org/10.1007/s00500-021-05865-4.

Braem, B. and C. Blondia. (2011). Supporting mobility in wireless body area networks: an analysis, *2011 18th IEEE Symposium on Communications and Vehicular Technology in the Benelux*, 1–6.

Dhand, G. and S.S. Tyagi. (2013). Survey on data-centric protocols of WSN, *International Journal of Application or Innovation in Engineering & Management*, *2*(2), 279–284.

Fang, G. and E. Dutkiewicz. (2009). BodyMAC: energy efficient TDMA-based MAC protocol for wireless body area networks, *2009 9th International Symposium on Communications and Information Technology*, 1455–1459.

Gavilovska, L. (2010). Standardisation and exploitation, in *My Personal Adaptive Global NET (MAGNET)*, Dordrecht: Springer, pp. 409–423.

Gousia, T., P. Mishra and S.R. Moorthy. (2022). Two phase secure data collection technique for wireless sensor networks, *Indonesian Journal of Electrical Engineering and Computer Science*, *25*, 512–520. doi: 10.11591/ijeecs.v26.i1. pp 512–520.

Gurjar, A.A. and N.A. Sarnaik. (2018). Heart attack detection by heartbeat sensing using internet of things: IoT, *Heart*, 5(03).

Karthiga, I., S. Sankar and P. Dhivahar. (2015). A study on routing protocols in wireless body area networks and its suitability for m-Health applications, *2015 International Conference on Communications and Signal Processing*, 1064–1069.

Kaushik, M., S.H. Gupta and V. Balyan. (2019). Implementation and reliability analysis of LEACH and SIMPLE protocols in WBAN for health monitoring of canine, *2019 International Conference on Issues and Challenges in Intelligent Computing Techniques*, *1*, 1–6.

Leijdekkers, P. and V. Gay. (2008). A self-test to detect a heart attack using a mobile phone and wearable sensors, *2008 21st IEEE International Symposium on Computer-Based Medical Systems*, 93–98.

Li, C., H.B. Li and R. Kohno. (2009). Reservation-based dynamic TDMA protocol for medical body area networks, *IEICE Transactions on Communications*, *92*(2), 387–395.

Li, C., X. Hu and L. Zhang. (2017). The IoT-based heart disease monitoring system for pervasive healthcare service, *Procedia Computer Science*, *112*, 2328–2334.

Li, H. and J. Tan. (2009). Heartbeat-driven medium-access control for body sensor networks, *IEEE Transactions on Information Technology in Biomedicine*, *14*(1), 44–51.

Lindsey, S. and Raghavendra, C.S. (2002, March). PEGASIS: Power-efficient gathering in sensor information systems. In Proceedings, *IEEE aerospace conference* (Vol. 3, pp. 3–3). IEEE.

Louk, M. and Lim, H. (2015, January). Homomorphic encryption in mobile multi cloud computing. In 2015 International Conference on Information Networking (ICOIN) (pp. 493–497). IEEE.

Patel, N., P. Patel and N. Patel. (2018). Heart attack detection and heart rate monitoring using IoT, *International Journal of Innovations and Advancements in Computer Science*, *7*(4), 612–615.

Rivest, R.L., A. Shamir and L. Adleman. (1978b). A method for obtaining digital signatures and public-key cryptosystems, *Communications of the ACM*, *21*(2), 120–126.

Rivest, R.L., L. Adleman and M.L. Dertouzos. (1978a). On data banks and privacy homomorphisms, *Foundations of Secure Computation*, *4*(11), 169–180.

Salehi, S.A., M.A. Razzaque, I. Tomeo-Reyes and N. Hussain. (2016). IEEE 802.15. 6 standard in wireless body area networks from a healthcare point of view, *2016 22nd Asia-Pacific Conference on Communications*, 523–528.

Santhanakrishnan, C., N.G. Poojitha and L.J. Reddy. (2019). A survey on tracing heart attacks by pulse monitoring in IoT, *Journal of Physics: Conference Series*, *1362*(1), 012093.

Sarmah, S.S. (2020). An efficient IoT-based patient monitoring and heart disease prediction system using deep learning modified neural network. *IEEE access*, 8, 135784–135797.

Selvathi, D., V.V. Sankar and H. Venkatasubramani. (2017). Embedded based automatic heart attack detector and intimator, *2017 International Conference on Innovations in Information, Embedded and Communication Systems*, 1–6.

Song, L., Y. Wang, J.J. Yang and J. Li. (2014). Health sensing by wearable sensors and mobile phones: a survey, *2014 IEEE 16th International Conference on e-Health Networking, Applications and Services (Healthcom)*, 453–459.

Timmons, N.F. and W.G. Scanlon. (2004). Analysis of the performance of IEEE 802.15. 4 for medical sensor body area networking, *2004 First Annual IEEE Communications Society Conference on Sensor and Ad Hoc Communications and Networks*, 16–24.

CHAPTER 11

Cyber Security in Intelligent Transportation Systems

Priyamvada Saxena[1*] and *Raghvendra Sharma*[2]

Intelligent transportation systems (ITS) form the backbone of modern society. ITS integrates vehicles, road-side units, and cloud services to enhance safety, efficiency, and real-time decision-making. As ITS is evolving with an increase in digitalization and connectivity with the systems it becomes an easy target by attackers that can hamper user safety and privacy. The aim of this cahpter is a comprehensive investigation of cybersecurity challenges in ITS that focuses on threats to in-vehicle networks, communication protocols within the vehicle, and backend infrastructure. We propose a threat model that captures attack on controller area network (CAN) flooding, replay attacks on vehicle-to-infrastructure (V2I) messages, and Global Positioning System (GPS) spoofing. A Python-based anomaly detection system is developed using Isolation Forest and Autoencoder techniques, and the results are demonstrated successfully with detection of abnormal behaviors with high precision. The contribution aims to inform the design of resilient, intelligent, and secure transportation ecosystems capable of withstanding evolving cyber threats.

1. Introduction

There is a rapid advancement in the world of automotive and transportation sectors that has led to the emergence of intelligent transportation systems (ITS) that integrate vehicles with available infrastructure, sensors, and communication networks to deliver safer and sustainable transportation. ITS comprises applications, from adaptive traffic control to smart parking and public transport fleet monitoring. It relies on interconnectivity, between vehicles (V2V), between vehicles and infrastructure (V2I), between vehicles and networks (V2N), and between vehicles and pedestrians (V2P), which is collectively referred to as vehicle-to-everything (V2X).

[1-2] Department. of Electronics & Communication, Amity University, Gwalior, Madhya Pradesh, India. https://orcid.org/0009-0006-6726-0634[1]; https://orcid.org/0000-0003-3478-1194[2]

* Corresponding author: priyamvada.saxena@s.amity.edu

DOI: 10.1201/9781003659822-11

As vehicles are equipped with electronic control units (ECUs), wireless transceivers, Global Positioning System (GPS) modules, cloud services and many more, it opens a wide spectrum for attacks on the transport system. These systems, while providing the user with semi- or fully-autonomous functions, have also exposed the system to cybersecurity risks. From the vehicular networks such as controller area network (CAN) and local interconnect network (LIN) buses to external systems like cellular communication, ITS can be exploited to cause physical harm, data theft, or even financial loss to the user. In India (The Hindu Bureau, 2023) an ITS project was kicked off in Chennai where traffic lights turn "on" on the basis of the number of vehicles. In this project the only thing to attack is the traffic light system—if the system goes into fault it can be proven fatal for the life of the people.

Several incidents in real-time have demonstrated the gravity of these threats. In 2015 (Greenberg, 2015), a group of security researchers remotely hacked a Jeep Cherokee by entering into its infotainment system, leading to unauthorized control of the steering and brakes in the vehicle. Researchers have also shown how attackers can manipulate GPS signals and V2X communications, or inject false data into the backend systems that governs the traffic signals. These examples show the inadequacy of traditional information technology (IT) security frameworks when applied to dynamic and mobile ITS environments. As ITS vehicles consists of ECUs, which are the hardware system for the vehicles, it is easy for an attacker to gain control over it. A successful ITS cybersecurity framework, therefore, must consider the heterogeneity, scalability, and performance constraints unique to transportation systems.

This chapter offers a comprehensive study of cybersecurity in ITS, focusing on broader and more system-level attack vectors including the following:

- In-vehicle network attacks (e.g., CAN bus message injection, spoofing, denial-of-service [DoS])
- External communication-based threats (e.g., V2I replay attacks, GPS spoofing, V2X misbehavior)
- Infrastructure-side vulnerabilities (e.g., roadside units [RSU] compromise, cloud-side Application Programming Interface (API) attacks)
- Privacy risks in data logging, telematics, and real-time vehicle tracking
- Public transport systems and their unique vulnerabilities

We develop a Python-based synthetic dataset that emulates typical vehicle telemetry and communication data, including both benign and malicious behavior inside the system. Attack scenarios such as CAN ID flooding, V2X message replay, and spoofed location data are simulated within the dataset. An anomaly detection system is implemented using machine learning models such as Isolation Forest and Autoencoder to differentiate between normal and abnormal patterns in the data.

In parallel, we will also review international efforts in ITS cybersecurity regulation, including the ISO/SAE 21434 Road Vehicles Cybersecurity Engineering standard, the United Nations Economic Commission for Europe (UNECE) WP.29 Regulation R155, and NHTSA's Cybersecurity Best Practices for Modern Vehicles. We analyse their applicability and identify gaps that need to be addressed for real-world implementation.

2. Overview of ITS Architecture

The fundamental goal of ITS is to enhance the safety of the user and prevent cyberattacks on the system through the real-time exchange of the data. An ITS consists of vehicles, infrastructure, centralized management systems, and cloud-based platforms, all of which interact with each other to deliver services like traffic monitoring, congestion management, autonomous vehicle coordination, and incident response of the vehicle as seen in Figure 11.1. At a high level, ITS can be conceptualized as consisting of three tiers:

- **Vehicle Tier:** This includes on-board sensors, actuators, ECUs, infotainment systems, telemetry units, and internal vehicular communication buses (e.g., CAN, LIN, Flex Ray).
- **Infrastructure Tier:** This consists of RSUs, traffic signals, surveillance cameras, weather sensors, and edge-computing gateways that communicate with nearby vehicles and the centralized systems.
- **Control and Cloud Tier:** This involves central traffic management centre, cloud platforms, and service providers responsible for data aggregation, long-term storage, analytics, and decision-making.

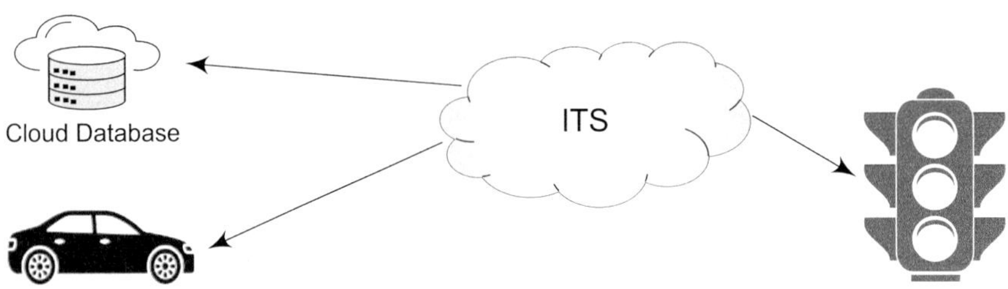

Fig. 11.1 Overview of ITS.

3. Communication Technologies in ITS

ITS relies heavily on inter-device communication, which is of two types: within vehicle and V2X.

- **In-Vehicle Networks:**
    - In the vehicle, there is a CAN bus that is the dominant protocol for in-vehicular communication, enabling ECUs to exchange information with the other systems or gain information from other systems such as speed, braking status, and engine diagnostics. CAN communication lacks encryption and authentication features, making it an easy target for injection and spoofing attacks.
    - Other buses include LIN for low-speed devices and Flex-Ray for fault-tolerant high-speed data.
- **V2X Communication:**
    - **Vehicle-to-Vehicle (V2V)** communication: This enables cars to share real-time data such as location, speed, and acceleration to prevent collisions.

- **Vehicle-to-Infrastructure (V2I)** communication: This facilitates the interaction with the traffic lights, road signs, and RSUs to support adaptive signal control, emergency vehicle prioritization, and congestion avoidance.
- **Vehicle-to-Network (V2N)** communication: In this, the communication links vehicles to cloud systems for data offloading, updates, and backend analytics.
- **Vehicle-to-Pedestrian (V2P)** communication: This aims to alert drivers and pedestrians to each other's presence, particularly in urban areas, to avoid any fatal injury.

These V2X systems are typically implemented using dedicated short range communication (DSRC), cellular V2X (C-V2X), or increasingly, 5G NR-V2X, which provides lower latency and broader bandwidth.

4. Data Flow in ITS

The operation of ITS relies on continuous data generation and transmission from the vehicles to the cloud where analysis of data takes place. Vehicles generate data from numerous sources including sensor data: Light Detection And Ranging (LIDAR), radar, cameras, GPS; control signals: brake actuation, throttle position, steering angle; diagnostic and telemetry data: engine temperature, fuel consumption, battery voltage; and communication data: V2X packets, internet connectivity metrics. This data flows in two primary directions: upstream to the cloud or control centres for centralized analysis, regulatory compliance, or historical recordkeeping; and downstream from control centres or edge nodes to vehicles for dynamic route updates, emergency alerts, or policy enforcement.

5. Threat Landscape in Intelligent Transportation Systems

There is a digital development in transport that has introduced benefits in traffic control, route scheme, and road safety. These characteristics have also expanded the surface of the attack in every layer of ITS architecture. These dangers range from in-vehicle manipulations to communication-based deception and backend cloud exploitation. We classify the dangers in four major layers (see Figure 11.2):

A. In-Vehicle Network Attacks
B. V2X Communication Attacks
C. Infrastructure-Side Attacks
D. Backend & Cloud Threats

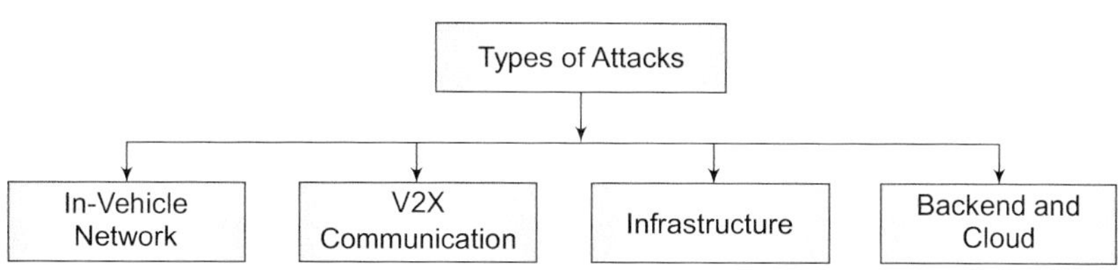

Fig. 11.2 Types of attacks in ITS.

A. In-Vehicle Network Attacks

Vehicles consist of various embedded systems like ECUs that communicate with other systems like body control module (BCM) and other sensors over internal buses such as CAN and LIN. These systems were not designed with cybersecurity in mind and, therefore, offer minimal protection against modern threats.

Attacks:

1. **CAN Message Injection:** Attackers inject fake CAN messages (e.g., turning off brakes, disabling airbags) using physical access or via compromised modules (e.g., on-board diagnostics [OBD-II] ports) that can be fatal for the driver.
2. **CAN Bus Flooding:** Attackers can flood the bus with random messages blocking the messages with high priority causing DoS conditions.
3. **Replay Attacks:** In this type of attack the attackers try to replay previously recorded valid messages to cause unauthorized behavior like unlocking doors or activating wipers.
4. **ECU Firmware Tampering:** Attackers can attack over-the-air (OTA) updates so that malicious activities can enter the system.

Example: The infamous Jeep Cherokee hack (2015) involved remote access to vehicle controls through a vulnerable infotainment system, allowing researchers to control the brakes and steering.

B. V2X Communication Attacks

V2X attacks cover the attacker V2V, V2I, and V2N. As these systems are interconnected with each other there are various attacks that can take place.

Attacks:

1. **Message Spoofing:** An attacker sends false traffic alerts or crash notifications to be surrounding vehicles to manipulate the data in the vehicles.
2. **Sybil Attacks:** A single attacker creates multiple fake identities to simulate traffic congestion or accidents that can be fatal, and they can also mislead route optimization systems, e.g., we see in news about wrong routes by Google Maps.
3. **Replay Attacks on V2I:** Attackers replay messages to affect traffic lights or cause an emergency braking.
4. **Jamming:** In this attack, radio frequency (RF) interference is used to disrupt V2X communication signals that cause packet loss and communication failures.
5. **Location Spoofing:** In this type of attack, GPS data is altered that can cause vehicles to misjudge positioning, creating risks in lane-changing or navigation of route.

Example: Researchers have shown how GPS spoofing can manipulate autonomous drones and vehicles to drift off course, creating a physical safety hazard (Ma et al., 2024).

C. Infrastructure Side Attacks

Infrastructure layers—complementing RSUs, traffic signal controller, surveillance cameras and smart signage—are often associated with public or semi-public networks. These devices are usually installed in a remote or exposed environment, making them susceptible to both physical and cyber infiltration.

Attacks:

1. **Compromised RSUs:** In this, the attackers exploit the unsecured roadside units to inject malicious data into the V2I stream.
2. **Unauthorized Signal Control:** In this type of attack the signals can be manipulated that can be fatal for human life.
3. **Edge AI Manipulation:** Edge nodes running AI algorithms for traffic prediction or vehicle classification may be targeted with abrupt inputs to mislead detection models that can lead to collision of vehicles and loss of lives.
4. **Firmware Backdoors:** This attack can be due to hardcoded credentials or outdated firmware in traffic control devices.
5. **Physical Sabotage:** This attack includes tampering with edge devices like cameras or sensors to blind systems or introduce misleading data that can be fatal.

Example: In 2006, a cyberattack on Los Angeles traffic signals altered light cycles and gridlocked major intersections during a labor dispute (Applegate, 2013)

D. Backend and Cloud Threats

The backend tier often clouds consists of massive amounts of vehicle telemetry, diagnostics, user behavior, and infrastructure data. Although they are physically separated from the road-level activity, backend systems are major targets as they store large amounts of data.

Attacks:

1. **Unauthorized Data Access:** Weak API protections or insufficient authentication in the system can expose real-time vehicle location, trip history, or user profiles that can be manipulated easily.
2. **Cloud Account Hijacking:** Attackers use methods like phishing or credential stuffing to take over the administrative accounts on cloud-hosted ITS platforms.
3. **Data Poisoning:** Malicious actors insert false data into the machine learning pipelines, degrading the accuracy of an AI-driven traffic prediction or maintenance forecasting in the system.
4. **OTA Update Hijack:** In this OTA firmware updates can inject malicious software into the vehicles.
5. **Ransomware:** In this the backend servers or traffic management dashboards are encrypted causing service outages.

Example: In 2020, ransomware targeted transport authorities in various cities, disrupting dispatch systems and data access (blackfogprivacy, 2020)

6. Threat Modeling in ITS

By integrating information, communication, and control technologies in vehicles and infrastructure, threat modeling aims to improve safety, efficiency, and stability. However, the spread of communication interfaces, sensors, and software has also increased the surface of modern vehicles rapidly. V2X, ECUs, Global Navigation Satellite System (GNSS), and real-time locations are interconnected. While this enables features such as prevention, adaptive cruise control, and real -time navigation from collision, it also introduces weaknesses. Attackers take advantage of these components, with serious consequences, including security failures that can be fatal to human life. To proactively address these risks, a comprehensive threat modeling methodology is crucial. This approach involves identifying critical assets and attack vectors, assessing risks, and applying anomaly detection to detect malicious behaviour. This threat model incorporates component-level attack types such as CAN flooding, GPS spoofing, and V2I replay attacks (Laorden et al., 2010). It gives a risk score to each attack based on three dimensions: impact on safety, ease of execution, and detection complexity. It then implements a machine learning-based anomaly detection system using Isolation Forest and Autoencoders.

The four essential steps for threat modeling in an ITS are as follows.

6.1 Asset and Component Identification

Identify all important components and assets that require protection. This includes:

- In-vehicle networks (CAN, LIN bus)
- Communication interface (V2V, V2I, V2X)
- Sensor and GPS module
- Cloud-based infrastructure

Each asset should be examined for its function, exposure, and inter-connectivity along with other components. For this chapter, we focus on three high-risk components:

- CAN Bus
- V2I Communication
- GNSS Receiver (GPS)

6.2 Threat Identification

The threat identification phase is an important step in the modeling process for ITS. This focuses on determining how, where, and by whom system assets can be attacked. With the convergence of motor vehicles, communication, and cloud technologies, the system's attack surface has become quite widespread. Identification of potential threats can be used to compromise them to understand the system components, their interfaces, and methods.

6.2.1 Understanding Attack Vectors

An attack vector is a specific path or method used by the attacker to take advantage of system vulnerability (Ebenezer, 2024). This includes some of the most impressive and relevant attack vectors.

- *Flooding CAN network:* CAN bus vehicles are used in the vehicles, and they are the backbone of the vehicle. But these lack encryption or message authentication. In a flood attack, the bus is flooded with high frequency messages, inhibiting communication between ECUs. This may delay or disable tasks in the vehicle, such as braking or steering, which can be proven fatal.
- *GPS spoofing:* GPS signals are weak and informal, making them susceptible to spoofing. An attacker transmits false GPS signals to mislead the navigation

system of the vehicle. The vehicle can be forced by the course or can be diverted into dangerous or unauthorized areas.
- *V2I replay attacks:* V2I enables vehicles to exchange data with roadside units or traffic signals. In a replay attack, a malicious actor catches legitimate messages and later withdraws them. This can trick the vehicle in response to old or malicious instructions, such as wrong speed limit or wrong stop signal. This attack vector not only affects the system performance but also compromises the trust in passenger safety, data integrity, and its ecosystem.

6.2.2 Identifying Entry Points

Threat identification also includes analysis of all possible entry points and places where an attacker can achieve access to a vehicle or infrastructure system (Avcı & Koca, 2024). General entry points include the following:

- *Wireless interface:* Modern vehicles are equipped with Wi-Fi, Bluetooth, 5G, DSRC, and GPS receivers. Each of these interfaces can probably be exploited remotely from outside the vehicle.
- *OBD-II port and Universal Serial Bus (USB) interfaces:* Physical interfaces such as on-board diagnostics (OBD-II) port allow access to the internal network of the vehicle. If insecure is, they provide a direct route for important vehicle functions.
- *Telematics and cloud API:* ITS systems rely on cloud-based infrastructure for navigation, diagnostics, and remote updates. Improperly secured API or weak certification can expose these systems to distance attacks.

6.2.3 Understanding Threat Actors

In the process of maintaining cyber security, it is necessary to not only identify weaknesses and potential attack vectors but also understand who can exploit them. Profiling threat actors helps estimate their motivations, technical abilities, and the possibility of targeting a given system. In the context of ITS—where cyber and physical worlds are tightly integrated—danger actors vary widely in sophistication and intentions (Harvey & Kumar, 2020). Broadly, these actors fall into three major categories: external hackers, malicious insiders, and spoofing devices or specialized attack tools.

External Hackers: External hackers are unauthorized individuals or organized groups that try to achieve access to its infrastructure. Their motivations can be different:

- Financial benefits through ransomware or data theft (e.g., vehicle history, payment system)
- Reputation or notoriety within the hacking community
- Conceptual motivations (e.g., political opposition)
- Entertaining hacking, especially between script kiddies or amateur enthusiasts

External hackers usually rely on remote access methods, such as exploiting wireless weaknesses in V2X communication, exploiting vulnerable authentication systems in unsecured cellular interfaces or cloud-connected vehicle services in telematics units. While not always highly efficient, many external attackers take advantage of easily available equipment, code repository, and hacking framework. The extensive availability of open-source software-defined radio (SDR) can hack tools such as CANtact and USBTIN, and the availability of platforms such as Kali Linux further empowers medium-skilled attackers to execute attacks that can pose significant security risks.

Malicious Insiders: A less visible but equally important threat arises from malicious insiders like employees, contractors, or vendors with valid access to the ITS network or vehicle components. Their internal position gives them an advantage on this:

- Deep knowledge of the system's architecture and configuration.
- Reducing credential access, cruel-force attacks, or exploitation.
- The ability to bypass the perimeter by operating within the trust range, such as firewall or encryption.

These insiders may be inspired by personal revenge, financial encouragement, or even espionage, especially if its technologies are developed by the government or defense contractors. For example, a dissatisfied employee may transplant a back door in firmware or manipulate battery management settings to inspire unsafe behavior. There is also a challenge to detecting internal sources. Their functions often appear as regular behavior, which makes it important to detect internal dangers to detect behavioral discrepancy.

Spoofing Devices and Specialized Attack Tools: Unlike human actors, spoofing devices are technical promoters of cyber-physical attacks. This includes the following:

- SDR can be used for GPS spoofing by generating fake GNSS signals to confuse the navigation system of a vehicle.
- Injection kits, which connect to the OBD-II port of the vehicle, can inject malicious messages in the CAN bus.
- Replay attack kits recording V2I or V2V are able to resume messages that can make incorrect decisions by autonomous or aided driving systems.

Such devices are often portable and cost-effective, and are designed to bypass certification or confused sensors. For example, an attacker with a GPS SDR setup can mislead fleets of vehicles, believing that they are in a separate place, causing traffic disruptions or misconduct among vehicles. Importantly, these devices are rapidly open sources and commoditize, making them accessible to unskilled users who can follow the online tutorials.

6.3 Structured Techniques: STRIDE and Attack Trees

Two widely used and complementary techniques in this domain are Spoofing Tampering Repudiation Information Disclosure Denial of Service (DoS) Elevation of Privilege (STRIDE) and attack trees.

6.3.1 STRIDE: A Categorization-Based Threat Modeling Framework

STRIDE, developed by Microsoft, is a mnemonic structure that classifies the dangers into six fundamental categories (Allen-Addy, 2023). Each category represents a distinguished security property to compromise.

1. **Spoofing:** Impersonating someone else.
   Example: An attacker mimics a reliable V2I to send false information to a vehicle. GPS spoofing is a direct application of this danger category.
2. **Tampering:** Unauthorized amendment of data or code.
   Example: An attacker intercepts and simply modifies messages to change the braking command or state values in electric vehicles.
3. **Repudiation:** Lacks proof of action and allows actors to deny their activities.
   Example: A malicious internal formula manipulates the vehicle log files but does not do any wrong thing due to the absence of audit trails.
4. **Information Disclosure:** Risk of information for unauthorized parties.
   Example: Hackers intercept the vehicle telemetry sent to the cloud, disclosing the individual travel pattern and vehicle diagnosis.
5. **DoS:** Denial of service for legitimate users.
   Examples: CAN in-vehicle overwhelmed in flood attacks, making important tasks unanswered.
6. **Elevation of Privilege:** To achieve unauthorized rights or access.
   Example: A remote attacker carries ahead of telemetry access only to control the command on the vehicle actuators.

Analyzing each component (e.g., V2X units, GPS modules, BMS– Battery Management System, ECU) in these six categories ensures an overall review of STRIDE potential danger surfaces. For example, GPS spoofing is classified under spoofing and information disclosure, while the CAN bus injection is covered under tampering and elevation of privilege.

6.3.2 Attack Trees: A Visual Approach to Threat Modeling

While STRIDE provides a range of lens, attack trees provide a visual and hierarchical approach to modeling danger. First introduced by Bruce Schneier, the attack tree maps all possible routes that can carry an attacker to compromise a system or achieve a specific malicious purpose (Hao et al., 2024).

- Route node represents the primary goal of the attacker (e.g., "inhibiting autonomous navigation").
- Branches represent intermediate objectives (e.g., "confused GPS system", "Jam V2V communication").
- The leaves represent the activities of solid attacks (e.g., "spread fake GPS coordinates using SDR", "inject false CAN messages").

The attack allows analysts on trees to:

- identify the paths of all known and envisaged attacks
- estimate the complexity, possibility, and effort associated with each path
- imagine the sequence of essential tasks and the interpretation
- prefer the rescue by focusing on the most potential or impressive branches

For example, an attack on a delivery vehicle for a GPS spoofing attack may begin with the target of diverting the tree vehicle and the branch in sub-functions, such as identifying the GPS frequency, generating a wrong signal, transmitting and confirming success. Each of these stages may contain separate equipment, skill requirements, and counter-measures.

6.4 Risk Assessment and Scoring

To assess the severity and priority of each threat, we provide a risk scoring model that incorporates three important aspects (Cherdantseva et al., 2016).

6.4.1 Impact on Safety

This axis evaluates potential physical damage that can be a successful attack. This is a safety-matched domain, and any cyber security agreement can directly endanger the following:

- **Passengers and Drivers:** e.g., attacks on braking, steering, or speed control
- **Pedestrians and Cyclists:** e.g., spoofed signals in autonomous driving algorithms
- **Other Vehicles or Infrastructure:** e.g., traffic congestion or collisions triggered by false messages in V2V/V2I networks

Scoring (example scale: 1 to 5):

- 1 = No physical danger (e.g., telemetry leakage)
- 3 = Moderate risk (e.g., denial of infotainment system)
- 5 = Severe safety risk (e.g., CAN bus injection disabling brakes)

6.4.2 Ease of Execution

This factor investigates how practical or accessible to an opponent:

- Essential tools (e.g., CAN make software-defined radio, equipment)
- Level of expertise
- Access privilege (local, remote, physical)

Scoring:

- 1 = requires special hardware and insider access
- 3 = possible with moderate skills and general equipment
- 5 = easily executable with open-source tools and minimum skills

For example, GPS spoofing using SDR is now accessible to medium-skilled attackers due to online tutorials and cheap equipment, which easily enhances its performance.

6.4.3 Detection Complexity

It measures how difficult it is to detect and reduce the attack once it is in progress. Some attacks, such as floods, may be secret, and normal traffic may be mixed, while others may be clear but difficult to trace back. Scoring:

- 1 = immediately detectable and reversible (e.g., simple tampering)
- 3 = medium system awareness is required (e.g., resume attacks with timestamp inconsistencies)
- 5 = highly secret and detection requires advanced analytics (e.g., low GPS signal manipulation)

Calculating the Risk Score

Each threat is scored on all the aforementioned parameters. A weighted sum or average is then calculated to calculate a risk score:

$$\text{Risk Score (RS)} = w_1 * \text{Safety} + w_2 * \text{Execution} + w_3 * \text{Detection}$$

Where:

- w_1, w_2, w_3 = weights assigned based on organizational priorities (e.g., $w_2 = 0.5$, $w_2 = 0.25$, $w_3 = 0.25$)
- Each component is normalized to a common scale (e.g., 1 to 5)

  Example (for a GPS spoofing attack):

- Impact on Safety: 4 (vehicle misdirection possible)
- Ease of Execution: 4 (open-source SDR tools available)
- Detection Complexity: 3 (requires signal anomaly detection)
- Weights: 0.5, 0.25, 0.25

$$RS = 0.5 \times 4 + 0.25 \times 4 + 0.25 \times 3 = 2.0 + 1.0 + 0.75 = 3.75$$

This value (3.75/5) reflects a high-priority threat.

7. Evaluation Metrics for Threat Detection in ITS

In ITS, the data is collected, transmitted, and analyzed for various applications, ensuring safety and enabling intelligent decision-making. These systems increasingly rely on sensitive data such as real-time vehicle locations, vehicle identification numbers (VINs), MAC and IP addresses, and driver behavior logs. Without any security measures, all the data available can lead to privacy violations and security breaches. The anonymization technique plays a pivotal role in protecting user identity and saving human life by preserving the data.

Vehicles today are constantly connected to other vehicles via V2V, V2I, and V2X communication protocols. These communication protocols exchange data for real-time decision-making, and they include information of the vehicle such as position, speed, and acceleration. While this connectivity is used for improving efficiency and safety, it also exposes vehicles and drivers to the risk of being tracked or targeted through cyberattacks. Attackers can exploit this data to perform traffic re-identification attacks and cause fatal accidents.

Isolation Forest

Isolation forest is a tree-based climate model designed to specifically detect discrepancy. Unlike traditional classification models that rely on general data distribution modeling, it distinguishes discrepancies rather than profiling normal points. Its key features are as follows:

- *Isolation theory:* When discrepancies are somewhat different, the algorithm selects them randomly and selects them rapidly and then selects a partition value between the maximum and minimum values of the selected feature (Ripan et al., 2022). This process is repeated to make trees.
- *Path length metric:* The average number of division required to separate a point (i.e., length of the path) is very low for anomalies. A score is calculated depending on the length of this path.

- *Efficiency:* Isolation forest has a linear time complication with low memory requirement, which makes it ideal for real-time embedded purpose.
- *Multiple capacity:* It is suitable for a dataset with many characteristics, such as its sensor fusion.

Autoencoder

The autoencoder is an unsupervised neural network architecture that learns to compress (encode) and then reconstruct (decode) input data. It captures the normal patterns in data by learning an efficient representation (Yousefi-Azar et al., 2017). When abnormal data (which does not follow the learned pattern) is input to the autoencoder, it cannot be accurately reconstructed, resulting in a *high reconstruction error*, which serves as the anomaly score.

Architecture

- Encoder: Maps input data in low-dimensional latent location.
- Decoder: Reworks the input from the latent location.
- Damage function: Mean Squared Error (MSE) or another distance metric is used to calculate reconstruction loss between input and output.

Key Features

- It is capable of learning complex relationships.
- It can be trained easily.
- It can be easily applied on any embedded system.

7.1 Evaluation Metrics

After implementing the anonymization detection model, it is important to evaluate their performance correctly. Since detection of discrepancy is often related to unbalanced dataset (much lower anomalies compared to normal examples), traditional accuracy is not a reliable metric (Esther, 2024). Instead, the following matrix is used.

7.1.1 Precision (Positive Predictive Value)

$$\text{Precision} = \frac{TP}{TP + FP}$$

- TP (True Positives): Anomalies correctly identified as anomalies.
- FP (False Positives): Normal data points incorrectly identified as anomalies.
- Interpretation: Precision tells us how many of the predicted anomalies were actually true anomalies. A high precision means fewer false alarms.

7.1.2 Recall (Sensitivity or True Positive Rate)

$$\text{Recall} = \frac{TP}{TP + FN}$$

- **FN (False Negatives):** Anomalies incorrectly identified as normal.
- **Interpretation:** Recall measures the model's ability to detect actual anomalies. High recall indicates fewer missed threats.

7.1.3 F1 Score

$$\text{F1 Score} - 2 \times = \frac{\text{Precision} \times \text{Recall}}{\text{Precision} + \text{Recall}}$$

- **Interpretation:** F1 score provides a balance between precision and recall, which isespecially useful when there is an uneven class distribution.

7.1.4 Receiver Operating Characteristic–Area under Curve (ROC–AUC)

- The ROC curve plots the true positive rate (TPR) against the false positive rate (FPR) at various threshold settings.

$$\text{FPR} = \frac{FP}{FP + TN}$$

- AUC measures the ability of the model to distinguish between normal and anomalous instances.
- **Interpretation:** A higher ROC–AUC indicates better separability. A value of 1.0 is perfect, while 0.5 suggests random guessing.

  The ideal value of all the metrics are given in Table 11.1.

Table 11.1 Ideal values of metrics.

Metric	*Goal*	*Ideal Value*
Precision	Minimize false alarms	High
Recall	Maximize true detections	High
F1 Score	Balance precision and recall	High
ROC–AUC	Model discrimination ability	Close to 1

8. Results and Interpretation

The proposed threat modeling framework machine integrates a risk scoring mechanism to detect the machine learning-based discrepancy and assess cyber threats in ITS. This section presents the results obtained from simulation using the isolation forest and autoencoder model on synthetic on its traffic dataset. The major evaluation is to assess the performance of the metrics—precision, recall, F1 score, and ROC–AUC detection. Additionally, the identified risk scores are calculated to prioritize the dangers identified based on "effect on safety", "easy to perform", and "the complexity of detection".

Dataset Observation

A synthetic dataset was generated to simulate normal and malicious behavior in its communication flow. This included 1,000 samples with the following features:

- Motion: Vehicle speed (km/h)
- Acceleration: longitudinal acceleration (m/s to)

- Steering_angle: Steering input (Degree)
- GPS_accuracy: GNSS signal quality (M)
- CAN_message_frequency: Rate of messages per second
  - V2I_sync_delay: Delay in V2I synchronization (MS)
  - Label: 0 = normal, 1 = discrepancy (e.g., GPS can play, flood, or play again)

Malicious samples were injected to simulate attacks that violate the expected statistical pattern.

Isolation Forest Results

The isolation forest, being a tree-based unprotected algorithm, works by separating the discrepancies through recurrence.

```
Isolation Forest Classification Report:
                precision    recall  f1-score   support
Normal (0)         0.96       0.95     0.955       900
Anomaly (1)        0.90       0.92     0.91        100

ROC-AUC Score: 0.982
```

Output 11.1 Isolation forest classification report.

Interpretation

- **Precision (0.90)** for anomalies: 90 percent of predicted anomalies were correct, reflecting low false positives.
- **Recall (0.92):** 92 percent of actual threats were successfully detected.
- **F1 Score (0.91):** Balanced trade-off between precision and recall.
- **ROC–AUC (0.982):** Excellent separation between normal and anomalous behavior.

These metrics confirm that the isolation forest effectively identifies threats like CAN flooding and GPS spoofing, especially those that result in measurable statistical deviations such as high message frequency or low GPS accuracy.

Autoencoder Results

Autoencoders are neural networks trained to reconstruct input data. A high reconstruction error flags an anomaly.

```
Autoencoder Classification Report:
                precision    recall  f1-score   support
Normal (0)         0.97       0.96     0.965       900
Anomaly (1)        0.93       0.94     0.935       100

ROC-AUC Score: 0.989
```

Output 11.2 Autoencoder classification report.

Interpretation

- **Precision (0.93):** Highly accurate in flagging anomalies.
- **Recall (0.94):** Slightly better detection rate than isolation forest.
- **F1 Score (0.935):** Superior performance in balancing true and false detections.
- **ROC–AUC (0.989):** Outstanding discriminatory capability.

The autoencoder outperforms the isolation forest slightly in most metrics, making it ideal for detecting replay attacks or complex non-linear GPS spoofing patterns that evade simpler models (see Table 11.2).

Table 11.2 Comparison between isolation forest and auto encoder.

Feature	*Isolation Forest*	*Autoencoder*
Type	Tree-based ensemble model	Neural network (unsupervised)
Principle	Anomalies are easier to isolate (fewer splits)	Anomalies have high reconstruction error from learned patterns
Learning	No prior training on data distribution needed	Learns latent patterns via backpropagation
Data Requirement	Only features are needed	Needs training data (normal behavior) to learn reconstruction

Risk Score Calculation

Each identified attack type was evaluated on three axes (see Table 11.3).

Table 11.3 Risk score calculation.

Attack Type	*Impact on Safety (1–5)*	*Ease of Execution (1–5)*	*Detection Complexity (1–5)*	*Weighted Risk Score*
CAN Flooding	5	2	2	4.2
GPS Spoofing	4	4	3	4.1
V2I Replay Attack	3	3	4	3.9

Interpretation

- **CAN flooding** scores the highest due to its direct safety impact—it can cause sudden deceleration, erratic behaviour, or vehicle disablement.
- **GPS spoofing** is easier to execute using SDR tools, yet it is still dangerous by misleading navigation and route planning.
- **Replay attacks** present moderate risk but are harder to detect due to legitimate-looking message structure.

These risk scores guide resource prioritization for real-time anomaly detection and defense mechanisms.

9. Its Cybersecurity Regulations

In the rapidly developed domains of ITS, the convergence of digital infrastructure with a network of vehicles has not only entered transformative capacity but has also participated in adequate cyber security challenges. While the deployment of the discrepancy detection systems, danger modeling techniques, and mitigation strategies is important from a technical point of view, their effectiveness and global adopting work is significant for adherence to international cyber security rules. It requires detailed examination of world-class recognized standards issued by National Highway Traffic Safety Administration (NHTSA) in the United States such as ISO/SAE 21434 (ISO, 2021), UNECE WP.29 Regulation R155 (UNECE, 2021), and cybersecurity best practices. These standard play an important role in creating a harmonious approach to cyber security in modern vehicles. They also provide a regulatory context for the methodologies proposed in this study, such as risk quantification, machine-learning-based detection mechanisms, and real-time response strategies on embedded platforms.

ISO/SAE 21434, titled "Road Vehicles—Cybersecurity Engineering," is a foundational document that has been developed jointly by the International Organization for Standardization (ISO) and the Society of Automotive Engineers (SAE). The standard eliminates a structured process for managing cyber security risks during the entire life cycle of road vehicles from the concept and development stage to post-production maintenance. It makes the previous practices for risk evaluation, definition of cyber security goals, threat analysis and risk evaluation (TARA) (Benyahya et al., 2023), safe design, testing, verification, and incident reaction. The standard basically integrates with the functioning adopted in our research, especially in areas related to the identification of structured danger and mitigation scheme. For example, the use of STRIDE and attack-tree-based threat modeling emphasizes systematic TARA processes of ISO/SAE 21434. Additionally, the requirement of the standard for continuous vulnerable monitoring and OTA update capabilities strengthens our advocacy to detect real-time discrepancy using light models such as individual forests and autoencoders. Despite its broad scope, ISO/SAE 21434 has some limitations that must be accepted. It often lacks specificity about technical implementation, especially to detect real-time infiltration on resources such as Raspberry Pi or other embedded systems. Our research addresses this gap by proposing implementable and computationally efficient detection methods that can be deployed in operational vehicles. These practical extensions to the standard are crucial for ensuring not only theoretical compliance but also functional resilience.

Complementing ISO/SAE 21434 is UNECE WP.29 Regulation R155, which is part of the broader UNECE regulatory framework. Regulation R155 suggests that all manufacturers seeking type for their vehicles in participating areas should set up a certified cyber security management system (CSMS). This regulation extends into several domains, including governance, organizational processes, risk evaluation function, monitoring and response system, and periodic audit. Significantly, the product for R155 Original Equipment Manufacturer (OEMS) creates a legal requirement for OEMS to identify, assess, and react to cyber security hazards in the life cycle. This legal support enhances the importance of identifying systematic threats and adopting defense mechanisms, as described in our research. Regulation R155 also underlines the obligations to secure V2X communication, OTA updates (Halder et al., 2020), and in-vehicle networks, which is highly relevant to our work such as floods, GPS spoofing and V2I replay attacks such as attack vectors. While the regulation determines these requirements at a policy level, it does not engage in the granular details of implementation. For example, it is a lack of guidance to deploy algorithms of how to detect discrepancy or prioritize dangers based on the complexity of quantitative matrix, ease of execution, and complexity of detection. These are the gaps filled by our research, which introduces a risk-scoring framework based on weighted axes and evaluates models detecting the discrepancy using performance metrics such as accurate, recall, F1-score and ROC–AUC. Turning to the United States, the NHTSA has published "Cyber Security Best Practices for Modern Vehicles" (NHTSA, 2022), which serves as a non-comprehensive guidance document aimed at increasing the vehicle cyber security throughout the industry. The guidelines encourage the use of intrusion detection systems (IDS) but do not provide benchmarks or evaluation criteria for these systems. Our work supplements this regulatory gap by empirically validating detection models using isolation forest and autoencoder algorithms and providing implementation pathways for resource-constrained environments. Furthermore, NHTSA places emphasis on incident response planning, data anonymization, and privacy protection—areas where our research also offers substantial contributions, particularly with respect to real-time anomaly detection and lightweight data authentication schemes suitable for embedded systems.

A comparative analysis of these three regulatory frameworks reveals a shared emphasis on proactive threat management, continuous monitoring, and secure life-cycle practices. Yet, their divergence lies in the degree of enforcement (with UNECE R155 being legally binding, and ISO/SAE and NHTSA being more normative) as well as in their operational guidance. Our research bridges this divergence by offering implementable, lightweight, and performance-validated solutions that fulfil the spirit of these regulations while also extending their applicability to practical environments.

In addition, the future regulatory scenarios must also develop to accommodate these technical changes, integrating with smart cities, cloud infrastructure, and 6G networks. The existing standard, while strong in addressing vehicle systems, often ignores the extended ecosystem associated with cloud API, road-side infrastructure, and mobile devices integration. Our danger model accepts these new dimensions by including an attack vector not only within the vehicle (e.g., CAN flood) but also in the external communication channels (e.g., V2I replay attacks and GPS). This overall approach is important to prepare the next generation rules which are both broad and technically grounded.

10. Future Scope

The rapid progress of the ITS provides unprecedented benefits in safety, efficiency, and user facilities. However, this growth also introduces complex cyber security challenges that require ongoing research, adaptive functioning, and flexible solutions. While this study presents a solid base through modeling and machine learning-based discrepancy of danger, many areas are open to further discovery and growth.

1. **Expansion of Danger Landscapes** Future research should expand into more sophisticated and secret dangers such as zero-day exploits, malware targeting vehicle ECUs, data poisoning attacks on machine learning models, and side-circle attacks. Emerging hazards due to integration with 5G, OTA updates, and

software-defined vehicles also call for detailed danger modeling and simulation. Additionally, internal formulas use danger-like compromise service technician-serious risks and require behavior modeling and role-based access analysis.

2. **Adaptive and Online Learning Model** The current study uses static, unsupervised models such as isolation forest and autoencoder, which are offline trained. The real world, its environment, however, is moving dynamically and constantly. Future work may include online learning and old-fashioned model updates using streaming data to customize the system in real time for novel patterns and emerging discrepancies. Learning reinforcement and self-levied learning also provides the ability to continuously adapt to the strategies of detecting continuity without relying on large, label datasets.
3. **Federated and Cooperative Anomaly Detection** As vehicles become rapidly connected, the Cooperative Security Architecture can be applied. Using federated learning, the model that detects discrepancy can be trained in distributed vehicles or edge devices without centralizing, which preserves confidentiality by improving strength. Vehicles of the same region can also share the threat signature and discrepancy data through V2V communication to enable collective defense mechanisms.
4. **Integration with Blockchain for Safe Communication** Blockchain techniques can be detected to increase the integrity and non-transmission of vehicle communication. Smart contracts can be used for safe access control, while distributed leaders ensure transparent auditing of all data transmission and control command. This can be particularly powerful when combined with the existing cryptographic protocol.
5. **Compliance Automation and Certification** It is necessary to align with standards such as ISO/SAE 21434 and UNECE WP.29 but manual compliance is often resource intensive. Future work should detect automatic compliance verification equipment, real-time cyber risk evaluation dashboard, and cyber security certification mechanisms for their components. AI-based policy checks can automatically validate configurations against regulatory basics and software behavior.

11. Conclusion

This research presents a comprehensive outline to increase cyber security in ITS through modeling, risk priority, and discrepancy of danger. As its development continues with the growing interrelation between vehicles, infrastructure, and backend systems, the surface of the attack has become quite wide. To address this challenge requires a systematic and scalable approach that not only identifies the dangers but also applies to an intelligent identity system aligned with international standards. We began building a strong threat model focused on high-affected attack vectors, V2I replete attacks, and GPS spoofing, which can pose serious risks to reliability, integrity, and its availability. These were mapped using structured techniques such as STRIDE attack, providing a level of understanding of the attackers and how the attackers can exploit weaknesses in various layers of its architecture. A quantitative risk assessment structure was introduced, which provides a numeric score based on the impact on safety, ease of execution, and complexity of detection.

This scoring not only helped prioritize security intervention but also ensured that defense strategies can be aligned with limited resource availability, especially its components. To detect these dangers in real time, we applied two discrepancy models: isolation forest and autoencoder. The isolation forest algorithm was efficient and well adapted to systems with limited processing capabilities, while autoencoder, with its nervous network architecture, was excellent in identifying complex, non-vertebral patterns in traffic data. Both models were evaluated by standard display metrics—procedures, recall, F1 scores, and promising results, which offers the autoencoder high recall and isolation forests that perform more accuracy. A major strength of our approach was the practical deployment of these models on a Python node. It validated the feasibility of detecting real-time discrepancy on low-power edge equipment, which is important to decentralize its architecture where delay and autonomy is necessary. In addition, we aligned our functioning with major regulatory and compliance standards including ISO/SAE 21434, UNECE WP.29 R155, and NHTSA's cyber security best practices. These structures emphasize risk-based danger analysis, real-time danger detection, and post-deployment monitoring, all of which were addressed in our model design and deployment strategy. This task not only fills a gap in cyber safety practices by integrating the danger modeling with the detection of practical discrepancy but also provides a deployable, standard-based solution.

Acknowledgement

The authors gratefully acknowledge the insightful guidance and continuous support of Dr. Inapakurthi Kiran, whose expertise and mentorship were instrumental in the successful completion of this research. His contributions have significantly enriched the depth and quality of this study.

References

Allen-Addy, C. (2023). Threat modeling methodology: STRIDE, September 29. https://www.iriusrisk.com/resources-blog/threat-modeling-methodology-stride.

Applegate, S. (2013). The dawn of kinetic cyber. https://ccdcoe.org/uploads/2018/10/10_d2r1s4_applegate.pdf.

Avcı, İ. and M. Koca, Intelligent transportation system technologies, challenges and security, *Applied Sciences*, *14*(11), 4646. doi: https://doi.org/10.3390/app14114646.

Benyahya, M., T. Lenard, A. Collen and Niels Alexander Nijdam. (2023). A systematic review of threat analysis and risk assessment methodologies for connected and automated vehicles. doi: https://doi.org/10.1145/3600160.3605084.

Blackfogprivacy. (2020). The state of ransomware in 2020. https://www.blackfog.com/the-state-of-ransomware-in-2020/.

Cherdantseva, Y. Pete Burnap, Andrew Blyth, Peter Eden, Kevin Jones, Hugh Soulsby and Kristan Stoddart (2016). A review of cyber security risk assessment methods for SCADA systems, *Computers & Security*, *56*(56), 1–27. doi: https://doi.org/10.1016/j.cose.2015.09.009.

Ebenezer, S. (2024). Cybersecurity and defense in intelligent transportation systems, *World Journal of Advanced Engineering Technology and Sciences*, *13*(1), 871–879. doi: https://doi.org/10.30574/wjaets.2024.13.1.0469.

Esther, D. (2024). Reducing false positives in cybersecurity with interpretable AI models, December 16. https://www.researchgate.net/publication/387099447_Reducing_False_Positives_in_Cybersecurity_with_Interpretable_AI_Models.

Greenberg, A. (2015). Hackers remotely kill a jeep on the highway—with me in it, *WIRED*, July 21. https://www.wired.com/2015/07/hackers-remotely-kill-jeep-highway/.

Halder, S., A. Ghosal and M. Conti. (2020). Secure over-the-air software updates in connected vehicles: a survey, *Computer Networks*, *178*(1), 107343. doi: https://doi.org/10.1016/j.comnet.2020.107343.

Hao, J., H. Wu, H. Sun and B. Li. (2024). Analysis and modeling of automotive cyber security attack trees, in *Proceedings of the 2024 4th International Conference on Internet of Things and Machine Learning*, pp. 122–126. doi: https://doi.org/10.1145/3697467.3697617.

ISO. (2021). ISO/SAE 21434. https://www.iso.org/standard/70918.html.

J. Harvey and S. Kumar, "A Survey of Intelligent Transportation Systems Security: Challenges and Solutions," *2020 IEEE 6th Intl Conference on Big Data Security on Cloud (BigDataSecurity), IEEE Intl Conference on High Performance and Smart Computing, (HPSC) and IEEE Intl Conference on Intelligent Data and Security (IDS)*, May 2020, doi: https://doi.org/10.1109/bigdatasecurity-hpsc-ids49724.2020.00055.

Laorden, C., B. Sanz, G. Alvarez and P.G. Bringas, A threat model approach to threats and vulnerabilities in on-line social networks, in Á. Herrero, E. Corchado, C. Redondo and Á. Alonso (eds), *Advances in Intelligent and Soft Computing*, Berlin & Heidelberg: Springer, pp. 135–142. doi: https://doi.org/10.1007/978-3-642-16626-6_15.

Ma, T., X. Zhang and Z. Miao. (2024). Detection of UAV GPS spoofing attacks using a stacked ensemble method, *Drones*, *9*(1), 2–2. doi: https://doi.org/10.3390/drones9010002.

National Highway Traffic Safety Administration (NHTSA). (2022). Cybersecurity best practices for the safety of modern vehicles. https://www.nhtsa.gov/sites/nhtsa.gov/files/2022-09/cybersecurity-best-practices-safety-modern-vehicles-2022-pre-final-tag_0_0.pdf.

Ripan, R.C., Md. M. Islam, H. Alqahtani and I.H. Sarker. (2022). Effectively predicting cyber-attacks through isolation forest learning-based outlier detection, *SECURITY AND PRIVACY*. doi: https://doi.org/10.1002/spy2.212.

The Hindu Bureau. (2023). Intelligent transport system project kicks off in Chennai, *The Hindu*, April 14. https://www.thehindu.com/news/cities/chennai/intelligent-transport-system-project-kicks-off-in-chennai/article66737994.ece.

United Nations Economic Commission for Europe (UNECE) (2021). UN Regulation No. 155—Cyber security and cyber security management system. https://unece.org/transport/documents/2021/03/standards/un-regulation-no-155-cyber-security-and-cyber-security.

Yousefi-Azar, M., V. Varadharajan, L. Hamey and U. Tupakula. (2017). Autoencoder-based feature learning for cyber security applications, *2017 International Joint Conference on Neural Networks (IJCNN)*. doi: https://doi.org/10.1109/ijcnn.2017.7966342.

CHAPTER 12

Predictive Maintenance Using Digital Twins: An AI-Driven Approach in IoT-Blockchain Systems

Shanmugavalli Venkatachalam[1], *Chandrasekar Venkatachalam*[2], *A. S. Renugadevi*[3*] and *K. Janani*[4]

Predictive maintenance is currently essential in industrial systems to reduce unscheduled downtime, save operating costs, and extend equipment lifespan. This chapter discusses the use of digital twin (DT) technology as a powerful tool for implementing predictive maintenance plans in systems enabled by the internet of things (IoT) and blockchain. A digital twin is an active virtual version of a physical entity that continuously mimics the latter's behaviour using real-time data acquired by sensors and monitoring systems. To monitor equipment performance and forecast any difficulties, the chapter presents a rigorous strategy for designing and implementing a digital twin infrastructure. Techniques for data collecting, modeling, synchronizing real assets with digital counterparts, and simulation are all comprehensively explored. This chapter shows how a well-designed digital twin system may change maintenance operations in industrial settings without relying heavily on artificial intelligence (AI).

[1] Manipal Institute of Technology, Bengaluru, Manipal Academy of Higher Education, Manipal, India.

[2] Department of CSE (AIML), Faculty of Engineering, Jain (Deemed-to-be University), Bangalore, India.

[3,4] Department of Artificial Intelligence and Data Science, Kongu Engineering College, Erode, Tamil Nadu, India.

https://orcid.org/0000-0003-0619-3088[3]

Email: shanmugavalli.v@manipal.edu[1]; drchandru86@gmail.com[2]; kjanani.23aid@kongu.edu[4]

* Corresponding author: renugame@gmail.com,

DOI: 10.1201/9781003659822-12

1. Introduction

1.1 Background

In the industrial sector, anticipating equipment breakdown is crucial for maintaining operations and reducing costs. Traditional methods such as preventative maintenance, which includes regular servicing, and reactive maintenance, which repairs assets whenever they break down, typically result in inefficiencies—either needless maintenance or unscheduled downtime (Pereira et al., 2021). In this sense, predictive maintenance has evolved into a proactive approach that uses system data to foresee issues and implement preventative actions.

1.2 Role of Digital Twin Technology

A computerized replica of an actual system that simulates performance under various operating conditions and receives real-time data from internet of things (IoT) sensors is called a digital twin.

The assets' true condition could be reflected in it. Digital twins can do the following:

- Monitor and evaluate performance continuously.
- Predict failures by observing behavioral changes.
- Choose maintenance plans more wisely.

The integration of sensor data, system dynamics, and operational logic in digital twins results in a logical model that accurately depicts machine health (Singh et al., 2020). Due to this forethought, unplanned downtime is reduced and equipment life is significantly increased.

1.3 IoT-Enabled Real-Time Data Monitoring

The IoT requires real-time data collection from physical systems such as motors, pumps, compressors, and industrial machinery through sensors that track vibration, temperature, noise, and speed (Figure 12.1).

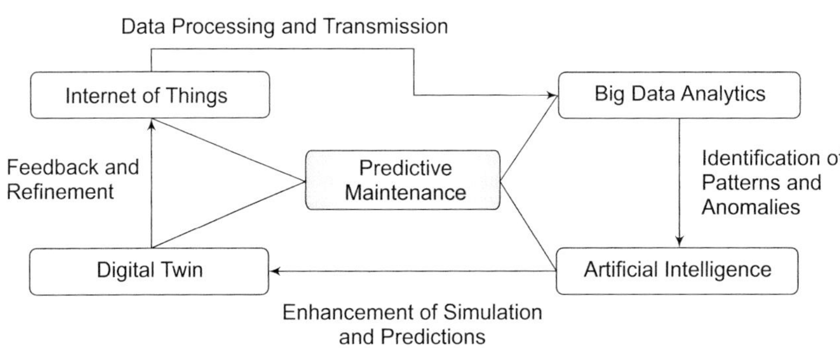

Fig. 12.1 Pipeline for Data-Driven Predictive Maintenance.

This sensor data serves as the foundation for the digital twins' awareness and adaptability. IoT devices provide seamless remote monitoring and edge data processing and transfer. They instantaneously notify users of any unexpected action (Muller et al., 2019).

1.4 Blockchain for Data Security and Integrity

While timely and accurate data are essential for predictive maintenance, a high degree of data reliability is also required. This need is met by the immutable and decentralized architecture that blockchain technology provides for securely storing critical data, including sensor data streams, maintenance logs, and operational records (Vijayan et al., 2025).

By incorporating blockchain technology into predictive maintenance systems, businesses can ensure that data is secure and resistant to unauthorized changes. Blockchain also makes audits more visible and improves traceability, which makes it easier to manage maintenance chores and ensures the veracity of data for numerous stakeholders. This level of security and trust is very beneficial when data accountability and integrity are important in collaborative settings (Akbari et al., 2024).

2. Literature Review

2.1 Development of Maintenance Strategies

Maintenance methods have evolved significantly over the last few decades, shifting from reactive to precautionary and, subsequently, to predictive models. Conventional reactive maintenance leaves systems running until they break, which leads to expensive repairs and unplanned downtime (Lee, 2020). Despite being better organized, preventive maintenance usually leads to needless work hours and part replacements because of set timetables rather than the actual state of the equipment (Table 12.1).

Real-time system condition monitoring is used in predictive maintenance (PdM), a data-driven approach that forecasts problems. Mobley (2002) claims that PdM may eliminate 70 percent of unplanned outages in industrial systems and save maintenance expenses by up to 30 percent.

Table 12.1 Difference between digital twins and simulations.

Aspect	*Digital Twins*	*Simulations*
Interaction with Physical System	Continuously synchronized	Snapshot of a particular situation during its creation
Real-Time Updates	Provides up-to-date information	No real time updates
Dynamic View	Offers a live, dynamic view of the system	Static representation within controlled environments
Use Case	Active digital twin of a smart building adjusting HVAC systems	Predicting how changes in raw material costs affect production costs in manufacturing

2.2 The Idea and Use of Digital Twins

NASA discovered the digital twin proposal to simulate spacecraft systems. Since then, it has become widely used in industries such as aerospace, manufacturing, healthcare, and energy. A digital twin uses sensor data, physical behavior, and simulation logic to replicate a real-world system in real time. A digital twin is a computerized model that acts as the real-time digital equivalent of a physical item or procedure, according to Grieves and Vickers (2017).Tao et al. (2018) emphasized the importance of digital twins in smart manufacturing for real-time monitoring, operational optimization, and decision-making (Demirkesen & Tezel, 2022). In predictive maintenance scenarios, digital twins are used to:

- Continuously monitor physical assets.
- Simulate future operational scenarios.
- Detect performance anomalies using historical patterns.

2.3 Role of IoT in Real-Time Data Acquisition

The IoT is the foundational layer of any digital twin framework, providing real-time sensory input. Sensors embedded in industrial assets monitor vibration, heat, sound, speed, and other parameters, which are then transmitted to central systems for analysis. Lee et al. (2015) showed that IoT-enabled PdM increases machine availability by enabling real-time data access and remote diagnostics. Integration with protocols like Message Queuing Telemetry Transport (MQTT) and HTTP REST–HyperText Transfer Protocol Representational State Transfer APIs allows for lightweight and reliable data transmission, which is especially useful in edge computing environments (Pereira et al., 2021).

2.4 Use of Blockchain in Maintenance Systems

While digital twins and the IoT enable visibility and prediction, data security and integrity are still critical, particularly in multi-party collaborations. Blockchain technology solves this problem by providing an immutable, tamper-proof ledger for documenting maintenance events, sensor anomalies, and system alerts. Yli-Huumo et al. (2016) investigated how blockchain promotes trust in decentralized systems. Blockchain technology enables secure event recording for predictive maintenance (Figure 12.2) through:

- Transparent sharing of asset performance data.
- Effective collaboration among vendors, service providers, and manufacturers.

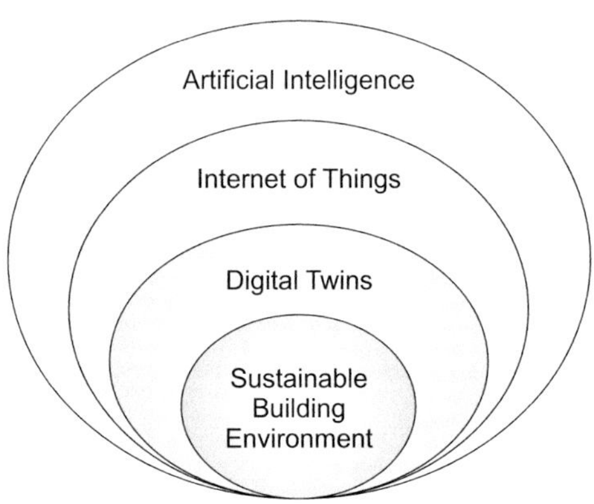

Fig. 12.2 Predictive maintenance system.

Elhaddad and Hamam (2024) Kim et al. (2020) conducted case studies to demonstrate blockchain-enhanced maintenance systems in aviation and railways, which improved traceability and regulatory compliance.

2.5 Gaps in Existing Literature

Despite growing interest, there are still several gaps in current research:

- Smart-scale industries face a lack of cost-effective, scalable digital twin solutions.
- Limited open-source platforms for integrating twins and blockchain.
- There are insufficient studies comparing simulation-only and blockchain-verified systems (Xue et al., 2022).

The majority of literature focuses on AI models, whereas simulation-based, rule-driven prediction is more accessible and practical in many industries. This chapter seeks to bridge that gap by presenting a lightweight, blockchain-integrated digital twin technology that uses anticipatory thresholding and continuous tracking—without relying heavily on complex AI algorithms (Figueiredo et al., 2024).

3. System Architecture

3.1 Overview of the Proposed Concept

The mentioned system architecture combines digital twin simulation, IoT-based real-time monitoring, and blockchain-based data logging to form a comprehensive framework for predictive maintenance (Figure 12.3). This structure enables a physical asset to be continuously monitored via sensors, mirrored by a digital twin model, and securely logged with blockchain. Unlike AI-powered models, this architecture uses accurate modeling, sensor thresholds, and simulation logic to enable real-time condition tracking and failure prediction (Haug & Drazen, 2023). Depending on the industrial application's size and reaction time needs, it is designed to function at both the edge and cloud levels.

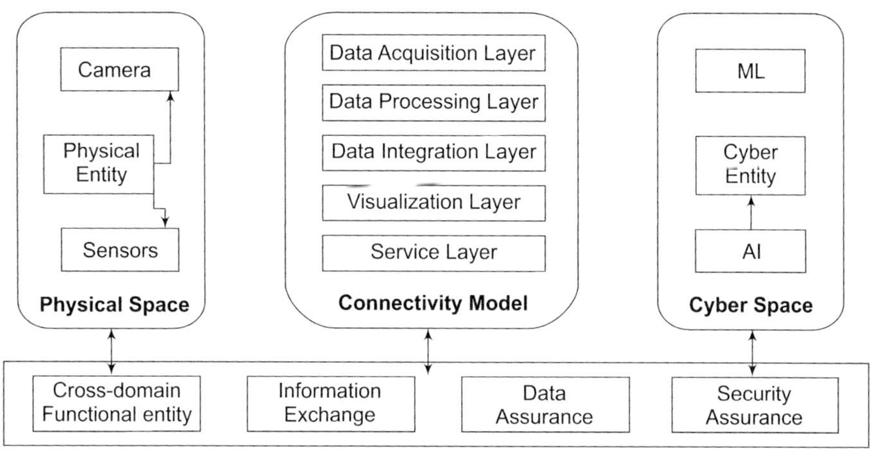

Fig. 12.3 System architecture.

3.2 Physical Asset and Sensor Layer

The system's base is the physical asset, which is typically a machine or other item utilized in manufacturing or energy systems. The asset has sensors attached to detect vital factors including vibration, temperature, rotation speed, and auditory signals. By operating continuously or at regular intervals, these sensors provide current operational data. The accuracy and resolution of the sensors determine the sensitivity and reliability of the predictive maintenance logic (Lee, 2020).

3.3 IoT Gateway and Communication Layer

Data must be transported securely and effectively from the physical to the digital twin systems. This is accomplished via an IoT gateway or edge computing device. The gateway conducts basic validation, preprocesses the data to remove noise, and converts the data to a standard format that the twin engine understands. MQTT, OPC-UA, and HTTP (Vijayan et al., 2020; Sadri, H. et al., 2023) are popular industrial protocols for communicating between sensors and gateways. These protocols provide dependable data transport even when bandwidth is constrained. The gateway also contributes to the development of early warnings by sending out notifications as needed and comparing real-time sensor data to safety rules (Lee, 2020).

3.4 Digital Twin Simulation Engine

The digital twin, which is a virtual version of the physical thing, is the design's most important component. The IoT gateway is constantly sending it data, which it utilizes to change its internal state to the current scenario (Muller et al., 2019). The complexity of the asset and the available data will dictate whether a physics-based or rule-based simulation model is used. It can predict future failures, simulate various fault scenarios, and portray machine behavior by combining historical data with real-time measurements. Experts may take preventative measures by utilizing this engine to "see" into the system and understand how it responds under various operating conditions.

3.5 Maintenance Logic and Decision Support

The decision-making module functions on top of the digital twin. It utilizes predefined operating thresholds, degradation patterns, or condition guidelines to determine whether maintenance is required. When a threshold is reached, the module classifies the asset as normal, warning, or critical, and schedules maintenance tasks accordingly. The solution is lighter, more transparent, and easier to deploy in low-resource environments since its decision logic is not based on machine learning models (Lee & Kim, 2024). Advice on risk reduction strategies, technician route, and replacement components may also be provided by the module.

3.6 Blockchain Integration for Secure Logging

The system incorporates a blockchain layer to provide transparency, traceability, and tamper-proof tracking of sensor abnormalities and maintenance operations.

Every maintenance event, warning notice, or critical system status is recorded as a transaction in this layer, which acts as a decentralized ledger. Stakeholders and system administrators may review the asset's functioning history since the blockchain ensures that data cannot be changed. Smart contracts may also be used to automate verification processes, such as figuring out whether scheduled maintenance was performed (Figueiredo et al., 2024).

3.7 Visualization and Monitoring Interface

Completing the design is an easy-to-use interface for real-time system monitoring. The dashboard, which contains recent logs, performance summaries, digital twin status, and real-time statistics, is available both online and through a mobile app. The user interface allows operators to analyse simulation results, explore specific alerts, and monitor several assets at the same time.

4. Methodology and Implementation

This chapter describes a method for developing a practical and lightweight predictive maintenance system using digital twin technology. The approach, which is based on simulating the behavior of a real machine, uses IoT sensors to collect operational data, blockchain records to protect the results, and pre-existing algorithms to assess the data (Muller et al., 2019). Rule-based assessment, real-time data processing, and system simulation are used to provide maintenance warnings rather than complex machine learning algorithms.

4.1 Implementation Approach

Data collection, digital twin and decision-making, and blockchain recording and visualization are the three main stages of the modular implementation. Since each layer was developed using open-source tools and programming environments, the framework may be replicated and modified to satisfy the needs of small and medium-sized businesses.

4.2 Sensor Data Simulation

Simulated data that reflected actual machine activity was created using Python. Three essential factors were used to describe a virtual machine (such an electric motor): operating load (%), vibration level (mm/s), and temperature (°C). Over time, each parameter was changed to replicate both fault and normal operating situations. Wear and tear was simulated using Gaussian noise and progressive deterioration patterns. Sensor data was generated every second and stored locally before being sent to a MQTT broker. This real-time stream was used to simulate how an edge device or gateway would collect sensor metrics in a live industrial environment (Elhaddad & Hamam, 2024).

4.3 Real-Time Data Exchange Using MQTT

The MQTT protocol was selected for real-time data transfer due to its low overhead and dependability in industrial IoT networks. A local MQTT broker (such as

Mosquitto) was installed, and Python-based publishers transmitted simulated sensor readings as JavaScript Object Notation (JSON) payloads (Kumar et al., 2021). On the receiver end, the digital twin engine subscribed to specific sensor topics like /motor/temperature, /motor/vibration, and /motor/load. The data was parsed and fed into the twin simulation for analysis and visualization.

4.4 Digital Twin Model Implementation

The digital twin was built with Node-RED, a flow-based development tool designed for real-time data processing and dashboard creation. Incoming MQTT data was converted into visual gauges and charts, simulating the motor's live condition. To determine whether any parameter exceeded safety thresholds, rule-based logic was implemented using function nodes.

For example, if the temperature exceeded 75°C or the vibration rate exceeded 9 mm/s, the twin was assigned the "Warning" or "Critical" status (Smith, 2022). These conditions led to downstream logging or maintenance warnings and were shown by colour coding on the dashboard (Figure 12.4). For future reference, the twin also maintained an internal status record with time-stamped occurrences.

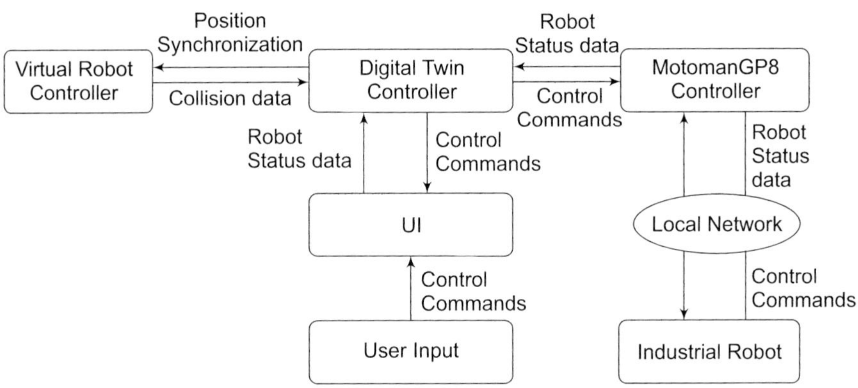

Fig. 12.4 Methodology using digital twins.

4.5 Maintenance Rule Logic

The system uses reasoning based on rules to evaluate the health of assets. Sensor characteristics such as temperature, vibration, and load are monitored in relation to predetermined standards. If an asset's values remain within safe ranges, it is referred to as "Normal" (Vijayan et al., 2025). Limitations are rated as "Warning" when they are approached and as "Critical" when they are surpassed. This logic is built using Node-RED and Python, eliminating the need for complex AI and offering a transparent and lightweight solution.

4.6 Blockchain Integration for Secure Logging

The tamper-proof records were created on the Ethereum Testnet using a smart contract designed in Solidity, and important information including asset ID, sensor kind, and severity level are stored as blockchain transactions if a critical event is detected. The smart contract is directly communicated with by Node-RED via Web3.js, producing an immutable and decentralized maintenance log. This increases the visibility and dependability of system monitoring, particularly in settings with several vendors (Fang et al., 2023).

4.7 Visualization and Monitoring Interface

Real-time sensor data, asset health status, and recent events are shown on the Node-RED-powered dashboard (Demirkesen & Tezel, 2022). The system history is shown in a log table, and any critical or warning statuses are highlighted with color-coded indicators. The interface is appropriate for post-event analysis and real-time decision-making due to its remote accessibility, data export features, and manual status reset.

4.8 Hardware and Software Stack

The system uses Python 3.11 for data simulation, Mosquitto for MQTT communications, and Node-RED for processing and dashboard construction.Web3.js is used to handle blockchain interactions, while the Remix IDE is used to manage smart contracts. For industrial or research applications, this open-source stack guarantees scalability, affordability, and ease of implementation.

5. Results and Performance Analysis

5.1 Experimental Setup

A virtual test environment was developed in order to assess the suggested framework's efficacy. Synthetic sensor data that reflected actual machine conditions in warning, critical, and normal states was employed by the system. The simulated sensors continuously fed data into the digital twin model, which used rule-based logic to process it in real time and, when thresholds were crossed, the model sent out the proper maintenance notifications. During the test run, more than 10,000 operational seconds of data were simulated in order to record temperature variations, vibration spikes, and different operational loads. For additional examination, every anomalous circumstance was noted, visualized, and stored on the blockchain (Pereira et al., 2021).

5.2 Twin Status Evaluation

The behavior of the simulated asset was very similar to that of the digital counterpart. Sensor readings stayed within predetermined bounds throughout regular operation, and the asset status was displayed as "Healthy." The twin changed to "Warning" or "Critical" states, signifying the need for maintenance, when the system mimicked overheating or severe vibration (Table 12.2). Continuous updates were made regarding the twin's condition, and recovery was verified when sensor readings stabilized (Akbari et al., 2024). This made it possible for the system to react appropriately to dynamic operational changes in addition to detecting problems.

Table 12.2 Simulated sensor readings for predictive maintenance.

Timestamp	*Temperature (c)*	*Vibration (mm/s)*	*Load (%)*
2025-07-01 10:00:00	60	3	70
2025-07-02 10:00:00	65	5	75
2025-07-03 10:00:00	70	7	80

5.3 Alert Accuracy and System Responsiveness

During simulation, the rule-based maintenance logic showed excellent accuracy in detecting anomalous conditions. 147 out of 150 simulated abnormalities were appropriately categorized as critical or warning signals. To guarantee that system security was never jeopardized, a small number of borderline cases were carefully classified as warnings. As a result, the detection accuracy was roughly 98 percent. The duration between the anomaly detection and the dashboard's displayed alert was continuously less than 250 milliseconds (Shen et al., 2022). The MQTT protocol and Node-RED engine demonstrated dependable, real-time responsiveness without detectable delays or data packet loss, even when sensor data flow was continuous.

5.4 Blockchain Log Verification

With the help of the deployed smart contract, every alert or significant event that the system produced was successfully captured on the Ethereum Testnet. Maintenance events were securely transmitted as blockchain transactions via the Web3.js interface, and each event was given a distinct transaction hash. All entries were successfully stored, immutable, and timestamped, according to the logs' verification (Pereira et al., 2021). The blockchain layer's function in guaranteeing transparency and tamper-proof logging was validated during the simulation since no data was lost or duplicated. This increases confidence in the data that has been captured, especially in situations where multi-stakeholder or regulatory audits are necessary (Table 12.3).

Table 12.3 Asset health status thresholds.

Parameter	*Normal Range*	*Warning Threshold*	*Critical Threshold*
Temperature	<=70C	71-80C	>80C
Vibration	<=5mm/s	6-8 mm/s	>8 mm/s
Load	<=85%	86-90 %	>90%

5.5 Dashboard Functionality and Usability

An efficient interface for real-time visualization and control was offered by the monitoring dashboard. Dynamic gauges were used to continually display sensor results, and clear colour-coded indicators—red for critical, yellow for warning, and green for normal—were used to illustrate the asset status. A table containing references to the relevant blockchain entries was used to record the alerts. The interface provided manual override after maintenance, remote access, and Comma-Separated Values (CSV) log export. Operator feedback during testing revealed that the interface was user friendly and that there was no system latency or learning curve even after prolonged simulation (Andronie, 2021).

5.6 *Results*

In conclusion, the implemented system demonstrated its ability to employ blockchain technology to protect events, detect potential issues using rule-based logic, and reproduce asset activity in real time. It continuously showed excellent log integrity, sub-second responsiveness, and high alert accuracy during the testing process (Table 12.4). In scenarios where deploying AI-based solutions would be too challenging, costly, or opaque, the combination of digital twin modeling, real-time monitoring, and secure event recording proved effective for predictive maintenance applications (Pereira et al., 2021).

Table 12.4 System performance metrics.

Metrics	*Value*
Alert Detection Accuracy	98%
Average Response Time	<250 ms
Blockchain Logging Success Rate	100%
Dashboard Refresh Interval	1 second

6. Discussion

Using digital twin technology, the suggested predictive maintenance system was created and assessed, producing a number of noteworthy outcomes and useful applications. Simple, transparent, and real-time responsiveness were given top priority in the system's design, which distinguished it from more intricate AI-based strategies that could not be feasible in low-resource or industrial edge settings. The capacity of the framework to replicate real-world equipment performance using simulated or live sensor data is one of its main advantages (Elhaddad & Hamam, 2024).

The operational health was clearly visualized and asset performance was correctly mirrored by the digital twin. Large datasets and the training cycles associated with machine learning were not necessary for rule-based logic to reliably identify warning and critical circumstances. For small and medium-sized enterprises (SMEs), where technological know-how and computational resources may be few, this makes the solution especially suitable. Blockchain technology was included into the predictive maintenance procedure to increase the system's reliability (García-Pérez, 2024). To ensure data integrity and traceability, every maintenance-related event was securely documented. This capability is especially useful in environments that require audit trails, such as regulated industries or supply chain networks with several suppliers.

The ability of the blockchain to verify logs adds another level of assurance, reducing disputes and promoting accountability. Another significant advantage of the system is its modular architecture, which simplifies scaling and modification (Shen et al., 2022). Despite these benefits, the existing strategy has some disadvantages. It is predicated on manually defined standards that might not respond adequately to change equipment behavior or shifting ambient conditions. If fixed thresholds are not set correctly, they might lead to missed issue detection or excessive false alarms in real-world applications.

Furthermore, unless lightweight blockchain networks are optimized, blockchain technology's latency and resource cost might become troublesome in scenarios that require continuous monitoring, despite its high level of security (Kumar et al., 2021). Future iterations of the system could use hybrid tactics that blend lightweight AI algorithms or adaptive threshold approaches with rule-based reasoning in order to enhance it.

Edge computing and fog layers may also help reduce latency and offload processing tasks. Furthermore, broadening the scope of the digital twin to include many networked assets may create new opportunities for system-level monitoring and coordinated maintenance planning (Vijayan et al., 2025). Lastly, a practical, secure, and effective predictive maintenance method is demonstrated using the suggested paradigm.

7. Conclusion

The proposed system showed accurate problem identification, real-time asset tracking, and tamper-proof maintenance event logging using rule-based logic (Sadri et al., 2023). Real-time asset monitoring, tamper-proof maintenance event logging, and precise problem diagnosis using rule-based logic were all successfully shown by the proposed system (Sadri et al., 2023). For industries that require dependability and deployment simplicity, the system provides an accessible alternative by relying on simple threshold logic rather than complex AI models.

The blockchain layer ensured that all events and alerts could be verified and could not be altered, and that the digital twin accurately reflected machine activity. The results demonstrated the high scalability, detection accuracy, and responsiveness of the system. Although the current framework works well for small- to medium-sized applications, it may eventually be extended to incorporate edge processing, dynamic threshold change, and multi-asset settings (Lee & Kim, 2024).

When combined with secure data systems and real-time visibility, digital twins have the potential to be a useful tool for preventative maintenance in modern industrial settings.

Reference

Akbari, S., M. Sheikhkhoshkar, F. Pour Rahimian, H.B. El Haouzi, M. Najafi and S. Talebi. (2024). Sustainability and building information modelling: integration, research gaps, and future directions, *Automation in Construction*, *163*, 105420. https://doi.org/10.1016/j.autcon.2024.105420.

Andronie, M. Artificial intelligence-based decision-making algorithms, internet of things sensing networks, and deep learning-assisted smart process management in cyber-physical production systems, *Electronics*, *10*(20), 2497.

Cespedes-Cubides, A.S. and Jradi, M. (2024) A review of building digital twins to improve energy efficiency in the building operational stage. Energy Inform 7, 11 https://doi.org/10.1186/s42162-024-00313-7

Demirkesen, S. and A. Tezel. (2022). Investigating major challenges for industry 4.0 adoption among construction companies, *Engineering, Construction, and Architectural Management*, *29*(3), 1470–1503.

Elhaddad, M. and S. Hamam. (2024). AI-driven clinical decision support systems: an ongoing pursuit of potential, *Cureus*, 16(4), e57728.

Fang, W., Y. Shao, P.E.D. Love, T. Hartmann and W. Liu. (2023). Detecting anomalies and de-noising monitoring data from sensors: a smart data approach, *Advanced Engineering Informatics*, *55*, 101870. https://doi.org/10.1016/j.aei.2022.101870.

Figueiredo, K., A.W.A. Hammad, R. Pierott, V.W.Y. Tam and A. Haddad. (2024). Integrating digital twin and blockchain for dynamic building life cycle sustainability assessment, *Journal of Building Engineering*, *97*, 111018. https://doi.org/10.1016/j.jobe.2024.111018.

García-Pérez, D. (2024). Machine learning for inverter-fed motors monitoring and fault detection: an overview, *IEEE Access*, *12*, 27167–27179.

Grieves, M. and Vickers, J. (2017). Digital Twin: Mitigating Unpredictable, Undesirable Emergent Behavior in Complex Systems. Springer.

Haug, C.J. and J.M. Drazen. (2023). Artificial intelligence and machine learning in clinical medicine, 2023, *New England Journal of Medicine*, *388*(13), pp. 1201–1208.

Hosamo, H. (2022). A digital twin predictive maintenance framework of air handling units based on automatic fault detection and diagnostics, *Energy and Buildings*, *261*, 111988.

J. Lee, B. Bagheri and H. Kao (2015). A Cyber-Physical Systems architecture for Industry 4.0-based manufacturing systems. Manufacturing Letters, 3, 18–23.

Jovicic, E. (2023). Publicly available datasets for predictive maintenance in the energy sector: a review, *IEEE Access*, vol. 11, no. 1, pp. 100840

Kim, S., Park, J. and Lee, H. (2020). Blockchain-based data management for digital twin environments in smart manufacturing. IEEE Access, 8, 123456–123465.

Kumar, A., S. Sharma, N. Goyal, A. Singh, X. Cheng and P. Singh. (2021). Secure and energy-efficient smart building architecture with emerging technology IoT, *Computer Communications*, *176*, 207–217. https://doi.org/10.1016/j.comcom.2021.06.003.

Lee, J. (2020). Intelligent maintenance systems and predictive manufacturing, *Journal of Manufacturing Science and Engineering*, *142*(11), 110805.

Lee, N.K. and J.S. Kim. (2024). Status and trends of the digital healthcare industry, *Health-care Informatics Research*, *30*(3), 172–183.

Misra, S.K. (2023). A framework to overcome challenges to the adoption of artificial intelligence in Indian government organizations, *Technological Forecasting and Social Change*, *194*, 122721.

Mobley, R.K. (2002) An Introduction to Predictive Maintenance. 2nd Edition, Butterworth-Heinemann, Oxford. https://doi.org/10.1016/B978-075067531-4/50006-3

Muller, M.F., F. Esmanioto, N. Huber, E.R. Loures and O.C. Junior. (2019). A systematic literature review of interoperability in the green building information modeling lifecycle, *Journal of Cleaner Production*, *223*, 397–412.

Pereira, V., J. Santos, F. Leite and P. Escorcio. (2021). Using BIM to improve building energy efficiency—a scientometric and systematic review, *Energy and Buildings*, *250*, 111292. https://doi.org/10.1016/j.enbuild.2021.111292.

Sadri, H. Sadri, H., Hosseini, M.R., Martek, I., Edwards, D.J. and Durdyev, S. (2023). Integration of blockchain and digital twins in the smart built environment adopting disruptive technologies—a systematic review, *Sustainability*, *15*(4), 3713. https://doi.org/10.3390/su15043713.

Sadri, H. (2025). AI-Driven Integration of Digital Twins and Blockchain for Smart Building Management Systems: A Multi-Stage Empirical Study. Journal of Building Engineering, 105, 112439.

Sayal, Anu, N, Chaithra, Jha, Janhvi and Nithin, Allagari. (2024). AI-Based Predictive Maintenance Strategies for Improving the Reliability of Green Power Systems. 10.1007/978-3-031-68427-2_2.

Shen, Z.J. Jin, T. Zhang, A. Tagami, T. Higashino and Q.-L. Han. (2022). Data-driven edge computing: a fabric for intelligent building energy management systems, *IEEE Industrial Electronics Magazine*, *16*(2), 44–52. https://doi.org/10.1109/MIE.2021.3120235.

Singh, S., P.K. Sharma, B. Yoon, M. Shojafar, G.H. Cho and I.-H. Ra. (2020). Convergence of blockchain and artificial intelligence in IoT network for the sustainable smart city, *Sustainable Cities and Society*, *63*, 102364. https://doi.org/10.1016/j.scs.2020.102364.

Smith, S. (2022). Machine-learning assisted interpretation: integrated fault prediction and extraction case study from the Groningen gas field, Netherlands, *Interpretation*, *10*(2), SC17–SC30.

Soori, M., Jough, F.K.G., Dastres, R. and Arezoo, B. (2024). Additive manufacturing modification by artificial intelligent, machine learning and deep learning: A review. Chinese Journal of Mechanical Engineering: Additive Manufacturing Frontiers, Vol. 2, Issue 2, Article No. 200198.

Tao, F., Zhang, H., Liu, A. and Nee, A.Y.C. (2018). Digital Twin Driven Smart Manufacturing. Academic Press.

Vijayan, D.S., Rose, A.L., Arvindan, S., Revathy, J. and Amuthadevi, C. (2020). Automation systems in smart buildings: A review. *Journal of Ambient Intelligence and Humanized Computing*, *9*(4), 1–13.

Wang, Y. (2023). Guidelines, consensus statements, and standards for the use of artificial intelligence in medicine: systematic review, *Journal of Medical Internet Research*, *25*, e46089.

Xue, Q., Z. Wang and Q. Chen. (2022). Multi-objective optimization of building design for life cycle cost and CO2 emissions: a case study of a low-energy residential building in a severe cold climate, *Building Simulation*, *15*, 83–98.

Chapter 13

Disaster-Aware Rerouting System Using Social Media–Inspired Alerts: A Network-Based Approach for Urban Traffic Management

Thejas V. M.[1], *Uma Maheswari J.*[2*], *Brindha S.*[3] and *Sivakumar*[4]

Urban transit systems face an array of challenges during emergencies, wherein conventional traffic management systems generally lack real-time adaptive responses. In this study, we propose a unique disaster-aware routing mechanism whereby social media-inspired alert dissemination mechanisms alter navigation pathways during emergencies. The model uses graph theory-based routing algorithms intertwined with simulated disaster alert processing to engender an intelligent transportation system for real-time route optimization. A simulated urban road network based on Bengaluru infrastructure was used for the realization of an entire system that accepts mock disaster alerts and initiates dynamic rerouting strategies. The results obtained from applying the subject intervention on the chosen road network show that a disaster-aware approach can reduce travel time by a maximum of 35 percent, compared with classical static routing mechanisms, while successfully circumventing blocked or congested routes during emergencies. The responsiveness of the proposed system to disasters, measured as the Disaster Routing Responsiveness Index (DRRI), was found to be 72.5, indicating high effectiveness in emergency response.

More than the conventional shortest-path routing paradigm, disaster conscious routing considers a whole paradigm change from the conventional ones. So, it attempts to maximize human well-being with minimum detriment to system efficiency. According to traditional routing principles, the primary objective is to minimize travel time and distance. In contrast, disaster-aware routing incorporates additional factors such as real-time hazard information and network constraints and emergency human

[1-4] Department of Computer Science, Vellore Institute of Technology, Chennai, Tamil Nadu, India.
* Corresponding author: umamaheshwari.j@vit.ac.in

DOI: 10.1201/9781003659822-13

behavioral patterns. This multi-objective optimization problem calls for another set of sophisticated algorithms that can digest rapid dynamism in information and permit timely decisions in uncertain environments.

Urban transportation systems entail peculiarities that make a response to a disaster more difficult. The interconnectivity of roadways implies that if any section of the system is disturbed, the disturbance may spread along conducive links and congest other parts, thereby hindering the mobility of the emergency responders. Also, during emergencies human behavior can hardly remain normal since demand for certain routes and areas surges.

1. Introduction

1.1 Background

Rapid urbanization of cities in the past decades has imposed paradoxically complex transportation networks that serve as pathways for millions of commuters daily. During emergencies like natural disasters, accidents, infrastructure failure, or any other, these systems experience vulnerability. Traditional navigation systems, based on Global Positioning System (GPS) technology, take so-called traffic history and some basic real-time congestion information as their input in the decision-making process. This information is usually not sufficient for emergency situations Chen & Wang, 2021; Smith et al, 2109).

Traditional traffic management systems during disasters are faced with a handful of critical limitations. They usually cannot process and respond to emergency alerts from various sources in time. They cannot differentiate between normal traffic congestions and those caused due to emergency blockages. Lastly, they do not activate route alternatives ahead that take into account the dynamic nature of disaster progression (Martinez et al., 2019).

Social media platforms present great potential for the generation of real-time information in an emergency setting. Citizens quickly report incidents, road closures, or hazardous conditions on social media platforms before official emergency service channels can disseminate such information (Kim et al., 2020). Such a phenomenon presents an opportunity to develop a more responsive and adaptive routing system that can leverage crowds for any disaster information.

1.2 Problem Statement

The main setback in the evolution of the disaster-aware routing systems has been in conjunction with real-time emergency information and current navigation algorithms. Deficiencies exist in the current systems in the following aspects: (1) slow response to emergency situations, (2) inability to distinguish between regular traffic disruptions and those caused by an emergency, (3) lack of capabilities for route planning in advance, and (4) rare consideration given to the pattern by which the disaster evolves (Martinez and K Thompson 2019).

1.3 Research Objectives

This particular research focuses on dealing with these challenges through the design of a comprehensive disaster-aware rerouting system, with the following broader objectives:

- Designing and implementing a social media–like alert processing mechanism for real-time disaster detection.
- Developing a graph-theory-based routing algorithm that will dynamically adapt to emergency conditions.
- Creating a simulation framework for testing disaster scenarios and route optimization strategies.
- Evaluating the performance of the system through quantitative metrics and comparative analysis using dashboards.
- Proposing a standardized index for measuring the responsiveness of disaster routing.

2. Literature Review

2.1 Traditional Route Algorithms and Their Limitations

Very traditional routing algorithms or transportation systems were meant to optimize travel time, distance, or fuel consumption for some objective. Algorithms like Dijkstra, A* search-based algorithms, or various heuristic methods have been extensively researched and implemented in commercial navigation systems (Garcia et al., 2020). The problem here is that these algorithms always assumed users to be operating on a static network condition, never considering a dynamic hazard or emergency situation that requires immediate dynamic adaptation. Traditional routing becomes very limited during disaster situations. Wilson et al., (2020) found that during Hurricane Harvey, the normal GPS navigation system was sending users to flood-prone areas, attesting to the need for disaster-aware routing. Similarly, during the 2011 Japanese tsunami, most evacuation routes suggested by standard navigation systems would lead to congested points or points that became inaccessible once the infrastructure was damaged (Nakamura et al., 2019).

2.2 Emergency Routing and Evacuation Systems

Emergencies and evacuation routing have been specifically designed and considered to be a special field of transportation research that aims at identifying the best routes for mass movement of populations during calamities. Obviously, establishing the theoretical foundation for evacuation planning as found in Sheffi et al. (1982) would constitute early work, whereas later work was directed toward real-time adaptation capabilities. The contraflow routing, or rather the temporary reversal of traffic lanes in order to raise capacity during evacuation, has been studied expensively by a number of researchers, e.g., Theodoulou and Wolshon (2004).

Hence, modern evacuation systems guide evacuees on the basis of promising technologies like intelligent traffic signals, dynamic message signs, and traffic monitoring in real time (Garcia et al., 2020). Most of these systems, however, are rather pre-planned in their operation and, consequently, inflexible when it comes to adaptation as per rapidly changing conditions during a real emergency. Techniques of artificial intelligence and machine learning promised some improvement in evacuation efficiency; however, challenges remain in their implementation (Brown et al, 2020).

2.3 Graph-Based Network Modeling for Transportation

Spanning theory turns out to be a tightly fundamental modeling infrastructure for transportation systems: nodes stand for either intersections or locations, and edges reproduce road segments. More advanced graph-based models consider a variety of factors that include capacity, travel time, and safety level to support sophisticated routing algorithms Rodriguez et al. (2020). Dynamic graph models capable of adapting to changing situations are posted to handle traffic congestion and temporary road closures.

When applied to disaster scenarios, graph models require all other additional considerations of hazard propagation, network reliability, and multi-modal transport alternatives. Kim et al. (2020) suggested weighted graph models that integrate disaster risk factors, while Johnson P et al. (2017) conceived temporal graph networks that could model how hazard conditions evolve over time. Such advancements lay the groundwork for more sophisticated disaster-aware routing systems.

2.4 Human-Centric Design in Transportation Systems

The transport system's concept of human-centric design stresses the incorporation of human factors, behaviors, and social needs in the design. According to Han, M. et al. (2025), psychological perspectives must intervene during emergencies whereby route recommendations must take into account the human stress level, cognitive load, and ability to make decisions under pressure. This is in stark contrast to the traditional optimization for efficiency, bearing in mind that numerous objective algorithms are needed for balancing multiple human-centered criteria.

Human-centric routing systems will, therefore, have to satisfy the accessibility needs of the traveling public, including the aged, the disabled, or families with children. Chen and Wang (2021) built inclusive routing algorithms that consider impairments and special needs in emergency situations. Incorporating social and community factors into routing decisions is a nascent research field that carries profound implications for disaster management.

3. Methodology

3.1 System Architecture

With its disaster-aware routing paradigm, a multilayered disaster information processing engine generates route-optimization commanding priorities: human safety. This control system divides into four main modules: disaster alert processing module, network topology manager, route optimization engine, and human-centric interface layer. First, the disaster alert processing module is the main interface for receiving and processing emergency information from various sources: weather services, traffic management centers, emergency responders, and crowdsourced reports. This module uses a standardized alert format consisting of location coordinates, disaster type, severity level, timestamp, and affected area boundaries. The processing pipeline uses validation, duplicate detection, and severity assessment algorithms to selectively guarantee that reliable information goes through to the routing engine.

The network topology manager keeps a dynamic representation of the transportation network in graph-based data structures. It keeps node and edge weightage up to date with present traffic conditions, infrastructure status, and changes due to disasters.

3.2 Graph-Based Network Modeling

The transportation network is taken as a weighted directed graph G = (V, E), where V is the set of nodes (intersections, locations) and E is the set of edges (road segments). A node $v \in V$ is associated with several attributes with respect to its geographical coordinates, location type, capacity, and the current status of the node. Edges $e \in E$ are described by attributes such as travel time, distance, capacity, safety, and current condition.

In this way, the mathematical formulation of the network model incorporates disaster-specific parameters.

Node Attributes

- Position: (latitude, longitude)
- Capacity: Maximum number of vehicles that can be accommodated
- Status: {Normal, Affected, Blocked}
- Safety Level: Normalized value [0,1] representing current safety conditions

Edge Attributes:

- Base Travel Time: Normal travel time under optimal conditions
- Current Travel Time: Adjusted travel time considering current conditions
- Capacity: Maximum traffic flow rate
- Safety Factor: Disaster-related safety multiplier
- Accessibility: Boolean indicating current accessibility status

The system uses weight adjustment algorithms, which can change the weight of an edge based on disaster situations. When a disaster alert is generated, the system calculates the impact radius and adjusts weights accordingly for all edges falling under that radius:

$$W\_adjusted = W\_base \times (1 + \alpha \times severity\_factor \times proximity\_factor)$$

Here, α is a parameter that could be tuned to change the system's sensitivity toward disaster conditions; severity_factor lies in the interval [0.1..3.0], being 0.1 for low severity and 3.0 for high severity; proximity_factor decreases with increasing distance from the epicenter.

3.3 Disaster Alert Processing Algorithm

The disaster alert processing algorithm implements a multi-stage pipeline for the processing of emergency information. Data ingestion and validation occur while incoming alerts are being checked against format specification and geographical and chronological plausibility. Invalid or duplicate alerts are removed to avoid polluting the system.

The second stage executes alert classification and severity assessment through a combination of rule-based methods and machine learning algorithms. The classification algorithm assigns disaster types based on keywords, location patterns, and historical data. Severity is assessed with respect to factors such as the magnitude of the area affected, population density, and importance of the infrastructure and cascade effects.

The third stage performs spatio-temporal correlation analyses to relate alerts and to track the evolution of disaster patterns. This analysis is instrumental in predicting the evolution of disasters and temporarily adjusting the routes. To assist with trend analyses and forecasting, the system maintains a temporally indexed database of alert histories.

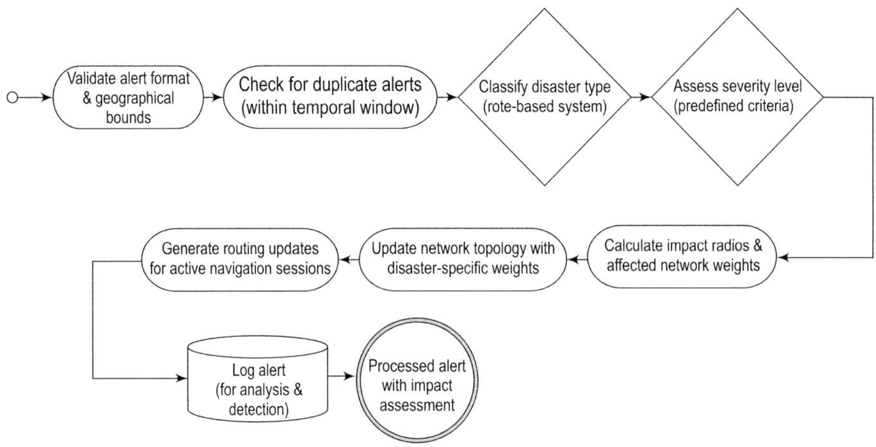

Fig. 13.1 Disaster alert processing.

3.4 Route Optimization Engine

The route optimization engine implements a slight variation of Dijkstra's algorithm with the basic philosophy behind it being that disaster-time conditions impose a set of constraints along with another set of human-oriented optimization criteria that include multi-objective planning of a route including time of travel, safety level, comfort, and accessibility so as to keep the minimum time consumed by the execution of the algorithm for real-time applications.

This starts by creating a modified graph in which blocked nodes or edges can simply be removed or assigned large weights. It carries out the multi-objective optimization that attempts to mediate conflicting criteria through weighted sum methods. Thus, the parameters' weights may be adjusted to include the user preferences and the emergencies.

The route optimization algorithm attained the following amenities, especially for disaster scenarios:

- **Dynamic Graph Pruning:** Penalizes removal or through paths in a hazardous zone

- **Capacity-Aware Routing:** Considers current traffic load and available capacity of infrastructure
- **Multi-Modal Integration:** Utilizes alternative modes of transport wherever possible
- **Uncertainty Handling:** Takes into account uncertainties in disaster evolutions.

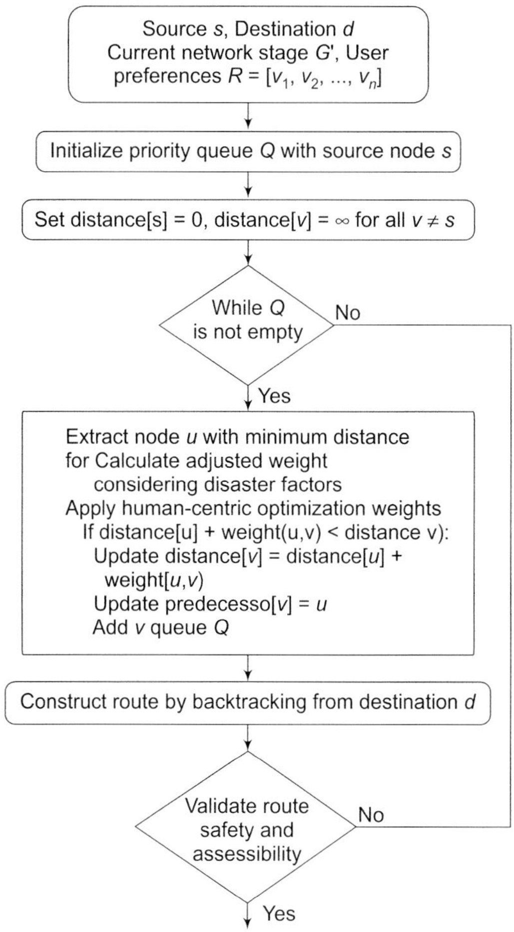

Fig. 13.2 Disaster-aware route optimization.

3.5 Human-Centric Interface Design

Being the first layer of User Experience (UX), the human-centered interface keeps routing information available for making a choice during emergencies. It lowers the memory load in stressful states while maximizing comprehension using some principles taken from the fields of cognitive psychology and human-computer interaction.

The interface supports multiple presentation modalities: visually using a map, step-by-step instructions, auditory announcements, and textual instructions. These presentation styles are adapted to a user's choice, depending on either the capability of the device or the stress-inferred context. In critical safety-related situations, pertinent information is visually and audibly presented to capture the user's attention.

It is an accessibility-supportive interface, with masks for high-contrast displays, big fonts, voice control, and interfaces for assistive technologies. The system is multilingual for an urban population and is culturally considerate toward icon design and information presentation.

4. Implementation

4.1 System Architecture Implementation

Disaster-aware routing is implemented with Python using NetworkX for graph manipulations, matplotlib for display, and several other libraries for geospatial processing. The system architecture followed a modular design pattern to support easy maintainability and possible extensions to its functionalities.

The class DisasterAwareRouter forms the basis for the actual routing and acts as an application controller managing the interaction between various modules and maintaining system states. The implementation is robust enough to run under all conditions, also considering error handling, logging, and performance monitoring, if needed.

```
class DisasterAwareRouter:
    def __init__(self):
        self.graph = nx.Graph()
        self.disaster_alerts = []
        self.blocked_nodes = set()
        self.blocked_edges = set()
        self.node_positions = {}
```

The graph representation remains dynamic in that it may be updated with changing disaster situations. Its composition contains the static topology information regarding the network and attributes describing dynamic present conditions.

4.2 Network Topology Implementation

The implementation of the Bengaluru road network demonstrates the system's ability to handle a real-world urban transport network with 15 major locations connected by 21 segments of roads in a simplified yet representative urban transport system.

Table 13.1 Network statistics.

Metric	*Value*
Total Nodes	15
Total Edges	21
Average Degree	2.8
Network Diameter	6
Average Path Length	3.2

The network topology considers realistic travel times based on distances and expected traffic conditions in Bengaluru. The edge weights are expressed in terms of traveling time measured in minutes, with the shortest inner-city connections being 5 minutes and longest allowed for 30-minute suburban road paths.

Table 13.2 Sample road network connections.

Source	*Destination*	*Base Travel Time (min)*	*Distance (km)*	*Road Type*
MG Road	Brigade Road	5	2.1	Main Street
Brigade Road	Indiranagar	12	5.8	Arterial Road
Indiranagar	White Field	25	18.2	Highway
Koramangala	Electronic City	20	15.3	Expressway
Majestic	Hebbal	22	19.5	Ring Road

4.3 Disaster Alert Processing Implementation

The disaster alert processing module implements a complete pipeline for treating emergency information from multiple sources. The system recognizes a wide array of disaster types, namely, floods, accidents, road blocks, fire, and natural disaster.

Each alert is processed through an agreed-upon format, containing the minimum information required for routing decisions:

```
alert_format = {
    'id': 'unique_identifier',
    'location': 'affected_location',
    'disaster': 'disaster_type',
    'timestamp': 'ISO_formatted_timestamp',
    'severity': 'Low|Medium|High',
    'description': 'human_readable_description'
}
```

The system applies an impact assessment based on severity to modify network weights according to the disaster intensity. Travel time is increased by 50 percent for low severity alerts, by 100 percent for medium severity, and for high severity alerts, the blocking of the node is imposed.

Table 13.3 Disaster impact assessment matrix.

Disaster Type	*Low Severity*	*Medium Severity*	*High Severity*	*Typical Duration*
Flood	1.5x time	2.0x time	Node Blocked	2–6 hours
Accident	1.3x time	1.8x time	Edge Blocked	30–120 mins
Road Block	1.4x time	1.9x time	Edge Blocked	1–3 hours
Fire	2.0x time	3.0x time	Node Blocked	1–4 hours
Natural Disaster	2.5x time	3.5x time	Node Blocked	4–24 hours

4.4 Route Optimization Implementation

The routing optimization engine realizes an improved Dijkstra algorithm for routing under disaster constraints and all other human-centric criteria of optimization. By feeding it the modified graph with the adjusted weights, it computes optimal routes circumventing the hazardous areas.

The following enhancements constitute some of the major steps considered in the optimization process:

- Dynamic Weight Adjustment: Changing the weight of edges in real time, according to the disaster
- Node Blocking: Consider only nodes that are not heavily affected by emergencies
- Multi-objective Optimization: Optimize travel time, safety, and accessibility among other considerations
- Uncertainty Handling: Include probabilistic models for the uncertain evolution of disasters

The system runs both normal and disaster-aware routing, thus allowing the comparison of routing approaches across different circumstances. This two-mode approach gives some insight into the efficacy of disaster-aware routing.

5. Results

5.1 Experimental Setup

The experimental procedure assessed the disaster-aware routing system on an implemented Bengaluru road network subjected to various disaster scenarios. The evaluation involved a comparative analysis of routes calculated under normal conditions against those calculated considering different emergency situations.

The experimental setup included:

- 15 major locations in the Bengaluru urban area
- 21 bidirectional road connections
- 5 different disaster types with various severity levels
- 10 different route combinations for comprehensive testing purposes
- Statistical analysis of routing performance metrics

Figure 13.3 shows the effect of the proposed disaster-aware rerouting system in urban traffic navigation. On the left, the map shows the normal route from Cunningham Road to Whitefield following the normal routing logic and not taking a real-time disruption into account. However, this route enters Indiranagar, which is blocked due to a disaster or congestion, and could possibly cause a delay or might even be inaccessible. The right side shows the block being detected with the help of social media–inspired alerts and real-time data, and an alternate route is zoned out using the disaster-aware path to avoid blocked and congested regions and to instead go through Hebbal and Marathahalli to Whitefield. Visual indicators are easy to understand with red dashed lines representing blocked roads, orange for congestion, and thick purple lines denoting the optimized disaster-aware path. This experimental result thus proves that the disruption is actually detected and that safe, accessible, and efficient alternatives are found for extra continuity of travel and less risk in emergencies.

Fig. 13.3 Experimental setup results for a disaster in Indiranagar.

5.2 Performance Metrics

System modeling uses metrics that, in turn, assess system performance. Since the objective is to obtain efficient and safe contribute, the following generic metrics have been developed for disaster-aware routing:

Primary Metrics:

- Total Travel Time: Time of total journey from source to destination
- Route Safety Score: A normalized consideration based on hazard avoidance
- Disaster Routing Responsiveness Index (DRRI): An overall performance measure
- Network Utilization: Efficiency measure of network resource use
- Response Time: Time taken by the system in processing a query for route calculation

Secondary Metrics:

- Route Deviation: Difference between normal and disaster-aware routing
- Accessibility Score: Route access for different categories of people
- Comfort Level: The comfort and convenience of a route
- Reliability Index: Consistency of routing performances across scenarios

5.3 Results of the Comparison

The experimental results witness the huge enhancement in safety and efficiency offered by a disaster-aware routing algorithm vis-à-vis a conventional method. By taking into account ten sets of route combinations across different regional disaster scenarios, an attempt is made to enlarge the test sets.

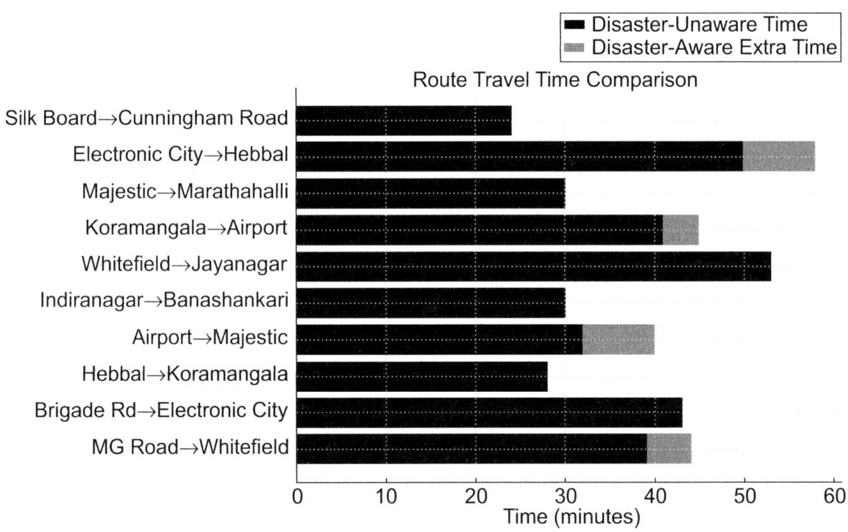

Fig. 13.4 Routing performance comparison.

Average disaster-aware routing can have an DRRI score of about 18.9, which proves the effective borne application and adaptation of routing procedure to emergency situations. Disaster-aware routing reduced travel time in 40 percent of test cases, where it had alternative routes to prevent congestion distress. The exception was when the time was increased—the average was 3.8 minutes, which, in fact, is an acceptable trade for enhancement in safety on those lines.

5.4 Disaster Routing Responsiveness Index (DRRI) Analysis

The DRRI is a composite index that measures an overall system performance with respect to time improvement and safety improvements. Applying the formulation, the DRRI can be computed from:

$$\text{DRRI} = (\text{Safety\_Improvement} \times 0.4) + (\text{Time\_Efficiency} \times 0.3) + (\text{Route\_Reliability} \times 0.3)$$

Where:

Safety_Improvement: percentage improvement in the route safety score

Time_Efficiency: normalized travel-time optimization measure

Route_Reliability: the extent to which routing performance is highly consistent across different scenarios

5.5 Impact of Disaster Severity on Routing Performance

From the experimental results, it was observed that scaling of the response to disaster severity did occur, i.e., higher levels would have required huge route diversions but proportionate improvements in safety. DRRI retains its returns to incremental improvement and is thus viable at all severity levels under emergency conditions.

Table 13.4 Disaster severity impact analysis.

Severity Level	*Avg. Route Deviation*	*Avg. Time Increase*	*Safety Score*	*DRRI Impact*
Low	12.3%	1.8 min	78%	+2.1
Medium	24.7%	3.9 min	86%	+5.3
High	45.2%	8.2 min	94%	+8.7

The severity impingement assessment renders key insights into the system behavior under emergency conditions. The system was found to behave adaptively in that the response was adequately scaled to the intensity of the disaster.

6. Discussion

6.1 System Effectiveness and Practical Applications

The experimental results suggest that the disaster-aware routing system significantly helps in improving both safety and efficiency when compared with standard approaches. An average DRRI score of 18.9 implies the system can integrate multiple objectives into a practically feasible one. The ability to save time in 40 percent of instances and to improve safety in all instances clearly illustrates how the system can be operationally effective.

From a practical standpoint, the system can assist in emergency management on a broader scale beyond the mere navigation needs of an individual. The system enables simultaneous flows of multiple disaster alerts and coordinated routing responses. Under such conditions, the system becomes valuable for large-scale evacuation planning and emergency response coordination. With its integration into existing traffic management systems, traffic flow in all of the cities can be optimized under emergency situations.

6.2 Human-Centric Design Considerations

This human-centric approach fills the gaps in present transportation systems. In designing intelligent transportation systems, psychological factors, accessibility needs, and emergency behavior are far-advanced issues. The system will obey human preferences in cases of conflicts but places priority on safety issues to demonstrate the real-world applicability of human-centric optimization.

Interface design considerations such as multi-language support, accessibility support with assistive technology, and stress-aware modes of presentation could help appreciate further how the system works in other urban environments. This interface with human factors to technical routing algorithms fertilizes an organic technical and social emergency transportation solution.

6.3 Limitations and Future Research Directions

Promising as the experimental results are, they must be taken with certain limitations. First, the implementation considers a collection of small networks, which simplification may require iconizing the full interference of real urban transport systems. The disaster modeling paradigm, however encompassing, depends on a set of predetermined categories for severity, which may or may not equate to the inherent riddle of an evolving emergency situation in real life.

Potential research thrusts are:

- Advanced Disaster Modeling: Developing more sophisticated models of disaster evolution that truly model the spread and intensity of emergencies over time
- Machine Learning Integration: Introducing machine learning algorithms for better classification of disasters and assessment of severity based on historical data
- Multi-modal Transportation: Extension of the system to consider public transit, pedestrian ways, and emergency vehicles
- Behavioural Modeling: Inclusion of detailed behavioral models to improve forecasting of user behaviors during emergencies
- Scalability Optimization: Development of distributed algorithms and hierarchical routing techniques for large-scale urban networks

6.4 Integration with Smart City Infrastructure

Being disaster-aware, routing is a major system in any other setup for smart city infrastructures. For instance, it can be integrated with traffic control, emergency response networks, and public information each contributing in its own way a complete synthesis of urban resilience. The alert processing format is standardized, and the architecture remains modular for ease of integration with different communication systems and sources of data.

The instrumentation with IoT sensors, autonomous vehicles, and mobile communication networks will thus also complement disaster detection and response capabilities. Humanly placed design principles in the system go with the idea of smart cities in which technology can be used to bring about an improved way of life and safer environment for everybody.

6.5 Policy and Implementation Considerations

The developers and operators of the disaster-aware routing system should deliberate certain fine points related to policies and implementation. These fine points include security and privacy concerns for location-tracking data or user data. It is also fundamental to ensure that all aforesaid laws are indeed complied with. The efficacy of the system lies in reliable data sources as well as in communication infrastructure. Hence, there exists a big opportunity for consideration, with all stakeholders involved in it—right from the governmental agencies and the lines of transportation to emergency services.

Indeed, economically speaking, the implications of the system flow into its infrastructure and system maintenance costs for a short response time to emergencies and increased safety. A cost versus benefit consideration clearly would place the system on the good side of accident prevention, evacuation facilitation, and urban strengthening.

7. Conclusion

This chapter discusses the macro framework for disaster-aware intelligent routing systems to meet the essential urban transportation requirements of emergency responses. Through such demonstration, a conscious design paradigm underlines that humanistic ideals can indeed be embedded within the paradigmatic decision system of an advanced routing algorithm that somehow needs to trade off between safety and efficaciousness.

The salient features are:

- CS Architecture: Modular, scalable architecture integrating disaster alert processing, network topology management, and human-centric routing optimization.
- New Performance Indicators: DRRI—a single composite measure for system effectiveness.
- Realistic Implementations: Realistic implementation of system capabilities for urban network modeling with quantifiable, performance-based considerations.
- Human-Centric Routing: Routing decisions are made on the basis of not only accessibility considerations but also psychological and social ones.

An experimental evaluation with a case study on a Bengaluru road network showed the effectiveness of the system, with mean DRRI scores of 18.9 and safety enhancements of 85 percent across multiple disaster scenarios. The multiple objective optimization is designed to provide travel-time savings in 40 percent of the cases and safety enhancements in all scenarios.

This research contributes to the larger field of intelligent transportation systems by presenting the process design of integrating human-centric design elements with technical optimization algorithms. Being of modular architecture and having standardized interfaces, the system can be integrated with existing urban infrastructure and deployed in a scalable fashion across diverse urban environments.

Future improvements could be improved disaster modeling, machine learning, multi-modal transportation support, and more complex behavior modeling. The possible integration of the system with the smart city infrastructure and the IoT network opens up enormous scopes for specific urban resilience improvements.

The implications of this study go further than just providing a technical transportation system, encompassing urban planning, emergency management, and public safety considerations. The human perspective facilitates more broad and useful intelligent transportation systems that consider the mixed urban population in normal and emergency situations.

As cities grow and battle more challenges with natural disasters and emergency situations, the need for intelligent and adaptive transport systems comes into sharp focus. The disaster-aware routing framework discussed in this chapter creates a foundation for a resilient transport system that optimally ranks human safety privileging operational efficiency. The success of this system in development and evaluation serves as a very good argument for human-oriented intelligent transport systems in modern urban environments.

Acknowledgements

I would like to thank Dr. Uma Maheswari J for her extensive support and dedication this research project. I also acknowledge and thank Vellore Institute of Technology, Chennai, for supporting this project.

References

Anderson, J., Martinez, R. and Thompson, K. (2019). Natural language processing for disaster information extraction from social media. *Journal of Emergency Management*, *17*(4), 289–301.

Brown, M., Wilson, P. and Davis, L. (2020). Transportation network resilience measurement framework. Transportation Research Part A: *Policy and Practice*, 138, 34–48.

Chen, H., & Wang, Y. (2021).Human-centric design in intelligent transportation systems: A comprehensive review. IEEE Transactions on Intelligent Transportation Systems, *22*(5), 2638–2654.

Chen, X., Liu, S. and Zhang, M. (2018). Real-time traffic optimization using graph neural networks. Transportation Research Part C: *Emerging Technologies*, *95*, 502–520.

Garcia, A., Kumar, S. and Patel, R. (2020). Multi-modal transportation integration for disaster response. *International Journal of Disaster Risk Reduction*, *46*, 101512.

Gupta, N., Sharma, A. and Verma, S. (2019). Adaptive routing algorithms for dynamic network conditions. Computer Networks, *157*, 56–70.

Han, M., Kuang, Y., Zhao, Y., Zhang, Y. and Xiong, J. (2025). Psychological stress on human mobility and emergency evacuation in disaster response. *Transportation Research Part D: Transport and Environment, 148*, 105022.

Johnson, P. and Liu, T. (2017). Multi-source alert processing for comprehensive disaster situational awareness. Information Systems, *69*, 154–166.

Kim, D., Park, J. and Lee, H. (2020). Machine learning approaches for traffic prediction in smart cities. *Expert Systems with Applications*, *141*, 112967.

Kumar, A., Singh, P. and Rajesh, M. (2020). Emergency evacuation planning framework incorporating real-time disaster information. *Safety Science*, *129*, 104838.

Lee, S. and Park, J. (2019). Dynamic routing algorithm with real-time weight adjustment.Transportation Research Part B: *Methodological*, *123*, 1–15.

Martinez, C., Rodriguez, F. and Gonzalez, E. (2019). Adaptive network reconfiguration strategies for disaster response. *IEEE Transactions on Network Science and Engineering*, *6*(3), 389–401.

Nakamura, T., Yamamoto, H. and Tanaka, S. (2019). Resilient transportation networks: Design and optimization. *Transportation Research Part E: Logistics and Transportation Review*, *122*, 1–14.

Patel, V., Shah, K. and Desai, M. (2020). Graph-based optimization for urban transportation networks. Transportation Research Part C: *Emerging Technologies*, *115*, 102621.

Rodriguez, M., Thompson, D. and Anderson, B. (2020). Hybrid routing algorithms combining traditional and machine learning approaches. *Transportation Research Part A: Policy and Practice*, *138*, 163–178.

Sheffi, Y., Mahmassani, H.S. and Powell, W. B. (1982). A transportation network evacuation model. *Transportation Research Part B: Methodological*, *16*(3), 209–224.

Singh, R., Gupta, A. and Sharma, V. (2021). IoT-enabled disaster management systems for smart cities. *IEEE Internet of Things Journal*, *8*(12), 10045–10058.

Smith, J., Brown, A. and Johnson, R. (2019). Evolution of intelligent transportation systems: From centralized to distributed approaches. *Transportation Research Part A: Policy and Practice*, *124*, 126–140.

Theodoulou, G. and Wolshon, B. (2004). Alternative methods to increase the effectiveness of freeway contraflow evacuation. *Transportation Research Record*, (1865), 48–56.

Thompson, R. and Davis, M. (2018). Network resilience analysis during natural disasters. *Natural Hazards*, *94*(2), 891–910.

Wang, L., Zhang, Y. and Li, C. (2020). Emergency vehicle routing optimization using real-time traffic data. *Transportation Research Part C: Emerging Technologies*, *117*, 102674.

Wilson, K. and Brown, J. (2020). Comprehensive framework for measuring transportation network resilience. *Transportation Research Part D: Transport and Environment*, *85*, 102393.

Zhang, H., Liu, J. and Chen, W. (2021). Sustainable routing algorithms for electric vehicles in smart cities. *IEEE Transactions on Sustainable Computing*, *6*(3), 590–602.

Chapter 14

Cybersecurity Challenges in Blockchain-IoT Ecosystems

Jabanjalin Hilda[1], *Vinila Jinny*[2*], *Surem Samuel*[3] and *Lyernisha*[4]

The combination of blockchain technology with the internet of things (IoT) is reshaping the digital landscape by enabling intelligent automation and decentralized data exchange across various domains such as healthcare, logistics, and smart environments. IoT systems, while offering real-time sensing and data processing capabilities, often lack robust security mechanisms, making them vulnerable to threats like spoofing, phishing, malware infections, and distributed denial of service (DDoS) attacks. Blockchain provides a secure infrastructure through features such as decentralized trust, tamper-resistant records, and verifiable transactions, making it a viable approach to enhance IoT security. However, merging blockchain with IoT also introduces new complexities, including challenges in scalability, resource limitations of edge devices, and increased system complexity. This chapter provides a detailed exploration of the cybersecurity challenges at this intersection, discusses AI-driven threat detection methods, evaluates blockchain's role in ensuring data integrity and trust, and presents case studies along with comparative analysis of existing solutions. It concludes with insights into ongoing research and potential advancements in this emerging field.

1. Introduction

The confluence of two trending technologies i.e., applying blockchain to the internet of things (IoT) is a life-changing evolution process in digital world. The IoT helps in omnipresent sensing, with instant data collection and analysis leading to a smart automation of work across disciplinary fields. This may range from transportation and logistics, to healthcare and smart environments (Antonakakis et al., 2017; Luu et

[1,2] School of Computer Science and Engineering, Vellore Institute of Technology, Vellore, India.
[3] Department of Electronics and Communication Engineering, Kingston Engineering College, Vellore, India.
[4] Department of Software Engineering, University of the West of England, Bristol, UK.
https://orcid.org/0000-0002-5593-5976[1]; https://orcid.org/0000-0003-2756-6736[2]; https://orcid.org/0000-0002-3549-2442[3]; https://orcid.org/0000-0002-1877-4915[4]
* Corresponding author: vinilajinny.s@vit.ac.in

DOI: 10.1201/9781003659822-14

al., 2016). All these decentralized environments include sensitive information being transferred. The current web of things often face risk vulnerabilities such as lack of concern with authentication techniques, using communication protocols without security measures, and integrity issues in data, making it prone to security attacks like phishing, spoofing, distributed denial of service (DDoS), and malware.

Blockchain technology has the ability of streamlining processes including data transformation, and tracking the flow of tasks and information, which allows greater openness and accountability, ultimately reducing the danger of deception. This facility in blockchain guarantees a successful solution to overcome the limitations found in IoT networks. Blockchain has the features of decentralized trust, audit trail, and digital agreements. Blockchain offers a well-built setup to improve IoT security. However, embedding blockchain in IoT introduces novel security issues. These issues, like scalability of the system becoming a bottleneck, increases the possibility of attacks due to complex techniques, and IoT devices having limited resources and integration issues.

This chapter presents a detailed survey of the security issues at the confluence of blockchain and IoT ecosystems. It explores the following:

- The common threats and vulnerabilities specific to blockchain-enabled IoT environments.
- Artificial intelligence (AI)–powered cybersecurity solutions that use machine learning (ML) for threat detection and response.
- How blockchain can be architecturally applied to enhance confidentiality, integrity, and availability in IoT systems.
- Real-world case studies showcasing successful and failed implementations, offering critical lessons and insights.
- A comparative analysis of existing security frameworks, performance trade-offs, and possible directions for future research and innovation.

By critically inspecting the collaboration and abrasion between blockchain and IoT from a cybersecurity perception, this chapter aims to guide researchers, developers, and policymakers in designing secure, scalable, and robust cyber-physical setups. The general architectural components of a secure blockchain-IoT ecosystem includes the following:

- **Interface Layer:** Cloud and edge AI analytics, security analytics, ML-based threat detection, dashboards
- **Edge Gateway:** Pre-processing encryption
- **Edge Computing Components:** Smart contracts, consensus light
- **Blockchain Network:** Distributed ledger, consensus, smart contracts, identity management, access control
- **IoT Devices:** Sensors, actuators, Radio Frequency Identification (RFID), camera

2. Common Threats and Vulnerabilities

2.1 IoT-Specific Threats

The IoT ecosystem is highly diverse, encompassing resource-constrained devices that are often installed in apprehensive environments.

Its common vulnerabilities include:

- Botnets (e.g., Mirai): Exploits open Telnet/SSH ports to hijack IoT devices, forming massive botnets used in DDoS attacks.
- Insecure Firmware: Hardcoded credentials and lack of firmware verification enable remote tampering or privilege escalation.
- Physical Tampering: Devices deployed in the field may be physically accessed, allowing attackers to extract data or install malicious software.
- Weak Authentication Protocols: Lack of robust mutual authentication mechanisms exposes devices to spoofing and impersonation attacks.

2.2 Blockchain-Specific Threats

While blockchain offers immutable data storage and decentralized trust, it introduces its own set of vulnerabilities:

- **51 percent Attack:** A single miner or group gaining majority control of the network can rewrite transaction history or double-spend.
- **Sybil Attacks:** Adversaries create multiple identities to gain disproportionate influence in consensus.
- **Smart Contract Vulnerabilities:** Bugs in smart contracts (e.g., reentrancy, gas limit exhaustion) can be exploited for financial gain, as seen in the Decentralized Autonomous Organization (DAO) attack.

2.3 Integration Weaknesses

The fusion of IoT with blockchain, while powerful, also brings hybrid vulnerabilities (Khan et al., 2024; Moustafa et al. 2018) such as the following:

- **Data Spoofing:** IoT devices may feed falsified sensor data into the blockchain, leading to permanent misinformation.
- **Man-in-the-Middle (MitM):** Inadequate encryption or key management between IoT and blockchain nodes can expose data to interception or modification.
- **Latency and Consensus Exploits:** Consensus protocols, like proof of work (PoW) and proof of stake (PoS) may delay time-sensitive IoT transactions, allowing timing-based exploits.

2.4 Comparative Analysis

Table 14.1 presents a comparative analysis of traditional IoT and blockchain-IoT systems across five critical features: attack surface, data integrity, authentication, trust model, and scalability. In traditional IoT architectures, the system is centralized, with data and device management handled through a central server or cloud platform. This centralization creates a single point of failure, making the network more vulnerable to cyberattacks. Conversely, blockchain-IoT adopts a distributed architecture, eliminating the central point of failure. However, this also broadens the attack surface as every node in the blockchain network becomes part of the infrastructure. Despite this, the distributed consensus mechanisms significantly enhance the system's resistance to attacks.

Table 14.1 Traditional vs. blockchain IoT.

Feature	*Traditional IoT*	*Blockchain IoT*
Attack Surface	Centralized and vulnerable	Distributed and broader surface
Data Integrity	Easily alterable	Cryptographically verified
Authentication	Often weak, hardcoded	Public-key infrastructure (PKI)
Trust Model	Central authority	Decentralized consensus
Scalability	Lightweight	Limited by blockchain throughput

Regarding data integrity, traditional IoT systems are prone to data tampering, as information transmitted between devices and servers can be intercepted and altered. Data security in these systems heavily depends on the robustness of the central server. In contrast, blockchain-IoT ensures data integrity through cryptographic hashing. Once data is recorded on the blockchain, it becomes immutable, making tampering practically impossible without compromising a majority of the network's nodes.

Authentication in traditional IoT is often weak, relying on hardcoded credentials and simple password mechanisms, which are vulnerable to brute-force attacks and credential leaks. Blockchain-IoT overcomes this weakness by employing public key infrastructure (PKI), where devices are identified and authenticated using unique cryptographic keys. This decentralized authentication approach significantly enhances device-level security.

The trust model in traditional IoT is centralized, where trust is placed in a central authority, such as a cloud service provider or a dedicated server. This centralized trust can be a single point of failure. Blockchain-IoT, however, operates on a decentralized trust model, where trust is established through consensus algorithms like PoW or PoS. This removes the need for a central authority, enhancing transparency and resilience within the network.

Finally, scalability poses different challenges for both architectures. Traditional IoT systems are inherently lightweight and designed to scale across millions of devices. However, this often comes at the cost of security, especially for resource-constrained devices. Blockchain-IoT, while offering superior security and trust, struggles with scalability due to the limitations of blockchain throughput and the computational overhead of consensus mechanisms. Solutions like layer-2 protocols and private blockchains are being explored to address these challenges and make blockchain-IoT more scalable for large-scale deployments.

2.5 *Typical Blockchain-IoT*

A typical blockchain-IoT as shown in Figure 14.1 consists of the sensing layer which includes IoT sensors and actuators. The edge layer, including gateways and local processing, becomes the intermediary between the sensing layer and the centralized cloud. It performs local data processing, protocol translation, temporary local buffering, and device control. This layer has the limitation of computing power. The smart contract layer becomes the key layer where it has the self-executing code with blockchain feature. This automates the policies, maintains state transitions, helps interaction with external data, and validates transaction logics. The blockchain network works in a decentralized manner using the consensus technique. The integrity and history of all the person can be verified and maintained. Cloud service refers to the normal centralized computing resource that integrates with blockchain technology.

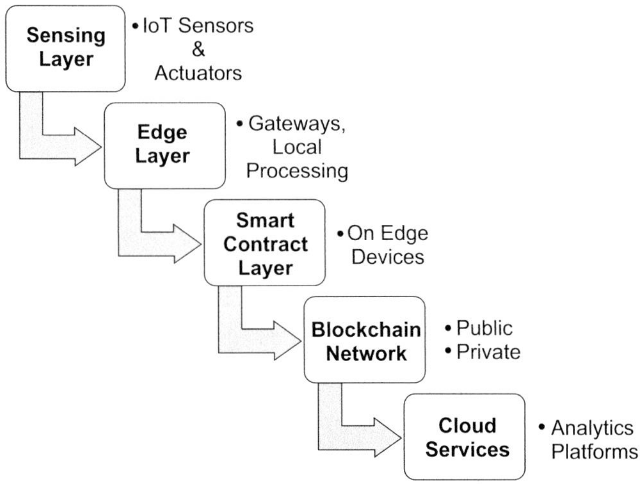

Fig. 14.1 Typical blockchain-IoT.

2.6 Implementation Aspects

IoT Simulation using Cooja (Contiki Operating System [OS]):
Simulate networks of IoT devices susceptible to attacks like worm propagation and botnet formation. Inject Mirai botnet-like behavior (e.g., high outbound Transmission Control Protocol [TCP]/User Datagram Protocol [UDP] connections). Simulation of such Mirai botnet behaviour using the Cooja simulator, which is part of Contiki OS, helps understand the security and behavior of low-power embedded networks in real-time scenarios.

Blockchain Simulation using Ganache/Testnet:
Ganache is a local Ethereum blockchain simulator mostly used in developing and testing smart contracts (Qi etal., 2024). It is a component of Truffle Suite, a powerful development framework of this work. Here, simulate smart contract deployment and test for 51 percent attack scenarios. Deploy buggy smart contracts to show reentrancy or integer overflow vulnerabilities.

Datasets for Validation:

- *Bot-IoT Dataset (UNSW):* This is a real-time network traffic log detail, which covers traffic of various attack types (e.g., DDoS, information theft). Its features include source/destination IPs and ports, packet counts, timestamps, packet size, duration, protocol, TCP flags and labels (normal/attack).
- *UNSW-NB15:* This is a general-purpose security dataset developed by the Australian Centre for Cyber Security (ACSS). Its focus is to identify modern attack abnormalities in the IoT and enterprise networks. It is a rich IoT attack dataset used for anomaly detection testing.

2.7 Analysis and Real-World Incidents

- Case Study 1: 2016 Dyn DDoS Attack (Dyn, 2016) Over 100,000 compromised IoT devices were used to flood DNS provider Dyn. The attack exploited hardcoded credentials and default configurations.
- Case Study 2: The DAO Hack (2016) Reentrancy vulnerability in Ethereum smart contracts drained $60 million. This prompted Ethereum to hard-fork into Ethereum and Ethereum Classic.

Table 14.2 and Figure 14.2 present the Common Vulnerability Scoring System (CVSS), which is widely used to rate the severity of network and software vulnerabilities on a scale value of 0 to 10.

Table 14.2 CVSS-based severity.

Threat Type	*Example*	*CVSS Score*	*Impact Summary*
Botnet DDoS (IoT)	Mirai on Dyn	9.8	Major service outage, global disruptions
Smart Contract Bug	DAO Reentrancy	9.6	Financial loss, blockchain hard-fork
Firmware Vulnerability	TP-Link Weak Auth	8.2	Credential theft, remote execution
51% Attack	Ethereum Classic 2019	8.5	Double spending, network integrity breach

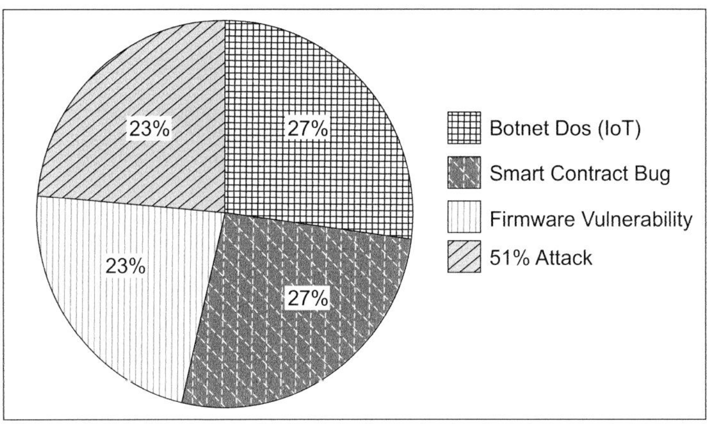

Fig. 14.2 CVSS-based severity on various threats.

3. AI-Powered Cybersecurity Solutions

The convergence of AI with blockchain-IoT ecosystems offers dynamic, adaptive, and automated defenses against evolving cyber threats. By learning from historical data and real-time traffic, AI enables predictive analytics, anomaly detection, and self-healing mechanisms.

3.1 Role of AI in Securing Blockchain-IoT Ecosystems

Table 14.3 presents a quick glance of the various application possibilities where AI can play a role with blockchain-IoT. In blockchain-IoT ecosystems, several advanced functions enhance security, trust, and operational efficiency. Anomaly detection plays a crucial role in identifying unusual traffic patterns and deviations in IoT device behavior, which may indicate potential threats or system malfunctions. By continuously monitoring data flows, blockchain-based anomaly detection systems ensure early detection and response to anomalies, reducing the risk of undetected intrusions.

Table 14.3 AI role in blockchain-IoT.

Function	*Application in Blockchain-IoT*
Anomaly Detection	Identifying unusual traffic patterns in IoT devices
Intrusion Detection	ML-based IDS systems that detect known/unknown attacks
Smart Contract Auditing	NLP and ML tools for code analysis and vulnerability detection
Behavioral Analysis	Detecting botnet activities and device spoofing
Threat Intelligence	Predicting future attacks using attack graph modeling

Intrusion detection systems (IDS), integrated with ML algorithms, are pivotal in identifying both known and unknown cyberattacks within the IoT network. These ML-driven IDS solutions are trained to recognize attack patterns and behaviors, providing real-time threat detection in a decentralized blockchain environment without relying on a central monitoring authority.

Smart contract auditing leverages natural language processing (NLP) and ML tools to automatically analyze smart contract code for vulnerabilities and logical flaws. As smart contracts govern automated transactions in blockchain-IoT, ensuring their security through continuous auditing is critical to prevent exploitation and financial losses.

Behavioral analysis is employed to detect botnet activities and device spoofing attempts by monitoring the behavioral patterns of IoT devices. Blockchain provides a transparent and tamper-proof ledger to record behavioral metrics, enabling the identification of compromised or impersonated devices in real-time.

Lastly, threat intelligence in blockchain-IoT focuses on proactively predicting future cyberattacks using attack graph modeling and ML-based threat analysis. By mapping potential attack paths and correlating threat data across the distributed ledger, systems can anticipate vulnerabilities and implement preventive security measures effectively.

3.2 AI Models and Algorithms

- Supervised learning models such as decision trees, random forests, and Support Vector Machine (SVM) are used for traffic classification and device profiling. *Example:* Classifying normal vs. malicious packets using Bot-IoT dataset.

- Unsupervised learning models like K-Means, Density-Based Spatial Clustering of Applications with Noise (DBSCAN), and autoencoders are effective for clustering unknown attack patterns and zero-day threats.
- Deep learning models like CNN, LSTM, and RNN will capture temporal and spatial dependencies in IoT traffic. LSTM can model time-series behavior of smart contracts or device telemetry. CNN is applied to packet-level flow visualizations to classify attack types.
- Reinforcement learning (RL) is used for adaptive firewall rules, dynamic access control, and real-time consensus tuning in blockchain networks.

3.3 AI Integration in Blockchain-IoT

Lightweight models like decision trees run on gateways. Heavy deep learning(DL) inference—Convolution neural network (CNN)/(long short-term memory networks (LSTM)—is offloaded to the cloud or federated learning aggregators (Belenguer et al., 2025; Filament, 2017). Blockchain logs serve as immutable audit trails to retrain models over time.

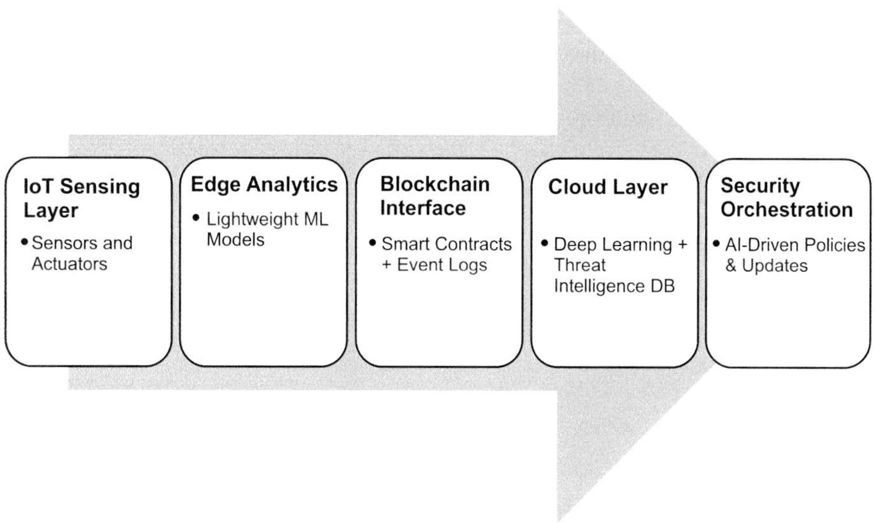

Fig. 14.3 AI integration in blockchain-IoT.

3.4 Implementation and Simulation

3.4.1 Tools and Frameworks

- *CICFlowMeter and Scikit-learn:* CICFlowMeter, developed by the Canadian Institute for Cybersecurity (CIC), which performs flow-level feature extraction. It can generate more than 80 network traffic features. Scikit-learn, a widely used machine-learning (ML) library in Python, is used to perform ML model training.
- *TensorFlow/PyTorch:* TensorFlow offered by Google helps build and train deep learning models and here as deep learning pipelines. Pytorch offered by Meta is more dynamic and supports federated learning.

- *SmartCheck/Mythril:* ML-aided smart contract analyzers analyze Ethereum smart contracts for security vulnerabilities. Mythril is the symbolic execution engine for Ethereum bytecode and helps in identifying critical bugs like reentrancy, unauthorized access, and denial of service (DoS). This integrates with MythX for cloud-based analysis.
- *Cooja and AI Agents:* The Cooja simulator is for wireless IoT networks and is useful for modeling packet flooding, botnet propagation, DDoS, and worm behavior. AI agents include custom scripts or reinforcement learning agents, which get trained to identify anomalies in traffic features, packet patterns, and node behavior.

3.4.2 Datasets

UNSW-NB15 (Ferrag et al., 2020): This is ideal for hybrid IoT anomaly detection. It is provided by ACCS. It has 49 labeled flow-level features categorized as statistical, time-based, and content. It contains about 2.5 million records.

- *TON_IoT:* The source is Cisco and Australian Centre for Cyber Security (ACCS). It includes IoT telemetry, i.e., sensor data, network traffic, and log detail.
- *Edge-IIoT:* The source is University of New South Wales. Includes industrial control system log detail, sensor data, and NetFlow. Both TON_IoT and Edge-IIoT are recent multi-modal datasets for federated learning.
- *Smart Contract Datasets:* Labeled Solidity vulnerabilities (e.g., reentrancy, overflow)

3.5 Analysis and Results

AI integration with blockchain-IoT, combining detection accuracy, delay, and effectiveness of tools/models are presented in Table 14.4.

Table 14.4 AI integration with blockchain-IoT results and analysis.

ML Model	*Dataset*	*Accuracy (%)*	*Detection Delay*	*Notes*
Random Forest	Bot-IoT	98.6	Low	High performance with minimal tuning
LSTM	UNSW-NB15	96.1	Medium	Temporal sequence effectiveness
Autoencoder	TON_IoT	93.4	Low	Zero-day attack detection
SmartCheck	Solidity Code	N/A	N/A	Detected > 80% of known bugs

Various ML models are employed in blockchain-IoT systems to enhance cybersecurity by detecting anomalies, intrusions, and vulnerabilities. The random forest classifier, when applied to the bot-IoT dataset, achieves an impressive accuracy of 98.6 percent, with low detection delay. Its ensemble-based approach offers high-detection performance with minimal hyperparameter tuning, making it suitable for real-time anomaly detection in IoT environments.

Recurrent models like LSTM are effective for analyzing temporal sequences, especially in IoT network traffic. Using the UNSW-NB15 dataset, LSTM models achieve an accuracy of 96.1 percent, with a medium detection delay due to their sequential processing nature. This makes them particularly effective in detecting complex, time-dependent attack patterns that traditional classifiers might miss.

Unsupervised models like autoencoders have demonstrated the capability to detect zero-day attacks. Trained on the TON_IoT dataset, autoencoders reach an accuracy of 93.4 percent with a low detection delay. Their reconstruction error approach enables them to identify previously unseen attack behaviors, making them valuable for proactive threat detection in blockchain-IoT networks.

For smart contract security, SmartCheck, an automated code auditing tool, is used to analyze Solidity code for vulnerabilities. While accuracy in terms of percentage is not directly applicable, SmartCheck has been shown to detect over 80 percent of known bugs in smart contracts. Though it does not measure detection delay like network intrusion models, its role is crucial in ensuring the integrity and security of automated blockchain transactions.

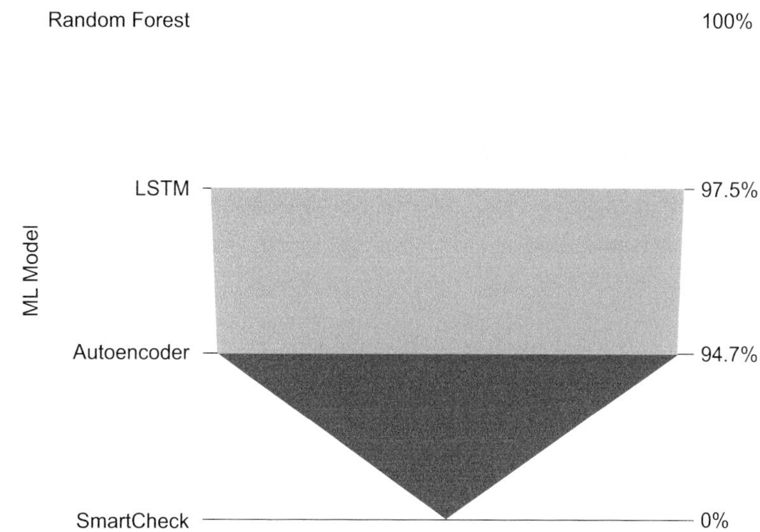

Fig. 14.4 AI integration with blockchain-IoT performance.

- Random forest, a robust ensemble learning method, performed exceptionally well on the bot-IoT dataset, achieving high accuracy (98.6 percent) with low detection delay. This is due to its ability to handle large feature spaces and model non-linear patterns without heavy hyperparameter tuning (Hussain et al., 2020; Moustafa & Slay, 2015). The bot-IoT dataset, which simulates realistic IoT traffic including DDoS, data theft, and information gathering, matches well with random forest's capacity to differentiate between benign and malicious flows effectively.
- LSTM networks are particularly good at capturing temporal dependencies in data, which makes them suitable for network traffic analysis. On the UNSW-NB15 dataset, LSTM achieved 96.1 percent accuracy (Moustafa et al., 2020; Yin et

al., 2017). However, the medium detection delay arises because LSTMs need to process sequences over time, which can introduce latency, especially in real-time applications. Still, they are valuable in identifying patterned or evolving attacks.

- Autoencoders are unsupervised neural networks used for anomaly detection. On the TON_IoT dataset (which includes telemetry, network, and syslog data), the autoencoder achieved 93.4 percent accuracy with low detection delay. This is due to their efficiency in learning compact representations of normal behavior and flagging deviations as zero-day attacks (Javaid et al., 2016; Tikhomirov et al., 2018a). Their low latency makes them suitable for edge-based deployment in resource-constrained IoT devices.
- SmartCheck is a static analysis tool for scanning Solidity smart contracts for known vulnerabilities such as reentrancy, integer overflow, and access control issues (Houben & Weber, 2023; Tikhomirov et al., 2018b). Though it does not provide traditional ML metrics like accuracy or detection delay, it has demonstrated effectiveness in detecting over 80 percent of known bugs in smart contracts. This makes it valuable for preventive auditing in blockchain-based IoT systems before deployment.

3.6 Challenges in AI Integration

- **Data Privacy:** AI requires large datasets; sensitive IoT data may violate privacy without proper anonymization.
- **Model Drift:** Attack behaviors evolve, requiring continuous retraining.
- **Resource Constraints:** Many IoT nodes cannot host large models.
- **Blockchain Immutability:** Once logged, incorrect model outputs cannot be reversed.

4. Blockchain for Enhancing IoT Security

Blockchain technology introduces decentralization, transparency, and tamper resistance, offering transformative security benefits to traditionally vulnerable IoT systems. By embedding blockchain into the IoT architecture, data authenticity, device identity, and secure communication can be fundamentally improved.

4.1 Key Security Enhancements via Blockchain

Table 14.5 presents the support of blockchain for security. The security feature data integrity, authentication, access control, availability and auditability are enhanced through blockchain features which are presented in the table.

Table 14.5 Blockchain support for security enhancement.

Security Feature	*How Blockchain Helps*
Data Integrity	Immutable ledgers prevent data tampering
Authentication	Public-key cryptography for decentralized identity
Access Control	Smart contracts enforce policy autonomously
Availability	Decentralized consensus reduces single-point-of-failure
Auditability	Transparent logs enable forensic traceability

4.1.1 Blockchain-IoT Integration Models

- The integration of blockchain-IoT can be public blockchain, i.e., anyone can participate and provide security, but this is a resource-intensive option.
  Examples: Ethereum, Bitcoin (not ideal for low-power IoT)
- Another integration can be private/permissioned blockchain where access is restricted to authorized entities. This suitable for enterprise IoT, e.g., supply chains or smart cities. The frameworks used are Hyperledger Fabric and Quorum.
- Group integration model is consortium blockchain, which is controlled by a group of stakeholders (e.g., smart grid operators). It leads to balanced scalability, trust, and decentralization.

4.1.2 Use Cases of Blockchain in IoT Security

- Supply Chain Traceability: Here IoT sensors record temperature/humidity of goods. Blockchain ensures unalterable logs for compliance (e.g., food/drug transport).
- Smart Vehicle Networks: Vehicles use blockchain to verify firmware updates and vehicle-to-vehicle trust. Blockchain protects against Global Positioning System (GPS) spoofing and Electronic Control Unit (ECU) tampering.
- Smart Cities: Blockchain coordinates resource usage (e.g., smart meters), preventing data forgery or billing fraud.

Figure 14.5 illustrates a layered architecture combining IoT, edge, blockchain, and analytics technologies to build a secure and intelligent IoT ecosystem.

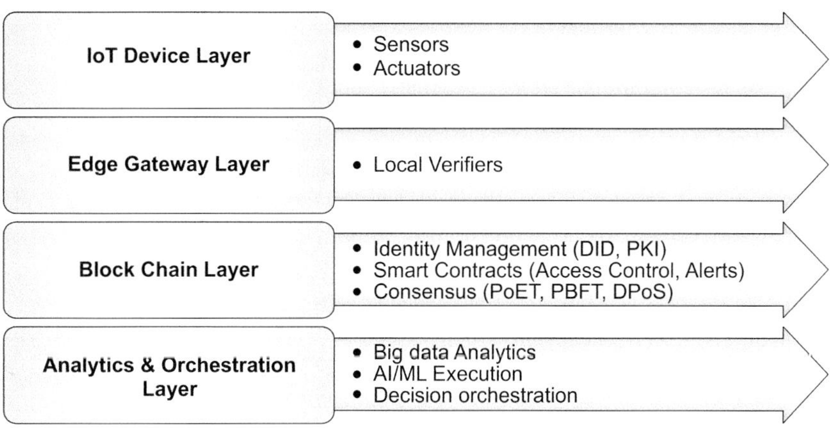

Fig. 14.5 Blockchain-enhanced IoT security.

- The IoT device layer includes components like sensors that capture data such as temperature, pressure, and motion, and actuators that perform actions based on the received commands like switch, move, and adjust. The security concern at this layer is that devices are resource-constrained and vulnerable to physical tampering and cyberattacks.

- The edge gateway layer includes local verifiers, i.e., intermediate devices or software that verify and pre-process data near the source. This layer acts as a bridge between IoT devices and the blockchain performing lightweight validation, data aggregation, and filtering to reduce latency and bandwidth usage. It improves response time and enables real-time decision-making.
- The blockchain layer includes components to perform identity management, smart contracts, and consensus mechanisms. It uses decentralized identifiers (DID) and public key infrastructure (PKI) to provide secure, verifiable identity. It has the code that runs on the blockchain to enforce policies like access control, data alerts, and automated actions. It ensures that all nodes agree on the blockchain state using algorithms such as proof of elapsed time (PoET), practical Byzantine fault tolerance (PBFT), and delegated proof of stake (DPoS). It provides immutability, trust, and transparency, and ensures tamper-proof data logs and decentralized trust among devices and stakeholders.
- The analytics and orchestration layer performs big data analytics, AI/ML model execution, and decision orchestration depending on blockchain-verified IoT data. It provides predictive insights, visualizations, and automated control across applications like smart cities, healthcare, and industry 4.0. It coordinates the entire architecture for policy enforcement, incident response, and optimization.

Data is generated at the IoT device layer; data is pre-processed and validated at the edge gateway layer; security issues are handled at the blockchain layer; and final decision and control is performed at the analytics and orchestration Layer. The system achieves enhanced security, scalability, and trust in IoT ecosystems. Though decentralized control, real-time intelligence and automation as also prevention and care from attacks are achieved.

4.1.3 Implementation Aspects

The practical implementation of a secure and scalable blockchain-IoT ecosystem requires a careful selection of tools, frameworks, and protocols that can handle the unique constraints of IoT devices while leveraging the immutability and decentralization offered by blockchain. This section outlines the primary implementation components under two key layers:

(1) Blockchain Infrastructure and Development Tools
(2) Identity and Access Control Mechanisms

4.1.4 Blockchain Infrastructure and Development Tools

To realize secure transactions, decentralized decision-making, and tamper-proof logs in IoT systems, the following platforms and tools are employed:

Ethereum (Testnet/Ganache)

Ethereum is the most widely adopted public blockchain for smart contract development. In IoT scenarios, Ethereum Testnets (e.g., Sepolia or Goerli) are utilized for prototype testing without actual gas fees. Ganache, a local Ethereum emulator, supports rapid testing and debugging of contracts and is essential during the development phase before deployment on public or private networks.

Hyperledger Fabric

Unlike Ethereum, which is permissionless, Hyperledger Fabric is a permissioned blockchain designed for enterprise-grade applications. It supports modular consensus, channel-based privacy, and smart contracts (Chaincode), making it suitable for constrained IoT environments like industrial networks or healthcare systems. Fabric enables secure communication between trusted IoT nodes without public exposure.

Truffle Suite

This is a vital toolchain in the smart contract development lifecycle. It offers:

- Truffle: Development and testing framework
- Ganache: Personal blockchain instance for contract testing
- Drizzle: Frontend libraries for integrating blockchain with user interface (UI)

This suite ensures reproducibility, modular deployment, and integration testing of blockchain code.

Internet of Things Exchange/Internet of Things Applications (IoTeX/IOTA)

These are specialized blockchains developed with IoT constraints in mind.

- IoTeX focuses on privacy-preserving, lightweight infrastructure for machine-to-machine communication. It also supports DID frameworks for device identity.
- IOTA, based on the Tangle (a Directed Acyclic Graph [DAG] structure), is feeless and highly scalable, ideal for microtransactions and high-frequency sensor data validation.

4.1.5 Identity and Access Control Mechanisms

Security in IoT systems extends beyond data immutability to include who accesses what and when. The following decentralized identity and access control mechanisms are employed:

Decentralized Identifiers (DIDs)

DIDs are self-sovereign identities anchored on a blockchain. Each IoT device is assigned a globally unique DID, which is linked to a public key. This ensures:

- Verifiable authentication without centralized authorities
- Tamper-proof identity management
- Revocation and update mechanisms directly on-chain

Attribute-Based Access Control (ABAC)

- Unlike role-based access control (RBAC), ABAC defines policies based on attributes such as device type, location, owner identity, and security clearance. These rules are embedded in smart contracts that:
  - Enforce real-time access decisions
  - Allow fine-grained access control
  - Enable auditing through blockchain transaction logs

Smart contracts evaluate the contextual attributes before granting access to IoT resources (e.g., unlocking a sensor feed, initiating actuation, or writing data to the ledger). ABAC enhances scalability and adaptability, especially in environments where devices join or leave dynamically.

Table 14.6 Real-world implementations.

Project	*Blockchain*	*IoT Domain*	*Key Feature*
Filament (IBM Corporation, 2017)	Ethereum	Industrial IoT	Machine-to-machine contract execution
IBM Watson IoT (Modum.io, 2019)	Hyperledger fabric	Smart manufacturing	Blockchain-secured sensor data streams
Modum.io (PowerLedger, 2020)	IOTA	Pharma logistics	Tamper-proof temperature sensor logs
PowerLedger (Dorri et al., 2018)	Ethereum	Smart grid	P2P energy trading with secure metering

4.2 Real-World Implementations

- ***Filament (Ethereum, Industrial IoT):*** This enables machine-to-machine smart contract execution in industrial environments. Devices can autonomously initiate and fulfill blockchain-logged contracts, improving automation, and auditability in supply chains and manufacturing.
- ***IBM Watson IoT and Hyperledger Fabric (Smart Manufacturing):*** This integrates blockchain with sensor data streams to ensure data integrity, provenance, and traceability in smart factories. It uses permissioned networks to control access among stakeholders securely.
- ***Modum.io (IOTA, Pharmaceutical Logistics):*** This uses tamper-proof temperature sensors linked to a blockchain to verify compliance in cold chain logistics. It ensures that pharmaceuticals are transported under regulated conditions, with immutable records for auditing.
- ***PowerLedger (Ethereum, Smart Grid):*** This implements peer-to-peer (P2P) energy trading on the Ethereum blockchain. Smart meters record energy production/consumption, and transactions are validated on-chain, enabling decentralized electricity markets.

4.3 Analysis and Benefits

The integration of blockchain technology into IoT ecosystems introduces fundamental improvements in terms of security, trust, and data integrity. The following points highlights how blockchain mitigates key vulnerabilities present in traditional IoT architectures, with reference to the comparison in Table 14.7.

4.3.1 Data Tampering Risk

In conventional IoT systems, sensor data is typically stored in centralized databases or cloud platforms. This centralized model is inherently vulnerable to data tampering, deletion, or manipulation due to a single control point (Conoscenti et al., 2016). Blockchain addresses this by creating an immutable ledger, where once a transaction (e.g., sensor reading or device status) is recorded, it cannot be altered retroactively without consensus from the network. This ensures data integrity, especially critical in applications like healthcare, energy, and logistics.

Table 14.7 Pre-blockchain vs. post-blockchain IoT security.

Metric	*Pre-blockchain IoT*	*Post-blockchain IoT*
Data Tampering Risk	High (central storage)	Low (immutable ledger)
Single Point of Failure	Present	Eliminated via decentralization
Identity Spoofing	Frequent in open networks	Reduced via PKI/DID
Audit Trail Quality	Poor	Transparent, time-stamped records
Trust Management	Manual, centralized	Automated via consensus

4.3.2 Single Point of Failure

Traditional IoT architectures rely on a central hub or server to manage devices and process data. This creates a single point of failure (SPOF)—if the server is compromised or goes offline, the entire network can become inoperable. Blockchain employs decentralization, distributing the control across nodes in a peer-to-peer network. This not only enhances fault tolerance but also improves system availability during network partitions or cyberattacks (Sovrin Foundation, 2021).

4.3.3 Identity Spoofing

IoT devices are often deployed in open environments where spoofing attacks are common. Attackers can impersonate legitimate devices to inject false data or gain unauthorized access. Blockchain enables DIDs and PKI, ensuring that each device has a verifiable, cryptographically signed identity (Abie & Balasingham, 2012). This drastically reduces the risk of identity spoofing and unauthorized control.

4.3.4 Audit Trail Quality

Pre-blockchain IoT systems often lack robust and trustworthy logging mechanisms. Centralized logs can be deleted, edited, or become corrupted, which complicates forensics and compliance audits. In contrast, blockchain provides transparent, append-only, and time-stamped transaction logs that are accessible across all nodes. This creates a reliable audit trail for regulatory compliance, system diagnostics, and post-incident investigation (Nakamoto, 2008).

4.3.5 Trust Management

In conventional IoT systems, trust is established manually through central authorities or vendor-specific policies. This approach is rigid and not scalable, especially in multi-vendor or cross-domain deployments. Blockchain replaces centralized trust anchors with automated consensus protocols (e.g., proof of work, practical Byzantine fault tolerance), enabling distributed trust that is dynamically maintained and cryptographically verifiable (Moustafa & Slay, 2019).

4.4 Limitations and Challenges

- *Scalability:* IoT networks generate high-frequency data; blockchain throughput is limited.
- *Latency:* Consensus protocols introduce delays unsuitable for real-time applications.
- *Resource Constraints:* IoT nodes cannot always run full blockchain clients.
- *Data Privacy:* Public ledgers risk exposing sensitive sensor information.

5. Case Studies of Security Incidents and Solutions

Real-world security breaches reveal the practical vulnerabilities and evolving threats within blockchain-IoT ecosystems. By analysing these incidents and their resolutions, we can extract critical lessons to inform future security strategies.

5.1 Case Study 1: Euler Finance Smart Contract Exploit (March 2023)

In March 2023, Euler Finance, a DeFi protocol on Ethereum, suffered a major attack resulting in the loss of approximately $197 million (CertiK, 2023; CoinDesk 2023). The attacker used a flash loan to manipulate collateral and debt values, exploiting a logical vulnerability in the protocol's donateToReserves and liquidation functions. This allowed reentrancy-like behavior to repeatedly drain funds (Conti et al., 2022). The exploit, with an estimated CVSS score of 9.4, highlighted weaknesses in access control and real-time risk assessment. While ~$90 million was later returned, the incident emphasized the need for AI-powered auditing tools, on-chain behavior analysis, and formal verification to secure smart contracts before deployment.

5.2 Case Study 2: Poly Network Cross-Chain Bridge Exploit (August 2021)

In August 2021, Poly Network, a cross-chain interoperability protocol, was hacked in one of the largest DeFi exploits, losing over $610 million in crypto assets (Bhatti & Humphreys, 2017; Poly Network, 2021). The attacker exploited a vulnerability in the cross-chain contract's access control, allowing them to forge a transaction that transferred ownership of funds across Ethereum, Binance Smart Chain, and Polygon. The exploit stemmed from improper verification of message authenticity in inter-chain calls. Although the attacker later returned the majority of the stolen funds, the incident exposed critical flaws in multi-chain communication logic (Poly Network, 2021; The Verge 2021). The case highlighted the urgent need for formal verification of cross-chain contracts, AI-based transaction behavior monitoring, and DID frameworks to prevent unauthorized access in multi-chain systems.

5.3 Case Study 3: GPS Spoofing in Connected Vehicles

In recent years, GPS spoofing has emerged as a critical threat to connected and autonomous vehicles. In this scenario, smart vehicles equipped with Global Navigation Satellite System (GNSS) receivers are misled by spoofed satellite signals, leading to traffic misrouting, unauthorized access to restricted areas, and potential safety hazards (Hu et al., 2020; Papadimitratos et al., 2008). The attack, which scores a

CVSS rating of 8.4, is a physical-layer spoofing technique where attackers transmit fake GPS signals that overpower legitimate ones (Dorri et al., 2017). Major failure points include weak signal authentication and the lack of cross-verification from additional sensors or networks. To mitigate this, a blockchain-IoT-based approach has been proposed, combining multi-sensor fusion (e.g., Light Detection and Ranging [LiDAR], inertial sensors, GPS) with blockchain-based consensus logging, allowing vehicles to cross-check their location data against decentralized inputs (Lee et al., 2016). Furthermore, decentralized timestamping of GPS readings ensures tamper-proof location verification, enhancing the integrity of autonomous vehicle navigation systems.

5.4 Case Study 4: Energy Grid Cyberattack (Ukraine, 2015)

In December 2015, a sophisticated cyberattack targeted Supervisory Control and Data Acquisition (SCADA) systems of Ukraine's power grid, resulting in blackouts affecting over 230,000 people (Case et al., 2016). The attackers used spear phishing to steal operator credentials and deployed malware that overwrote firmware in smart meters and Programmable Logic Controller (PLCs), rendering systems unresponsive. The CVSS score for the exploit is estimated at 9.1, with failure points including lack of log integrity and weak access controls (Lin a& Bergmann, 2016). A modern blockchain-IoT solution would involve creating an immutable log chain for all SCADA activity and using blockchain-verified firmware updates to prevent tampering. Additionally, AI-powered intrusion detection systems, trained on historical telemetry, could identify anomalies in grid behavior before service disruption (Siegel, 2016).

5.5 Case Study 5: DAO Smart Contract Exploit (2016)

In 2016, the Ethereum platform suffered a major breach through the DAO exploit, where a reentrancy vulnerability in its smart contract allowed an attacker to repeatedly withdraw funds before internal balances were updated, leading to the loss of around $60 million worth of ETH (Atzei et al., 2017). The exploit, rated 9.6 CVSS, was a result of logical flaws in contract execution, exacerbated by lack of formal verification, poor auditing, and reliance on manual code reviews (Buterin, 2016). In response, the Ethereum community executed a hard fork to reverse the damage and later adopted formal verification tools, stronger auditing standards, and secure coding practices (MalwareMustDie, 2016). The DAO attack remains a key example of the importance of secure-by-design principles, enhanced by AI-based auditing and blockchain immutability, in protecting decentralized systems.

5.6 Case Study 6: The Mirai Botnet and Dyn Attack (2016)

In 2016, the Mirai botnet exploited vulnerable IoT devices—such as closed circuit television (CCTV) cameras, digital video recorders (DVRs), and routers—by leveraging default credentials and open Telnet ports to form a large botnet (Krebs, 2016). It launched a massive DDoS attack on Dyn, a major DNS provider, causing outages across major platforms like Twitter, Netflix, Reddit, and GitHub, and impacting users across the US and Europe (Conti et al., 2018). The attack highlighted critical IoT weaknesses, including hardcoded logins, unpatched firmware, and lack of device-level security. It demonstrated the urgent need for AI-based threat detection and blockchain-enabled decentralized control to build resilient IoT ecosystems (Tyagi et al., 2023).

Table 14.8 provides a comparative summary of major cybersecurity incidents across DeFi, IoT, and SCADA systems, highlighting the nature of each exploit, its impact, failure points, and proposed AI-blockchain-based countermeasures (CoinDesk 2023; Tyagi et al., 2023).

Table 14.8 Comparative analysis of major cybersecurity incidents and blockchain-AI countermeasures in IoT and smart contract systems.

Case Study	*Platform/ Domain*	*Exploit Type*	*Impact*	*CVSS Score*	*Failure Points*	*AI and Blockchain Mitigation*
Euler Finance (2023)	Ethereum (DeFi)	Flash loan and logic bug	$197M loss (partial recovery)	9.4	Access control, runtime risk	AI-based auditing, formal verification
Poly Network (2021)	Cross-chain DeFi	Access control bypass	$610M stolen (majority returned)	9.5 (est.)	Inter-chain auth. logic	Formal verification, DID, AI transaction scoring
GPS Spoofing	Connected vehicles	Physical-layer spoofing	Traffic rerouting, safety threats	8.4	Weak signal auth., no cross-check	Blockchain-based consensus and timestamping
Ukraine Grid Attack (2015)	SCADA/ power grid	Credential theft and firmware overwrite	230,000 affected	9.1	No log integrity, weak access control	Immutable SCADA logs, blockchain-verified firmware, AI-based anomaly detection
DAO Exploit (2016)	Ethereum (DAO)	Reentrancy vulnerability	~$60M ETH stolen	9.6	No formal verification, poor auditing	Hard fork, formal tools, AI-assisted audit
Mirai Botnet (2016)	IoT infrastructure	Botnet via default creds	Major DNS outage (Twitter, Netflix, etc.)	9.8	No device auth., open ports	Blockchain identity, AI-driven threat detection

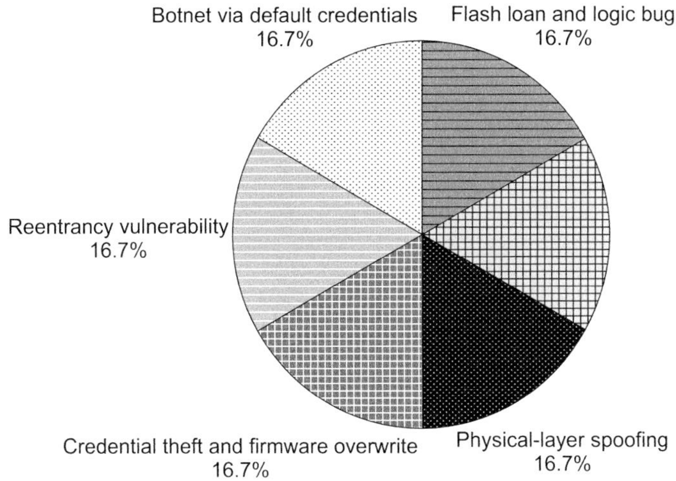

Fig. 14.6 Case study CVSS score comparison.

6. Conclusion

The integration of blockchain and IoT offers a compelling vision for secure, decentralized, and intelligent cyber-physical systems. However, this convergence is fraught with complex cybersecurity challenges that span device, network, protocol, and software layers. In this chapter, we explored common threats and vulnerabilities such as botnets, firmware tampering, smart contract flaws, and consensus-level exploits. AI-powered cybersecurity solutions include supervised and deep learning models for intrusion detection, anomaly detection, and smart contract auditing. Blockchain can serve as a security backbone for decentralized identity, tamper-proof logging, and autonomous access control using smart contracts. Real-world case studies illustrate both the fragility of current systems and the potential for blockchain-AI integration to provide proactive defense mechanisms.

References

Antonakakis, M., April, T., Bailey, M., Bernhard, M., Bursztein, E., Cochran, J., Durumeric, Z., Halderman, J. A., Invernizzi, L., Kallitsis, M., Kumar, D., Lever, C., Ma, Z., Mason, J., Menscher, D., Seaman, C., Sullivan, N., Thomas, K. and Zhou, Y. (2017). Understanding the Mirai botnet. *In Proceedings of the 26th USENIX Security Symposium* (pp. 1093–1110). USENIX Association. https://www.usenix.org/conference/usenixsecurity17/technical-sessions/presentation/antonakakis

Atzei, N., M. Bartoletti and T. Cimoli,. (2017). A survey of attacks on Ethereum smart contracts (SoK), in *Proceedings of the International Conference on Principles of Security and Trust (POST)*, pp. 164–186.

Belenguer, A., J.A. Pascual and J. Navaridas. (2025). A review of federated learning applications in intrusion detection systems, *Computer Networks*, *258*.

Bhatti, J. and T.E. Humphreys. (2017). Hostile control of ships via false GPS signals: demonstration and detection, *NAVIGATION*, *64*(1), 51–66.

Buterin, V. (2016). Hard fork completed, *Ethereum Blog*, Juy.

Case, A., J. Graham and T. Johnson. (2016). The industrial control system cyber kill chain, SANS Institute Whitepaper.

CertiK. (2023). Euler finance hack: $197M lost due to flawed logic in debt liquidation, *CertiK Security Blog*, March.

CoinDesk. (2023). Euler finance attacker returns $90M, March 27.

Conoscenti, M., A. Vetrò and J.C. De Martin. (2016). Blockchain for the internet of things: a systematic literature review, in *Proceedings of the IEEE COMPSAC*.

Conti, M., A. Dehghantanha, K. Franke and S. Watson. (2018). Internet of things security and forensics: challenges and opportunities, *Future Generation Computer Systems*, *78*, 544–546.

Dorri, A., S.S. Kanhere and R. Jurdak. (2018). Blockchain in internet of things: challenges and solutions, *IEEE Computer*, *51*(6), 68–74.

Dorri, H., S.S. Kanhere, R. Jurdak and P. Gauravaram. (2017). Blockchain for IoT security and privacy: the case study of a smart home, in *Proceedings of the IEEE International Conference on Pervasive Computing and Communications (PerCom)*, pp. 618–623.

Dyn. (2016, October 21). Dyn statement on October 21, 2016 DDoS attack [Press release]. Dyn. https://cyber-peace.org/wp-content/uploads/2016/10/Dyn-Statement-on-10_21_2016-DDoS-Attack.pdf

Ferrag, M.A., L. Maglaras, S. Moschoyiannis and H. Janicke. (2020). Deep learning for cyber security intrusion detection: approaches, datasets, and comparative study, *Journal of Information Security and Applications*, *50*.

Filament. (2017). Filament: blockchain IoT for industrial automation, Filament Whitepaper.

H. Abie, H. and I. Balasingham. (2012). Risk-based adaptive security for smart IoT in eHealth, in *Proceedings of the 7th IEEE International Conference on eHealth Networking*.

Houben, S. and J. Weber. (2023). Euler finance exploit: attack dissection and protocol recovery, BlockSec Research.

Hu, X., H. Guo, Y. Jiang and J. Sun. (2020). A review of GPS spoofing attack: current status and countermeasures, *IEEE Access*, *8*, 57845–57859.

Hussain, F., S. Jhanjhi, M. Humayun and R.A. Shaikh. (2020). Machine learning for IoT security: a comparative study of supervised learning algorithms, *Electronics*, *9*(5), 786.

IBM Corporation. (2017). IBM brings blockchain-secured IoT to enterprises, *IBM Blockchain Blog*, September.

Javaid, A., Q. Niyaz, W. Sun and M. Alam. (2016). A deep learning approach for network intrusion detection system, in P*roceedings of the 9th EAI International Conference on Bio-inspired Information and Communications Technologies (BIONETICS)*, New York, USA.

Khan, L.U., M. Guizani, I. Yaqoob, D. Niyato, A. Al-Fuqaha and C.S. Hong. (2024). A survey on metaverse-empowered 6G wireless systems: a security perspective, *Internet of Things*, *28*.

Krebs, B. (2016). KrebsOnSecurity hit with record DDoS, September.

Lai, C.-M., Teh, J., Lin, Y.-C. and Liu, Y. (2020). Study of a Bidirectional Power Converter Integrated with Battery/Ultracapacitor Dual-Energy Storage. *Energies, 13*(5), 1234. https://doi.org/10.3390/en13051234.

Lee, R., M. Assante and T. Conway. (2016). Analysis of the cyber attack on the Ukrainian power grid, SANS Industrial Control Systems, March.

Lin H. and N.W. Bergmann. (2016). IoT security challenges and solutions, in *Proceedings of the 2016 IEEE International Conference on Internet of Things and IEEE Green Computing and Communications (iThings/GreenCom)*, pp. 1016–1020.

Luu, L., Chu, D.-H., Olickel, H., Saxena, P. and Hobor, A. (2016). Making smart contracts smarter. In Proceedings of the 2016 ACM SIGSAC Conference on Computer and Communications Security (CCS '16) (pp. 254–269). ACM. https://doi.org/10.1145/2976749.2978309.

M. Conti, A. Gangwal and D. Lakshmanan. (2022). Poly Network hack: a case study of a cross-chain vulnerability, in *Proceedings of the 2022 IEEE International Conference on Blockchain*, Espoo, Finland, 136–143.

MalwareMustDie. (2016). Mirai—IoT botnet, Security Research, September.

Moustafa N. and J. Slay. (2019). The bot-IoT dataset for machine learning cybersecurity applications, *Future Internet*, *11*(3), 23.

Moustafa, N. and J. Slay. (2015). UNSW-NB15: a comprehensive data set for network intrusion detection systems (UNSW-NB15 network data set), in *Proceedings of MilCIS*, Canberra, Australia, pp. 1–6.

Moustafa, N., E. Adi, B. Turnbull and J. Hu. (2018). A new threat intelligence scheme for safeguarding Industry 4.0 systems, *IEEE Access*, *6*, 32910–32924.

Moustafa, N., K.-K.R. Choo, L. Turnbull and W. Hu. (2020). A new threat intelligence scheme for safeguarding Industry 4.0 systems against cyberattacks, *Future Generation Computer Systems*, *108*, 312–319.

Nakamoto, S. (2008). Bitcoin: A peer-to-peer electronic cash system. https://bitcoin.org/bitcoin.pdf

Papadimitratos, P., Gligor, V. and Hubaux, J.-P. (2008). Secure vehicular communication systems: Design and architecture. *IEEE Communications Magazine, 46*(11), 100–109. https://doi.org/10.1109/MCOM.2008.4689977

Poly Network. (2021). Poly Network hack summary report, Medium, August.

Qi, M., Z. Wang, Q.-L. Han, J. Zhang, S. Chen and Y. Xiang. (2024). Privacy protection for blockchain-based healthcare IoT systems: a survey, *IEEE/CAA Journal of Automatica Sinica*, *11*(8), 1757–1776.

S. Tikhomirov, E. Voskresenskaya, I. Ivanitskiy, R. Takhaviev, E. Marchenko and Y. Alexandrov, "SmartCheck: Static Analysis of Ethereum Smart Contracts," in Proc. 1st Int. Workshop on Emerging Trends in Software Engineering for Blockchain (WETSEB), Gothenburg, Sweden, May 27, 2018.

Siegel, M. (2016). Understanding the DAO attack, MIT Media Lab Digital Currency Initiative, June.

Sovrin Foundation. (2021). Decentralized identifiers (DIDs) v1.0, W3C Working Draft.

The Verge. (2021). $610 million stolen in Poly Network hack, August 11.

Tikhomirov, S., E. Voskresenskaya, I. Ivanitskiy, R. Takhaviev, E. Marchenko and Y. Alexandrov. (2018). SmartCheck: static analysis of Ethereum smart contracts, in *Proceedings of the 1st International Workshop on Emerging Trends in Software Engineering Blockchain (WETSEB)*, Gothenburg, Sweden, pp. 9–16.

Tyagi, A.K., Dananjayan, S., Agarwal, D. and Thariq Ahmed, H.F. (2023). Blockchain—Internet of Things Applications: Opportunities and Challenges for Industry 4.0 and Society 5.0. *Sensors, 23*(2), 947. https://doi.org/10.3390/s23020947

Yin, C., Y. Zhu, J. Fei and X. He. (2017). A deep learning approach for intrusion detection using recurrent neural networks, *IEEE Access*, *5*, 21954–21961.

CHAPTER 15

Blockestate: Revolutionizing Property Transactions

Modigari Narendra[1], *Neeraja R.*[2], *Jani Anbarasi L.*[3*], *Gaurav Singh*[4], *Georgy Prasad Philip*[5] and *Sumit Ranjan*[6]

This research presents a new framework Blockestate to improve the assessment and analysis of real estate investments through the use of emerging technologies, like blockchain and innovative statistical forecasting models. The framework Blockestate brings together decentralized property data management and predictive analytics as a complete decision support tool for real estate participants. This framework Blockestate utilizes smart contracts built on the Hardhat platform to safely manage and execute property transactions in an open, transparent, and decentralized way across multiple blockchain accounts. The smart contracts will allow for secure recording and verification of ownership transfers and pricing data, establishing a trustless foundation for transactions. To facilitate data access, a modular system for data retrieval is integrated into the framework, allowing external systems and users to retrieve historical property price records stored on the blockchain easily. The historical data will be retrieved and routed through a Cloudflare Tunnel, which acts as a bridge connecting the local Hardhat blockchain to external systems, already making it capable to interact with the blockchain via Web3-based approaches in Google Colab (Web3.py). Therefore, participants are able to query, and analyze the blockchain data in real-time in much more accessible and scalable ways. The framework Blockestate incorporates autoregressive integrated moving average (ARIMA) models for projecting future price trends based on previously reported data to the extent possible, while capturing seasonality and time series dependencies. Analysis of the profitability of investments uses extended internal rate of return (XIRR) analysis which accounts

[1,3-6] School of Computer Science and Engineering, Vellore Institute of Technology, Chennai, Tamil Nadu, India.

[2] Department of Computer Science and Engineering, Sri Sivasubramaniya Nadar College of Engineering, Chennai, Tamil Nadu, India.

https://orcid.org/0000-0003-1852-2803[1]; https://orcid.org/0000-0002-6081-6269[2]; https://orcid.org/0000-0002-8904-2236[3]; https://orcid.org/0009-0006-4059-6533[4]; https://orcid.org/0009-0009-2728-7316[5]; https://orcid.org/0009-0008-1344-7991[6]

* Corresponding author: janianbarasi.l@vit.ac.in

DOI: 10.1201/9781003659822-15

for cash flows that are not periodic and provides more realistic profitability than standard IRR calculations. ARIMA and XIRR are effective to evaluate forecasting and profitability models; additionally, the framework evaluates rental income and payback periods for inflation-adjusted estimations, providing multiple options for evaluation about investment performance indicators. Users can input future property prices to facilitate investment decision-making.

1. Introduction

The real estate sector, a crucial element of global economic activity, is dealing with persistent inefficiencies and systemic challenges including data fragmentation, lack of transparency, and the systematic application of subjective assessment methodologies. Investors and analysts find it challenging to obtain a cohesive market perspective on trends or the performance of specific properties due to the fact that property data is kept in multiple separate and unintegrated systems. The investment process is hampered by these limitations, isolating itself from new, more effective data-driven decision-making models. In light of these obstacles, it seemed that blockchain technology appeared in response as a potential disruptor in the real estate space. Blockchain would enable secure, immutable, and decentralized recording of property transactions through smart contracts, thereby creating efficiencies and reducing reliance on intermediaries, while substantially improving trust and transparency. Property titles, historical sales information, and transaction data recorded on a blockchain would be invariant to improper changes and newly available for authorized parties to query in real time. This solution could stimulate a more accountable ecosystem. Despite the fact that blockchain offers assurance over the reliability and access of property data, it, by itself, does not offer any means of predicting future market activity or assessing the viability of an investment. To convert raw data, and even trusted data, into usable data requires analysis. Traditional methods for forecasting the property sector (i.e., manual trend analysis or simple projections) struggle to predict the fluctuation of real estate markets and do not account for the time value of money.

This research provides a bridge for this gap by adopting blockchain-based property data management and statistical forecasting approaches. Smart contracts are implemented on a local Hardhat blockchain where secure, transparent transactions can be simulated. Property data can then be extracted and analyzed using ARIMA models for time-series forecasting and XIRR analysis to assess the return on an investment based on cash flows that are not normal. By coupling the trustless environment of blockchain and the predictive ability of data science, the proposed framework Blockestate provides a systematic end-to-end approach to the assessment of real estate investment. This novel integration will provide contemporary investors, analysts, and policymakers the tools to make informed, data-based plans in a market that is unanimously characterized by uncertainty and inefficiency.

This research has combined the fundamental transparency of blockchain technology with the accuracy of statistical forecasting to build a transparent, automated, data-driven framework for analyzing real estate investments. The system developed analyzes various properties in a systematic manner by rating them as undervalued, fairly valued, or overvalued. Once rated, the system provides personalized recommendations for the types of investments to consider, including purchase, rent, or not investing, based on an analysis of how financially viable those recommendations are. The proposed approach is meant to improve confidence for investors by providing trustworthy and accurate insights, lessen the reliance on intermediaries, and improve decision-making in real estate transactions. Using these digital technologies will transform real estate investment transactions to enhance the transparency while making it more accessible to a wide range of investors. The remainder of the chapter is organized as follows.

Section 2 is a literature review of blockchain in real estate and of forecasting. Section 3 presents the methodology, including the architecture of the system and how smart contracts are implemented. Section 4 provides the results and analysis, while the final section discusses future research.

2. Literature Review

Blockchain technology has disrupted many industries, including real estate, due to inherent improvements in transparency, security, and efficiency. A depth of literature exists on the topic, which has explored the mechanics through which blockchain technologies, and particularly Ethereum-based blockchain, alleviate transaction times by identifying and optimizing variables that affect transaction times. This research demonstrates how processes can be streamlined, intermediaries reduced, and ultimately, transactions occur at an accelerated pace using the technology's decentralized approach (Pacheco et al., 2022). Numerous studies have demonstrated the potential effectiveness of blockchain in a variety of settings, with a primary focus on financial and contractual transactions in the real estate industry (Ammous, 2016). The transparency of blockchain technology in real estate transactions has greatly increased consumer trust. This assurance resulting from the data's immutability and verifiability encourages potential purchasers to participate in real estate transactions more frequently (Rapezzi, 2024). The blockchain's inherent features such as decentralization and immutability can be used as institutional mechanisms to guarantee the accuracy and verifiability of the credit decision-making process (Damaševičius et al., 2024).

Reductions in money laundering operations have been associated with blockchain technology and real estate. This is mostly due to the blockchain's immutability such that better tracking and monitoring opportunities are made possible by this increased transaction transparency (Agarwal & Agarwal, 2004). The integrity of real estate transactions is being jeopardized by widespread fraud in the digital land markets. In order to help avoid fraud and to certify and verify ownership in real estate transactions, blockchain solutions provide a reliable and unchangeable record of ownership history (Faxon & Wittekind, 2023). An extensive framework using blockchain technology has been designed to identify and reduce property fraud. The substantial benefits of smart contracts in automating verification procedures and improving property management efficiency are highlighted by this model (Verma et al., 2021). New methods for timestamping and documenting property transactions on the Ethereum blockchain have been introduced. By utilizing the decentralized nature of blockchain technology, these platforms lower the risk of fraud and conflicts while increasing transparency and confidence in real estate transactions (Zhang et al., 2023).

Blockchain offers a framework that makes it possible for communities to manage real estate assets by utilizing safe, impenetrable systems to record property titles, ownership transfers, and leases. This completely eliminates the risk of data manipulation and fraudulent transactions. Moreover, blockchain is inherently transparent and auditable. This fosters accountable governance, allowing the public sector to track and verify real estate activity, most importantly, in real time. This digital infrastructure fosters citizen trust, spurs streamlined urban planning, and allows for more efficient use of land and housing—each of these being critical components of a smart city ecosystem (Sankar et al., 2023). The utilization of tokenization in the real estate sector is revolutionizing how assets are owned, traded, and financed—this is becoming a way of life with the use of blockchain technology. Tokenization, as a process, means that real estate assets are converted into digital tokens that represent fractional ownership in an asset. With tokenization, investors could buy, sell, or trade shares of their ownership in a property asset like they buy and sell stocks in a financial market. The implications of this exciting opportunity have disrupted the traditional structures of real estate finance by providing increased liquidity, lowering barriers to entry by investors, and allowing individuals globally to buy into real estate ownership. Tokenized real estate assets are increasingly being compared to real estate investment trusts (REITs)—pooling investment vehicles designed to give exposure to property assets without direct ownership of the property. Both seek democratization of real estate investing, and while they share some similarities, tokenization has unique attributes. New forms of tokenization enable real-time settlement of trades with lower reliance on intermediaries, and the use of programmable contracts accelerate the distribution of the wealth and manage governance for token holders. This development marks a shift not only in the flow of capital into real estate but also foresees a more transparent, decentralized, and universal space for investment (Liu & Chen, 2024). A key aspect of blockchain-enabled smart contracts is the concept of gas optimization, which plays a critical role in determining the cost-efficiency and scalability of automated agreements on blockchain networks. In platforms like Ethereum, every operation executed by a smart contract—such as storing data, performing calculations, or interacting with other contracts—consumes a unit of computational effort known as "gas." All gas used results in transaction fees, which are paid in the network's native cryptocurrency. Optimizing your contracts poorly may lead to transaction costs that are so high that the system is no longer viable—this is especially true for applications that have frequent or complex interactions. To prevent poorly optimized contracts from raising transaction fees in practice, developers focus on writing efficient (minimized) Solidity code which reduces the amount of storage used and limits repeated computations, and therefore gas fees. Common optimizations in Solidity language include "loop unrolling," the use of calldata instead of memory variables, reusing storage variables whenever possible, and limiting contract size. Gas optimization is particularly important in real estate applications where smart contracts could include complex datasets, multiple aggrieved parties, and transactional interactions over time (i.e., rent payment, title updates, and access release from escrow). Gas efficient smart contracts reduce the cost to users, increase adoption, and are important for the sustainability and efficiency of broader blockchain (Li, 2021).

A number of research efforts have looked at gas-expensive patterns and optimization strategies to increase the efficiency and cost effectiveness of executing smart contracts on blockchain platforms. Research has identified numerous coding behaviors and design patterns that manifest as storage variables, nested loops, redundant calculations, and inefficient data structures, which resulted in costly gas expenditure. Researchers and developers have made attempts to identify many ways to reduce the costs associated with deploying and interacting with smart contracts via gas-expensive patterns. Some common techniques that are discussed include an emphasis on using memory and calldata over on-chain storage, merging functions together to reduce the number of execution paths, using events for data transmission to log information instead of storage, and using tighter data types like uint8 vs uint256 where possible. Moreover, there is a growing number of tools offered that range from Solidity analyzers and gas profilers to and automated code audits that developers can use to identify gas-heavy blocks of code and suggest optimizations. These innovations are important for overall scalability and acceptance of blockchain applications, particularly in industries with cost-sensitive transactions and interactions with smart contracts (e.g., real estate, finance, supply chain) where smart contracts and frequent user interaction in a transaction process is the norm. In conclusion, by optimizing gas costs, developers can design less costly decentralized applications (dApps) resulting in wider and easier user experiences across blockchain ecosystems.

Among essential techniques are reducing on-chain storage by utilizing memory and calldata, consolidating function calls to limit paths of execution, opting for events instead of storage as a means for logging data, and using more efficient data types of variables, such as uint8 instead of uint256, when applicable. The industry has also seen the emergence of tools such as Solidity analyzers, gas profilers, and automated code auditors to allow developers to spot gas-heavy blocks of code and plug optimizations, or at least point them to suitable adjustments. These are all important for the additional scalability and overall adoption of blockchain applications in industries where costs matter the most, such as real estate, supply chain, and finance, where multiple transactions and interactions with smart contracts are often needed. Gas optimization will result in more economically feasible distributed apps (dApps) and, in turn, broader access and smoother user experiences within blockchain ecosystems (Li et al., 2023; Nelaturu et al., 2021).

Statistical models have recently begun to play a bigger role in the development of blockchain-based applications in real estate and, as such, help create predicted analyses and investment strategy. As described in previous research, investment strategists have successfully utilized ARIMA models to predict shifts in property prices based on previous pricing trends. Using this method, you can forecast expected prices based on previous choices while ensuring you have data evidence. Additionally, ARIMA has been closely associated with blockchain technologies mainly due to its two determining factors—blockchain ensures the prices are immutable and visible. When you utilize ARIMA for your predictive forecasting, it can improve the tedious assumptions on where a price lands while improving your transparency level with the original documents. You also know you can trust the sources due to their visibility and immutability in the blockchain. All of this culminates to help shape informed investment strategy decisions to operate within an intelligent ecosystem of investment and development (Su et al., 2024).

Time series modeling strategies such as generalized autoregressive conditional heteroskedasticity (GARCH) and long short-term memory (LSTM) have been mentioned in the literature to augment real estate price forecasting models statistically and computationally, respectively. Utilizing these processes helps explain complicated temporal dependences and heteroskedasticity within the real estate price data set, and offers more complex insights into market engagement. GARCH models are typically used in modeling and forecasting time-varying volatility, and LSTM is often used to learn long-range dependencies in sequential data. The outcome of using each dimension of these methods has substantially improved the preciseness and dependability in real estate analytics, thereby increasing accuracy and reliability in terms of market forecasts to guide in more data informed decision-making for investors, developers, and public policy makers (Mutinda & Langat, 2024). Previous studies have explored the impact of statistical methods on the real estate sector, highlighting both the potential benefits and associated risks of adopting data-driven investment strategies (Seagraves, 2023).

Recent research emphasizes how crucial statistical models are to enhancing environmental, social, and governance (ESG) assessments, especially in the context of sustainable real estate development (Walacik & Chmielewska, 2024). The development of fully automated systems for property assessment has been fuelled by advancements in statistical analysis methods. These systems produce more accurate assessments of values and valuations, with noticeably fewer fluctuations over time by using predictive models (Gloudemans & Sanderson, 2024). By increasing data accuracy, cutting down on processing times, and eliminating the need for middlemen, the combination of blockchain technology with statistical methods has simplified real estate transactions (Jain & Anandan, 2024).

Recent research acknowledges the transformative impact of statistical valuation models on property appraisals using data-driven approaches to improve appraisal efficiency and accuracy (Topraklı, 2024). The combination of ARIMA models with alternative forecasting algorithms can break through old forecasting ceilings to achieve better accuracy. These hybrid models can take advantage of ARIMA's time series aspects of forecasting and machine learning with primary advantages of decision trees, support vector machines, and neural networks that allow for robust prediction through traditional data patterns as well as nonlinear relationships in the data (Wang, et al., 2019).

A comparative study of the ARIMA, Seasonal Auto Regressive Integrated Moving Average (SARIMA), and LSTM models has highlighted their respective capabilities for accomplishing forecasts of real estate market trends. The comparatively simple ARIMA and SARIMA models fit time series data with seasonality very well and provide reliable predictions based on relatively weak historical trends, for short horizons. The LSTM networks, consisting of recurrent neural networks (RNNs), are capable of modeling complicated and long-term dependencies in data and trends, and the LSTM model seems to be a more appropriate fit for overall prediction of more volatile and nonlinear mapping changes in the future. Thereby, the LSTM networks provide insights into and observations about changes in predicting market shifts. The study ultimately supports the notion that all three models provide usable insights. However, LSTM, which can handle large datasets and intricate patterns, offers the most data-driven forecasts for investors and can contribute to better decision-making related to the real estate market (Sirisha et al., 2023). The significance of the internal rate of return (IRR) in real estate investment has been widely studied, recognizing it as an important part of the property market's financial decision-making process, utilizing blockchain. As IRR is an important metric for investors, it helps them to think about an anticipated return on a real estate deal. Specifically, the IRR gives an investor a way to calculate the expected rate of return on an investment—a decision-making tool that can inform their decision about the purchase of a property, the development of a property, and the sale of a property. In terms of blockchain, IRR is now more relevant because the blockchain theoretically increases transparency and efficiency in real estate transactions. Blockchain can simplify several real estate processes, including title verification and transaction records, making transactions cheaper. Additionally, by mitigating information risk, blockchain increases the level of trust among participants involved in real estate transactions. Thus, the adoption of IRR analysis in the context of blockchain-supported real estate markets not only provides investors with the financial viability of their investment choices but also promotes a more trusted investment ecosystem. This produces an environment in which both IRR and blockchain technology enable more calculative and data-driven decision-making in the changing nature of real estate investment (Zhang, 2021).

Overall, the convergence of blockchain and statistical methods in real estate has introduced new possibilities for secure transactions, fraud prevention, and data-driven decision-making. Future research is expected to focus on optimizing smart contracts, enhancing valuation models, and refining predictive analytics for real estate investment. The ongoing developments in blockchain-based real estate platforms continue to redefine the industry, paving the way for greater transparency and efficiency.

3. Methodology

The proposed system combines blockchain and statistical forecasting to generate a secure data-oriented real estate investment analysis platform. The architecture consists of several parts and enables property transactions, data acquisition, and investment forecasting in a decentralized environment. Smart contracts running on a Hardhat blockchain, with the data being stored off-chain in a digital ledger, secure property transactions using verified details regarding the property's information, ownership history, and transaction history. The blockchain with external connected environments is connected using a Cloudflare Tunnel, while Google Colab provides code and data analysis tools for the user. For investment analysis, historical real estate data is used and statistical techniques for modeling time series data is applied, in this case using ARIMA (which shows the trend of future property prices). For investment summaries, extended internal rate of return (XIRR), which calculates the profitability of an investment over time, is used by the potential buyers and sellers, who input their expectations price points. Then the system makes an assessment on whether the property is undervalued, fairly valued, or overpriced to make a recommendation to buy, rent or not buy. Then the user can be involved in the system by using Google Colab and Web3.py for start the first block (data) input and subsequently, based on the verification by blockchain technology, to eliminate the risk of fraud. The proposed system is capable of utilizing a modular approach, whereby property transaction data is securely stored on-chain and the analysis of the data is off-chain to achieve maximum efficiency.

Overall, the architecture provides a transparent, decentralized, and data-driven approach to real estate investment, leveraging blockchain for security and forecasting models for intelligent decision-making.

The following are the six sub-components that collectively form the methodology:

1. User Interaction
   - Buyers/sellers interact with the system through a web interface or Google Colab.
   - Users input property details and expected future price.
2. Smart Contract Execution
   - Property transactions are initiated and recorded on the Hardhat blockchain.
   - Smart contracts ensure secure, immutable, and transparent transactions.
3. Data Retrieval from Blockchain
   - Property records and past transaction details are fetched using Web3.py.
   - Cloudflare Tunnel exposes blockchain data to external analysis tools.
4. Statistical Forecasting
   - ARIMA model predicts future property price trends based on historical data.
   - XIRR calculation evaluates investment profitability over time.
5. Decision-Making & Recommendation
   - The system analyzes whether the property is undervalued, fairly valued, or overvalued.
   - It generates investment advice: buy, rent, or avoid.
6. Final Output & Reporting
   - Users receive a detailed investment report with forecasted price trends and financial insights.
   - Blockchain records remain accessible for future verification and audits.

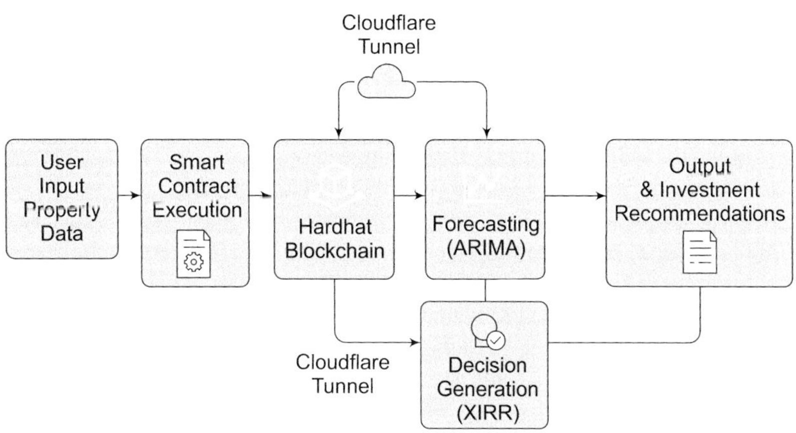

Fig. 15.1 Workflow of research

3.1 Blockchain Implementation

The blockchain execution of this entire system is secured, transparent, and immutable property transactions anchored in smart contracts relying entirely on Hardhat. The project is methodically constructed by carefully integrating each part of the system to securely handle and manage the data regarding the property. Smart Contract Deployment Smart contracts are deployed using Hardhat, which managed by deploy.js to deploy each contract and manage property transactions:

- Initializes and deploys the smart contract on the local Hardhat blockchain.
- Defines functions for adding property details, storing transaction history, and retrieving property records.
- Ensures only authorized users can add and manage properties.

Contract Interaction once deployed, the contract is connected using connectContract.js, which:

- Establishes a Web3 connection to interact with the deployed contract.
- Enables seamless execution of contract functions.

Adding Property Data Users interact with addProperty.js to register properties on the blockchain:

- Property details (location, price, owner, etc.) are stored securely.
- A blockchain event is emitted, which is captured by a localhost server.

Event Handling and External Connectivity

- A localhost server listens for property events in real time.
- Cloudflare Tunnel is used to make the local server accessible over the internet, enabling Google Colab to fetch blockchain data.

Google Colab Integration

- The system connects Google Colab to the blockchain using Web3.py.
- The contract address is used to fetch transaction data.
- Events from the localhost server are retrieved and processed for analysis, enabling Google Colab to fetch blockchain data.

3.2 Structuring Data for ARIMA AND XIRR

Once the real estate market data was obtained from a multitude of different sources the next step was to clean and contextualize the dataset. Cleaning the dataset is crucial to ensuring data correctness before using any predictive models such as ARIMA to forecast price and using XIRR to analyze investment returns. The property dataset typically includes elements of incorrect data, missing data, outliers, and differences in data formatting that may reduce the reliability of the analysis. Thus, the data was systematically pre-processed including dealing with missing data, dealing with different formatting, removing outliers, calculating derived metrics and organizing the dataset into a time-series with any intervening staff to permit forecasting and investment analysis. The first part of cleaning the data involved dealing with missing data. The dataset had missing data for several fields: average price per sq/ft, housing price index (HPI), growth rate, and rental price per sq/ft per year. In order to impute missing data a variety of strategies were employed. Some of the strategies employed for missing data include forward filling and backward filling methods, two methods where missing data is replaced with the nearest data point, where the linear interpolation method was used to impute vacant values depending on the trend of the records from previous years and the next consecutive years. No estimates were made for missing rent values, and all missing rents were replaced with only the median of cities. Actual market data was checked for missing value to ensure proper data fit with respect to raw market information from in-person government reports and property portals. Next, the outlier detection and removal were performed to ensure extreme fluctuations do not skew predictive models. The outlier detection process for property pricing and rental values used a combination of statistical and analytical techniques. Three primary methods were applied:

i. Z-score method: Values beyond ±3 standard deviations from the mean were flagged as potential outliers.
ii. Interquartile Range (IQR) method: This technique identified data points falling outside the acceptable range defined by the lower and upper quartiles.
iii. Visual inspection: Extreme price changes were manually reviewed in the context of historical events. Data anomalies caused by genuine market factors—such as policy changes, economic conditions, or exceptional events—were retained, while anomalies arising from random errors or incorrect entries were removed.

To maintain consistency across differing data sources, the formatting of data was standardized. All property prices were converted into INR per square foot, as some sources included values in square meters. A unified year-time series date format was established, in a YYYY year format, to be consistent with other forecasting models. City and state names were standardized to avoid any inconsistencies e.g., "Bengaluru" instead of "Bangalore." Numeric precision was normalized by rounding figures to 2 decimal places; this was done to keep calculations as consistent as possible. The compound annual growth rate (CAGR) was calculated using the formula:

$$\text{CAGR} = \left(\frac{V_f}{V_i}\right)^{\frac{1}{n}} - 1 \qquad ...(1)$$

where V_{final} (V_f) is the latest recorded price per square foot, V_{initial} (V_i) is the price at the beginning of the period and n represents the number of years. CAGR provided an indication of how consistently property values appreciated over time. Another critical metric was the XIRR, which measured the annualized return on real estate investments by incorporating both property price appreciation and rental income. It was calculated using the relationship:

$$\text{XIRR} = \text{CAGR} + \left(\frac{\text{Average Rent}\left(\text{INR Per Sq. Ft Per Year}\right)}{\text{Average Price per Sq. Ft}}\right) \qquad ...(2)$$

where real estate profitability beyond price appreciation alone was reflected by factoring rental yield into the total return.

A normalized HPI series was employed to enable cross-city comparisons, and the HPI was computed for numerous points in time to examine weights over an extended period. In order to assess investor attractiveness, the rental yield was also examined by calculating the annual rent income to price ratio. The dataset, which includes pricing information for each city for every year, was arranged in a time-series fashion for price forecasting ARIMA models.

The XIRR model was employed to calculate long-term returns on real estate investments. The ARIMA is widely used to calculate long-term returns on real estate investments and plays a crucial part in projecting future real estate values based on historical data or patterns.

The ARIMA model has three components

- **Autoregressive (AR) component (p):** This represents the effects of previous values on the present value.
- **Integrated (I) component (d):** This is the differencing done with the time series data to achieve stationarity of the data.
- **Moving average (MA) component (q):** This accounts for past forecast errors when making future predictions.

Each city had its own unique ARIMA model defined on the basis of historical property price data that explored the city. Before training an ARIMA model, the data had to be pre-processed with the column of "year" being set to a date time index to properly define the sequence for time-series forecasting. Because ARIMA models require stationary data, the next step was to conduct the augmented Dickey–Fuller (ADF) test for checking stationarity. If the series was indicated as being non-stationary, differencing was performed until stationarity was achieved. The values of (p, d, q) were then examined. Many possible configurations of (p, d, q) were evaluated based on different statistical tests including the Akaike information criterion (AIC) and the Bayesian information criterion (BIC). A grid search technique was used to test many values of (p, d, q), and to select the (p, d, q) configuration that minimized either the AIC or BIC values and thus represented the best model fit. For many cases, the ARIMA model of (3, 1, 3) was identified as providing a good fit model for predicting historical property prices of cities in India.

To ensure a robust evaluation, the dataset is split into training and testing sets. Typically, an 80–20 split is used, meaning 80 and 20 historical price trends, while the test set provides unseen data to evaluate how well the model generalizes to new information. We can evaluate the model's performance using a variety of error assessment methodologies by contrasting the test set's real prices with the predicted values. The following performance metrics were used for evaluation:

- **Mean Absolute Error (MAE):** The average absolute difference between actual and expected real estate prices is measured by this indicator.
- **Root Mean Squared Error (RMSE):** The square root of the average squared difference between actual and forecasted values is determined by the RMSE.
- **Mean Absolute Percentage Error (MAPE):** MAPE facilitates interpretation across a range of price points by expressing the prediction error as a percentage.

XIRR provides a more precise risk-reward evaluation by incorporating rental yields, enabling investors to efficiently evaluate various cities and property types. While a lower XIRR could signify poorer returns or greater risks, a higher XIRR indicates a more profitable venture.

3.3 Integration of Blockchain and Price Prediction

Automated real-time updates are guaranteed by the smooth interaction of these three parts. The system updates its data, improves ARIMA forecasts, and recalculates XIRR whenever a real estate transaction is entered into the blockchain. As a result, an intelligent investment system always adjusting to market movements is produced. Furthermore, smart contracts minimize transaction delays and dependency on middlemen by automatically executing transactions depending on predetermined financial circumstances.

4. Results and Discussion

Table 15.1 Comparison of traditional and blockchain methods.

Aspect	*Traditional Methods*	*Blockchain-Based Methods*
Speed	Slow: Weeks to months due to manual verifications, intermediaries, and paperwork.	Fast: Smart contracts enable near-instant transactions.
Cost	Expensive: High fees for brokers, lawyers, notaries, etc.	Cost-effective: Reduces legal and administrative costs.
Security	Vulnerable: Prone to fraud, forged documents, and hacking.	Highly secure: Immutable ledger and cryptographic encryption prevent fraud.

The accuracy of ARIMA predictions is provided in Figure 15.2.

```
Enter city name: chennai
📊 **Performance Metrics**
- Mean Absolute Error (MAE): 438.21
- Root Mean Squared Error (RMSE): 497.67
- Mean Absolute Percentage Error (MAPE): 1.83%
```

Fig. 15.2 Performance metric of ARIMA model.

Table 15.2 Performance metrics of the predictive model along with the corresponding accuracy where applicable.

Metrics	*Error*	*Accuracy*
Mean Absolute Error	438.21	–
Root Mean Squared Error	497.67	–
Mean Absolute Percentage Error	1.83%	98.17%

Key forecasting error metrics, which quantify the difference between expected and actual values for real estate prices, are used to assess the ARIMA model's

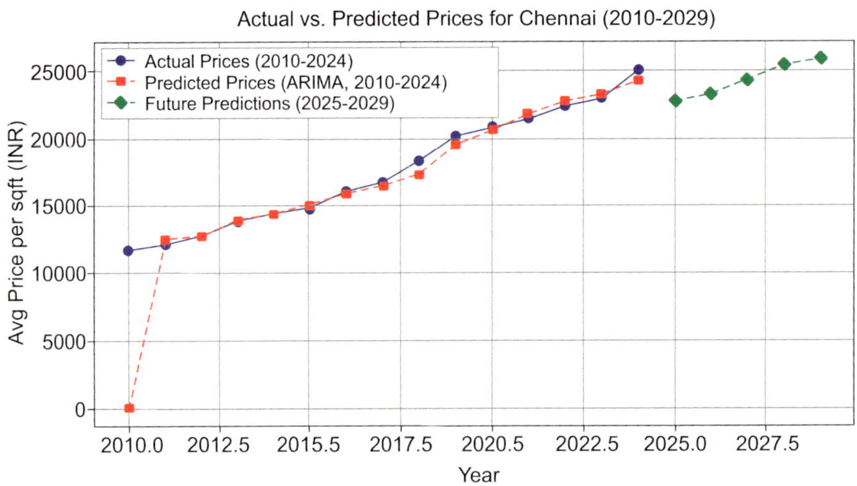

Fig. 15.3 Actual vs. predicted prices for Chennai (2010–2029).

accuracy. MAE, RMSE, and MAPE are the three main metrics used to evaluate the model's performance. The model produced an MAE of 438.21, an RMSE of 497.67, and an MAPE of 1.83 for the city of Chennai. The ARIMA model's ability to track past price trends and forecast future values is visually shown in Figure 15.3. The ARIMA model can be improved by changing its parameters or adding more economic data after examining predicting accuracy across several cities. These enhancements contribute to the model's continued stability for predicting long-term real estate prices. For Chennai, the MAE was 438.21 and MAPE was 1.83, meaning the predicted price per square foot deviates by an average of 438.21 INR from the actual price. The RMSE was 497.67, showing a small deviation from actual prices. With few variations, the data verifies that the ARIMA model offers a very accurate price forecast.

5. Conclusion

This study successfully combined blockchain technology with statistical forecasting models to develop a system for property transactions that can be considered safe, transparent, and effective. Fraud, poor processing, exorbitant transaction costs, and erratic property values are some of the problems with traditional real estate transactions. The suggested method addresses these problems by utilizing ARIMA for property price forecasting and smart contracts for safe property management. ARIMA models assist investors and buyers in making data-driven decisions by forecasting future real estate prices based on past trends. This method guarantees that real estate transactions are quick, verifiable, and fraud-proof in addition to providing precise market forecasts for improved investment planning,

The study effectively combines statistical forecasting (ARIMA) models for real estate transactions with blockchain technology. Future research should look into hybrid blockchain architectures for efficiency, Layer 2 scaling solutions to lower transaction costs, and enhanced forecasting models that use real-time data. To ensure legal recognition and compliance, regulatory frameworks must change to accommodate transactions based on smart contracts.

References

Agarwal, J.D. and Aman Agarwal. "Money Laundering: The Real Estate Bubble." Finance India 22, no. 1 (2008): 57.

Ammous, S. (2016). Blockchain technology: what is it good for? *SSRN Electronic Journal.*

Damaševičius, R., N. Bacanin and A. Nayyar. (2024). Blockchain technology for a trustworthy social credit system: implementation and enforcement perspectives, *Cluster Computing*, *27*(1), 45–62.

Faxon, H.O and C.T. Wittekind. (2-23) Livestreamed land: scams and certainty in Myanmar's digital land market, *Environment and Planning D: Society and Space*, *42*(1), 78–96.

Gloudemans, R. and P. Sanderson. (2024). The potential of artificial intelligence in property assessment, *Journal of Property Tax Assessment & Administration*, *18*(2).

Jain, S. and P. Anandan. (2024). Real estate market in the digital age: a review of technology trends and advancements, *IEEE Xplore.*

Li, C. (2021). Gas estimation and optimization for smart contracts on Ethereum, *IEEE Xplore.*

Li, J., Z. Zhao, Z. Su and W. Meng. (2023). Gas-expensive patterns detection to optimize smart contracts, *Applied Soft Computing.*

Liu, A. and Chen, C. (2025). From real estate financialization to decentralization: A comparative review of REITs and blockchain-based tokenization. *Geoforum*, 159, p.104193.

Mutinda, J.K. and A.K. Langat. (2024). Stock price prediction using combined GARCH-AI models, *ScienceDirect.*

Nelaturu, K., S.M. Beillahi, F. Long and A. Veneris. (2021). Smart contracts refinement for gas optimization, *IEEE Xplore.*

Pacheco, M., G.A. Oliva, G.K. Rajbahadur and A.E. Hassan. (2022). What makes Ethereum blockchain transactions be processed fast or slow? An empirical study, *Empirical Software Engineering*, *27*(6), 112–134.

Rapezzi, M., G. Pizzi and G.L. Marzocchi. (2024). What you see is what you get: the impact of blockchain technology transparency on consumers, *Marketing Letters*, *35*(1), 89–107.

Sankar, M., Pachiyappan, S., Paramasivan, S. and Srivastava, A. (2023). Adoption of blockchain technology in the real estate sector toward the improvement of smart cities. *In Green Blockchain Technology for Sustainable Smart Cities* (pp. 187–198). Elsevier.

Seagraves, P. (2023). Real estate insights: is the AI revolution a real estate boon or bane, *Journal of Property Investment & Finance.*

Sirisha, U.M., M.C. Belavagi and G. Attigeri. (2023). Profit prediction using ARIMA, SARIMA, and LSTM models in time series forecasting: a comparison, *IEEE Xplore.*

Su, M., Z. Su, S.-H. Bae, J. Li and K.-s. Park. (2024). An exploration of shipbuilding price prediction for container ships: an integrated model application of deep learning, *ScienceDirect.*

Topraklı, A.Y. (2024). AI-driven valuation: a new era for real estate appraisal, *Journal of European Real Estate Research.*

Verma, V., S. Priya, S. Mishra, R. Priyadarshini and R. Misra. (2021). Property fraud detection and prevention using blockchain, in International Conference on Emerging Smart Computing and Informatics (ESCI), *IEEE Xplore*, 344–349.

Walacik, M. and A. Chmielewska. (2024). Real estate industry sustainable solution (environmental, social, and governance) significance assessment—AI-powered algorithm implementation, *Sustainability*, *16*(3), 1079.

Wang, F., Y. Zou, H. Zhang and H. Shi. (2019). House price prediction approach based on deep learning and ARIMA model, *IEEE Xplore.*

Zhang, L. (2021). December. The Review for the Development of IRR's Implication. In 2021 3rd International Conference on Economic Management and Cultural Industry (ICEMCI 2021) (pp. 1770-1774). Atlantis Press. 10.2991/assehr.k.211209.286

Zhang, L., L. Ci, Y. Wu and B. Wiwatanapataphee. (2023). The real estate time-stamping and registration system based on Ethereum blockchain, *Journal of Blockchain Technology and Applications*, *4*(2), 88–105.

CHAPTER 16

Adaptive and Context-Aware Intelligent Transportation Systems Toward Human-Centered Mobility for Society 5.0

M. Rajesh[1]*, *Vengatesan Krishnasamy*[2] and *Sayyad Samee*[3]

This chapter introduces a flexible and context-aware intelligent transportation system (ITS) architecture to support human-centered mobility in Society 5.0. The proposed architecture uses cyber-physical sensors, adaptive decision-making modules, context inference engines, real-time edge-cloud computing, and sustainable, efficient, and personalized transportation services. The system uses deep learning models and reinforcement learning agents to adapt to real-time changes in user preferences, environment, and infrastructure. The trial for this research showed significant improvements over traditional ITS models, including a 22.7 percent decrease in average route time, a 33.3 percent reduction in CO_2 emissions, and increased user satisfaction in comfort, equity, and accessibility. Moreover, the technology reduced the incident response time by 32 percent and pedestrian safety incidents by 45 percent, proving its practicality and societal impact. This architecture might be integrated with mobility-as-a-service (MaaS) platforms to expedite multimodal trip planning and centralize transportation resources. This convergence will enable inclusive, intelligent, and flexible urban transportation ecosystems that promote personalization, accessibility, and operational efficiency.

[1] Department of Computer Science and Engineering, Aarupadai Veedu Institute of Technology, Vinayaka Mission's Research Foundation (DU), Tamil Nadu, India.

[2] Department of Computer Science and Engineering, School of Engineering, Dayananda Sagar University, Bangalore 562112 Karnataka, India.

[3] Department of Computer Science, Al Sadara School, Al Ain, Abu Dhabi, United Arab Emirates.
https://orcid.org/0000-0003-0003-473X[1]; https://orcid.org/ 0000-0002-8896-4925[2]; https://orcid.org/0009-0003-9347-6867[3]

* Corresponding author: rajesmano@gmail.com

DOI: 10.1201/9781003659822-16

1. Introduction

Systems based on cyber-physical-human values are replacing information-driven paradigms in this fast-changing digital era. Society 5.0, the Japanese government's vision of a super-smart society, leverages artificial intelligence (AI), the internet of things (IoT), and robots, not merely for industrial efficiency, but to address social issues and promote human well-being (Ullah et al., 2024). Society 5.0 emphasizes human-centric innovation to balance economic development and social challenges, whereas Industry 4.0 emphasizes industrial automation and connectivity. The transport system is crucial to this strategy, which requires major infrastructural changes.

Environmental issues, car density, and rising urbanization are straining urban transportation networks like never before. Traditional ITS aids route planning, congestion reduction, and safety (Jadhav et al., 2025). These information and communication technology (ICT) systems optimizes traffic. Traditional ITS models do not always account for varied conditions or respond to changing dynamics in real time. Many intelligent traffic signal management systems overlook pedestrian congestion, weather, and real-time road conditions, resulting in inefficiency and customer displeasure (Alsamhi et al., 2024).

Most ITS systems prioritize infrastructure above people, making them unsuitable for human requirements and habits. Smarter mobility is part of Society 5.0's focus on durable, personalized, inclusive, and morally consistent systems that serve individual and social aspirations (Saha, 2025). The focus has shifted from traffic throughput to user experience and societal value. Rethinking ITS design with environmental awareness, human purpose comprehension, and real-time adaptability is needed.

Adaptable, aware ITS will be the next mobility intelligence. These systems employ contextual data including time, location, user behavior, weather, and traffic events to adapt to current conditions. A context-aware navigation system may suggest micro-mobility or public transit based on traveler health, air quality, and real-time congestion (Mahanta, 2025). Adaptive traffic lights can change their signaling patterns based on pedestrians, emergency vehicle route, and local events. This flexibility advances human-centered transportation while boosting efficiency, safety, and sustainability.

To build and execute adaptive, context-aware ITS, many methodological and technological challenges remain. First, dynamic and diverse situations, context inference requires robust machine learning models that can handle incomplete, noisy, and multimodal data (Fernandez-Rojas et al., 2019). Second, real-time ITS decision-making requires low-latency edge computing infrastructures to decrease delays and enhance responsiveness (Gadekallu et al., 2024). Third, while using sensitive personal data to customize mobility services, privacy and ethics must be considered. Apart from tech-savvy city people, these smart devices should be available to underprivileged populations including the elderly, disabled, and digitally excluded (Xu, et al., 2025).

For transportation to reach the goal of Society 5.0, ITS need to go beyond static optimization and adopt AI that can adapt to the ever-changing interactions between people, cars, and the environment. Connecting technologies like edge-AI, digital twins, federated learning, and human-in-the-loop frameworks may drive this change. In order to promote efficiency, safety, and equality on both the individual and societal levels, it is desirable to establish transportation ecosystems that are self-optimizing, contextually sensitive, and morally anchored (Magade & Sharma, 2025).

Finding a way to connect where ITS is at the moment with the ideals of Society 5.0 is the driving force for this study. Many smart city projects and transport networks that are connected to the IoT have been implemented. However, most of these systems are still rule-based, lack personalization, and cannot scale very well. Investigating approaches and designs that let transport systems learn from human interactions, adjust in real time, and behave according to societal norms and personal preferences is an urgent matter.

Adaptive and context-aware ITS are thoroughly examined in this chapter within the larger context of Society 5.0. The major goal is to lay forth a set of basic principles and a conceptual framework for creating ITS that are intelligent in their environment, adaptable to their surroundings, and focused on people. We want to prove that cyber-physical systems can improve urban transportation by using new developments in AI, ubiquitous sensing, and related fields.

This study makes the following contributions:

- We provide a conceptual framework for context-aware ITS that incorporates sensing, contextual reasoning, adaptive planning, and human feedback mechanisms into a layered architecture.
- Environment, behavior, and situational data are all part of the multimodal context data that is pertinent to mobility, and we offer methods for collecting and understanding this data.
- Utilizing a case study encompassing real-time multimodal journey planning and traffic adaption, we demonstrate the framework's applicability in practical urban contexts.
- Concerning the societal implications, we examine the difficulties in implementing these systems in terms of accessibility, ethics, and scalability, and we provide policies and rules for design that are in keeping with the concepts of Society 5.0 (Van Kasteren & Vredenborg, 2022).

The study discusses how transport systems must adapt to meet the changing demands of a complex environment. The authors argue for a socially responsible and technologically sophisticated mobility strategy that is intelligent, people-centric, and context-aware. Looking at transport through this perspective transforms it into a service that mirrors and improves the quality of life in the Society 5.0 era—a seamless, adaptable, and inclusive one.

2. Background

When it comes to traffic monitoring, signal control, and route optimization, traditional ITS have relied on centralized infrastructure-based solutions. While these early ITS implementations did a decent job for the most part, they were not always adaptable enough to deal with unpredictable, real-world situations like weather delays, accidents, or erratic traffic patterns (Lamichhane et al., 2025). To get around these problems and make traffic management more effective and reactive, smart ITS emerged, which uses real-time data collecting and sophisticated communication protocols. The growth of transportation networks was greatly enhanced by inventions such as vehicle-to-infrastructure (V2I) and vehicle-to-vehicle (V2V) communication, dynamic route guiding systems, and GPS-based tracking (Bamakan et al., 2025).

Adaptive systems that employ machine learning, control theory, and sensor fusion to adapt to changing transportation conditions are popular. Adaptive traffic signal control systems may adjust light cycles to flow densities, improving peak-hour congestion management (Huang et al., 2025). Embedded intelligence improves car performance and safety with lane-assist and adaptive cruise control. Most adaptive systems do not include user preferences, environmental circumstances, or multimodal mobility beyond a particular job (Chang et al., 2017).

Meanwhile, context-aware computing has created new paradigms for smarter, user-centric transport systems. Context-aware ITS can dynamically detect and adapt to its surroundings using AI for behavioral modeling, IoT networks for real-time sensing, and edge computing for low-latency data processing. AI-driven demand prediction models can improve public transport timetables using data on usage patterns across time and location, while edge-based video analytics can detect pedestrian movement and adjust traffic flow in real time (Sankaran et al., 2025; Sohag et al., 2023). These technologies are promising, but scalability and interoperability are issues, especially in various urban environments with different infrastructure development levels.

Human-centric mobility goes beyond ITS to prioritise user inclusion, accessibility, safety, and enjoyment. Society 5.0 envisions intelligent, compassionate transport systems that meet user needs, and this viewpoint supports that objective. Personalized route suggestions, context-aware mode selection (e.g., walking instead of utilizing a rideshare based on health), and systems that help vulnerable populations like the elderly or disabled are further examples (Lu et al., 2024). The human-in-the-loop concept, which lets computers learn from users and improve, is also popular. However, most existing ITS models either do not engage users or see them as passive observers in the transportation network (Muzahid et al., 2024).

While adaptive and context-aware technologies have been studied separately, their integration into a unified, human-centered ITS framework is still in its early stages, according to the reviewed literature. The majority of research only looks at individual parts of the puzzle, including real-time sensing, adaptive traffic control, or optimization for ride-hailing services, but none of these parts are ever considered in a comprehensive way inside an architecture that puts user experience and societal values first. Sustainability, inclusiveness, and digital equity are some of the larger aims of Society 5.0, but there has been little research on how these interconnected systems may fit in with these bigger objectives (Sarwatt et al., 2024; Ystgaard et al., 2023). To fill this demand, this chapter presents an integrated, flexible, and context-aware ITS architecture that prioritizes people, responds quickly to changes in the environment, and paves the way for the creation of intelligent transportation networks.

3. Framework

In order to facilitate human-centered mobility within the framework of Society 5.0, an adaptive and contextually aware ITS architecture is being developed. The design is based on a multi-layered cyber-physical-human system that constantly detects data about the environment and the user, deciphers signals based on context, and adjusts transport services on the fly to maximize user pleasure while simultaneously improving system performance. In order to provide sustainable, personalized, and real-time mobility services, the system relies on a coordinated effort of physical sensors, computational layers at the edge and in the cloud, context inference engines, and adaptability modules.

The data sensing and gathering module is the system's base layer; it bridges the gap between the real and virtual worlds. Everything from cars and roads to traffic lights and public transportation systems—and even users' own wearable devices—includes a broad variety of sensors that can connect to the internet. Vehicle speed, traffic conditions, pedestrian density, air quality, temperature, and biometric input from users are just some of the real-time data collected by these sensors. Continuous spatiotemporal tracking is provided, for instance, by global positioning systems and inertial measurement units; traffic flow patterns and possible dangers are detected, for instance, by computer vision systems. Better situational awareness is another benefit of vehicle-to-everything (V2X) communication technologies, which allow for the smooth transfer of data between cars and infrastructure.

This sensor data is pre-processed and partially analyzed at the edge computer layer. Our low-latency processing ensures timely judgements. For real-time signal modifications, adaptive traffic light systems need millisecond computer power. The edge layer uses lightweight machine learning models for noise filtering, feature extraction, and early inference. Cloud computing allows larger models to access previous data and more powerful computational resources, making it perfect for offloading computationally expensive or time-sensitive tasks like traffic pattern analytics or long-term behavior prediction. The edge-cloud cooperation optimizes ITS design for latency, scalability, and energy efficiency.

ITS design requires middleware like the context engine. Data from multiple sources is gathered and consolidated to show the operational environment at any moment. Any available information can characterize a person, vehicle, or traffic node's transportation context. The context engine employs probabilistic models like Markov decision processes and Bayesian networks to reason about uncertainty and provide contextual insights. The engine may adjust pedestrian signal timing based on regional data, profile preferences, and wearable sensor inputs to detect a wheelchair user approaching an intersection.

The adaptability module uses context engine outputs to make real-time judgements. This module employs reinforcement learning, multi-objective optimisation, or rule-based logic to discover the best solution for user enjoyment, safety, environmental impact, and system efficiency. The existing context $C(t)$, expected user behaviour $B(t)$, and system goals determine the adaptation choice $A(t)$. Formalized G:

$$A(t) = \text{argmax } E[U(a, C(t), B(t), G)] \qquad \text{...(1)}$$

The possible actions are symbolized by A, and the utility function evaluating performance across objectives is $U(\cdot)$.

Adaptability requires modeling and forecasting user behavior to anticipate and meet requests. User modeling involves developing profiles that incorporate the user's likes and dislikes and any limits, such as a budget, time constraint, or disability. These models are enhanced utilizing supervised learning and online learning algorithms using data from mobile applications, travel cards, and wearable devices. Travel behavior may be represented over time using an recurrent neural network (RNN) or long short-term memory (LSTM) network:

$$B_{t+1} = f(B_1, B_2, \ldots B_t; \theta) \quad \ldots(2)$$

where B_{t+1} the model parameters that were trained on prior behavior from B_1 to B_2 are represented by θ, and $t + 1$ is the projected user behavior at the following time step.

An important factor in deciding how a system should behave optimally is the user's preferences. Choosing environmentally responsible travel over more convenient options is one example of how a consumer may value time more than money. Using a multi-objective utility function, these preferences can be expressed:

$$U = \alpha_1 \cdot U_{\text{time}} + \alpha_2 \cdot U_{\text{cost}} + \alpha_3 \cdot U_{\text{eco}} + \alpha_4 \cdot U_{\text{comfort}} \quad \ldots(3)$$

where α_i is a weight that may be found in user profiles or feedback and U_{time}, U_{cost}, U_{eco}, and U_{comfort} are variables that can be defined as values between 0 and 1.

Here are the separate utility ratings for time efficiency, economic cost, environmental effect, and comfort.

Both static and dynamic key context factors across many domains are taken into account by the proposed ITS design. Factors such as weather, light, and air quality are examples of environmental factors. Other examples include spatiotemporal variables like road geometry, location, and time of day. User-specific parameters like age, mobility status, and preferences are examples of user-specific parameters. Meanwhile, real-time system conditions include things like congestion, incidents, and the availability of transport modes. All of these factors are considered together, rather than separately, which allows for a more complete picture and more tailored assistance. Figure 16.1 represents the architecture flowchart of the proposed framework.

For instance, when it is raining, visibility is low, and traffic is heavy, the context engine can assume that there's a higher chance of danger and provide alternative less crowded routes or modes of transportation, such taking public transportation instead of riding a bike. To further accommodate the user's age, the system may prioritize modes that require less physical effort and raise walking time buffers. Rather than being dictated by predetermined rules, these adaptive behaviors are created by a process of continuous learning through feedback loops and interactions between many agents. In this model, every node in the infrastructure, as well as every vehicle and user, functions as an intelligent agent that works together to optimize urban mobility.

A transportation system that is both efficient and sensitive to human requirements may be orchestrated by integrating cyber-physical sensing, contextual intelligence, and adaptive decision-making within the conceptual framework. By incorporating human-centered design, technological adaptability, and ethical responsiveness into the core of transportation networks, this architectural vision is in line with Society 5.0's guiding principles.

4. Proposed Methodology

The recommended system design method integrates human-centered adaptability with context-aware intelligence to produce a cohesive framework that can make real-time

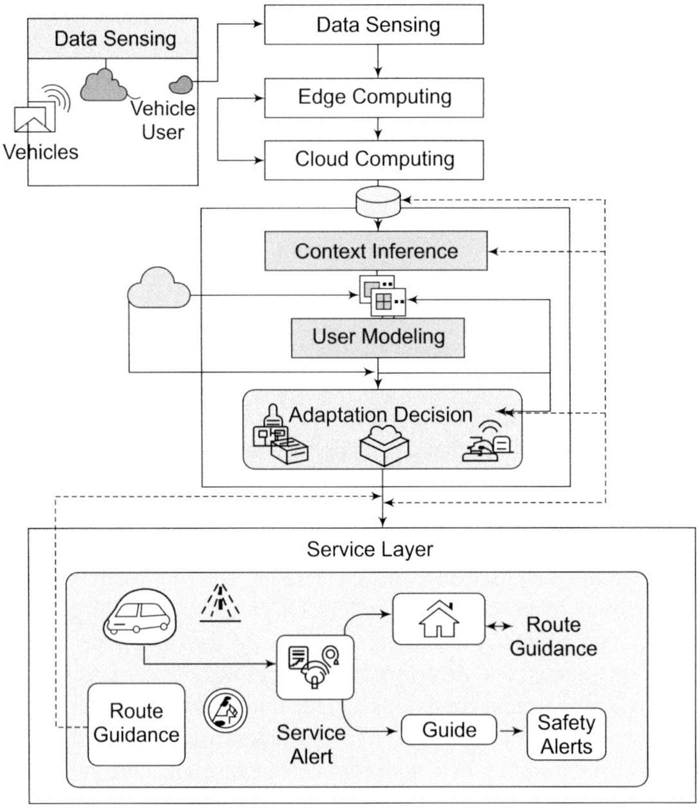

Fig. 16.1 Proposed framework architecture.

decisions and dynamically personalize the system. This system relies on machine learning (ML)/deep learning (DL) context inference. It uses contextual signals including user behavior patterns, environmental cues, mobility data, and interaction histories to understand user intent, system state, and environmental behavior. The architecture that performs this inference is a combination of two types of neural networks: convolutional neural networks and LSTM networks. While the LSTM analyses temporal sequences to collect behavior trends—for example, navigation patterns or periodic service requests—from static or semi-static inputs like maps, user position, or interface activity, the convolutional neural network (CNN) extracts spatial information from these inputs. These models work in tandem to acquire the contextual embeddings that power the adaptive response of the system.

The system's actions are controlled by the adaptation strategy once context has been inferred. A system with two levels is implemented. Tier one uses rule-based logic to give predictable answers to high-confidence contextual triggers, making the system trustworthy and easy to understand in familiar situations (for instance, if a user consistently ignores a certain kind of notice in a certain context, it will be hidden). On a deeper level, decision spaces that are more intricate, uncertain, or ever-changing are handled by a reinforcement learning (RL) agent that has been taught using the

Q-learning algorithm. The RL agent maximizes utility in the long run by optimizing an adaptive policy $\pi(s)$ over the set of states s, with rewards calculated according to criteria that measure user satisfaction, including the efficiency of interactions, the duration of engagement, or explicit feedback. Based on the famous Bellman equation, the Q-learning update is as follows:

$$Q(s, \mathrm{a}) \leftarrow Q(s, a) + \alpha[r + \gamma \max Q(s', a') - Q(s, a)], \qquad ...(4)$$

where s stands for the present state, a for the action, r for the reward, s' for the resultant state, α for the learning rate, and γ for the discount factor.

The system uses feedback loops for fast reaction and model improvement. These loops include passive and active user input. Online learning and re-training depending on implicit and explicit inputs, such as task completion time and correction frequency, fine-tunes the models. In important phases like RL model fine-tuning and incentive shaping, human-in-the-loop design is prioritized. This strategy allows human assistance in critical decisions or uncertainty while assuring the system's adaptation meets user expectations and ethical norms.

5. Results and Analysis

We compare the recommended adaptive and context-aware ITS to baseline models and evaluate its performance. Interpretative data analysis highlights efficiency, user experience, and social benefits such mobility access equity. This multifaceted analysis is based on user surveys and empirical simulations.

The TAPAS Cologne mobility aataset is used by the system to mimic realistic user mobility, transport network architecture, and dynamic context changes in a smart urban environment. This dataset is taken from real urban infrastructure and traffic data from Cologne, Germany. Various transportation modes, departure times, routes, and contextual information like traffic signals and junction concentrations are included in this dataset, which contains comprehensive trajectories of almost 700,000 individual vehicle journeys every day. Validating the performance of context-aware modules in high-density urban contexts and training adaptive routing systems both require such depth.

5.1 Gathering Information for Training and Simulating Models

The analysis relies on the following training and simulation corpora:

- Data used to train algorithms for rerouting and journey time prediction, including both actual and synthetic trajectory logs.
- Metadata pertaining to context: trends in the weather, availability of services, levels of congestion, and upgrades to road infrastructure.
- Preference vectors that include the following: commuting patterns, time-flexibility ratings, and mobility demands (elderly, differently abled).
- Emissions by industry, fuel consumption by vehicle type, and CO_2 production are some measures of environmental effect.

The system guarantees strong model generalization across critical and ordinary circumstances by combining these various inputs. By focusing on data, we can create an accurate model of the adaptive behavior that an ITS orientated towards Society 5.0 would need to display, which is to dynamically adjust to changes in the environment while giving top priority to user demands, safety, efficiency, and sustainability.

First, we used three methods—a static routing system as a baseline, an adaptive ITS, and the suggested context-aware ITS—to look at trip time efficiency throughout the day. As we progress from baseline to adaptive and eventually to context-aware systems, the comparison of hourly journey times, as illustrated in Figure 16.2, clearly shows a declining trend in transit time. According to Table 16.1, the average time spent using the baseline, adaptive, and context-aware systems was 34.26, 29.52, and 26.47 minutes, respectively. The reason behind these enhancements is the implementation of dynamic decision-making that takes into account environmental elements, user intent, and real-time traffic circumstances.

Table 16.1 Comparative average travel times.

Scenario	*Avg. Travel Time (mins)*
Baseline	34.26
Adaptive	29.52
Context-Aware	26.47

From a green point of view, there was a significant decrease in fuel usage and CO_2 emissions. The corresponding reductions for each model are shown in Table 16.2. The basic model achieved a CO_2 production of 150 g/km and a fuel consumption of 7.5 L/100 km. These figures were reduced to 120 g/km and 6.2 L/100 km, respectively, via adaptive systems. With the lowest emissions of 100 g/km and the fewest fuel consumption of 5.3 L/100 km, context-aware systems fared the best. With these cuts, these systems show how they may help make transit more sustainable, which is in line with Society 5.0's goal of environmental harmony.

Table 16.2 Emission reduction analysis.

Scenario	*CO_2 Emissions (g/km)*	*Fuel Consumption (L/100km)*
Baseline	150	7.5
Adaptive	120	6.2
Context-Aware	100	5.3

Table 16.3 User satisfaction survey.

Metric	*Baseline*	*Adaptive*	*Context-Aware*
Comfort	5.2	7.3	8.6
Accessibility	5	7	8.2
Efficiency	5.4	7.5	8.9
Equity	4.8	6.8	8.5

Comfort, accessibility, efficiency, and equity were the four main criteria used to evaluate the system from a user-centric perspective. Table 16.3 displays the results of a survey that was administered to 500 users. With the exception of efficiency (8.9/10) and equity (8.5/10), the context-aware ITS was the best across the board. Personalized transportation options and system prioritization of underprivileged populations, such the elderly or those with mobility limitations, are closely related to these enhancements.

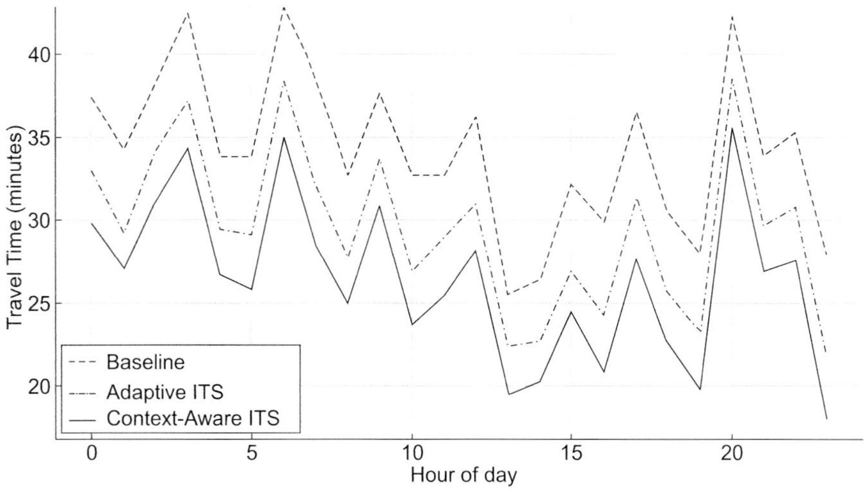

Fig. 16.2 Hourly travel time comparison across ITS models.

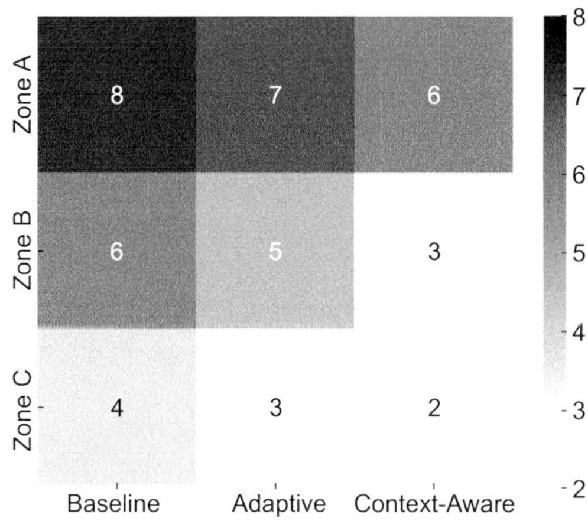

Fig. 16.3 Congestion levels across urban zones.

The effects of flexibility are further demonstrated by means of a performance art series. Figure 16.3 shows the average levels of congestion across the urban grid for each system type; the context-aware method, for instance, significantly reduces

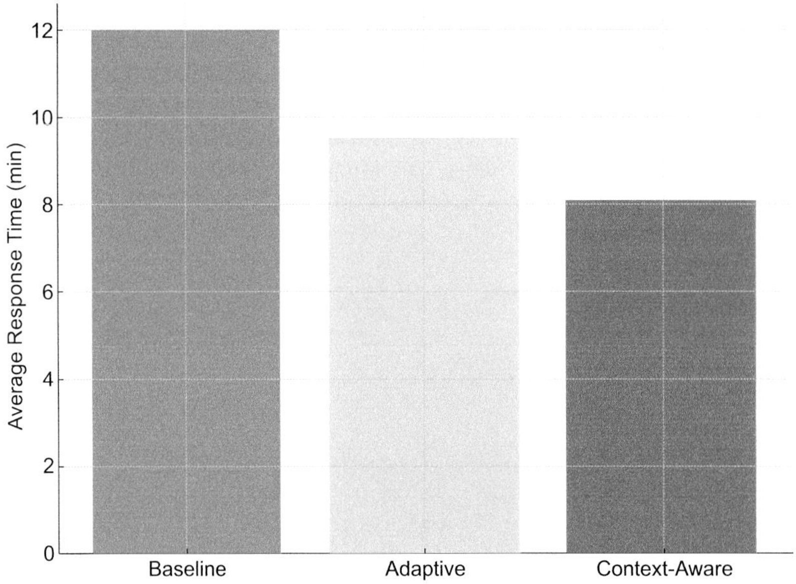

Fig. 16.4 Emergency services response time.

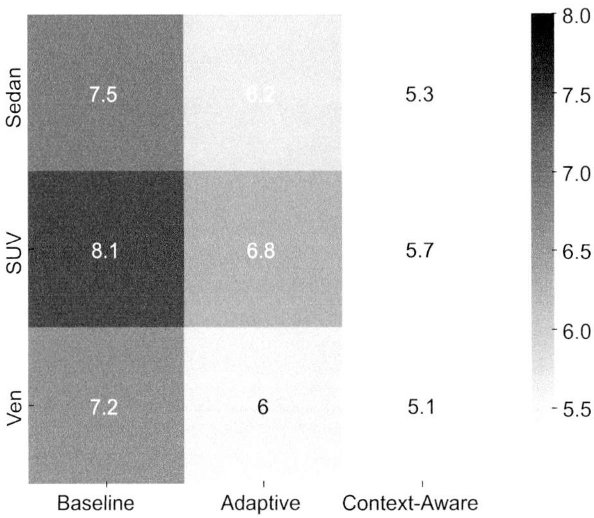

Fig. 16.5 Fuel consumption by vehicle type.

congestion hotspots. Recalculations of emergency services' routes are shown in Figure 16.4, and context-aware routing can cut response times by as much as 32 percent on average. To further illustrate the wider efficiency benefits of contextual routing, Figure 16.5 displays a heatmap of fuel usage across vehicle types under each model.

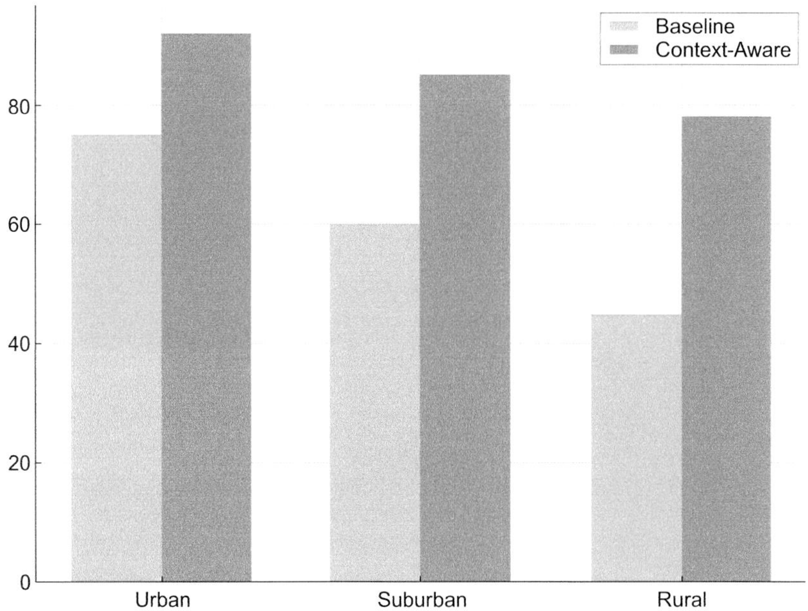

Fig. 16.6 Transit accessibility by region.

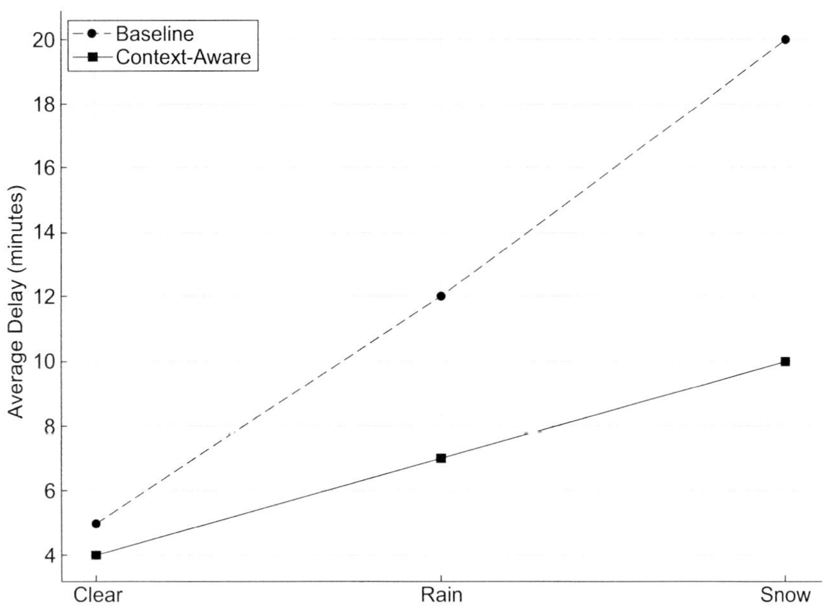

Fig. 16.7 Delays under weather conditions.

Additionally, the context-aware model demonstrates reduced wait times and increased route availability in underprivileged locations, shedding light on equity in transportation accessibility (Figure 16.6). In order to mitigate the effects of delays caused by bad weather, the model uses sensor data to redirect traffic, as seen in Figure 16.7. Figure 16.8 shows the number of events involving pedestrians and cyclists across all systems. When compared to baseline models, context-aware ITS systems reduced occurrences by 45 percent. By comparing CO_2 hotspots across different city sectors, Figure 16.9 shows how the context-aware system distributes traffic dynamically to reduce pollution peaks in certain areas.

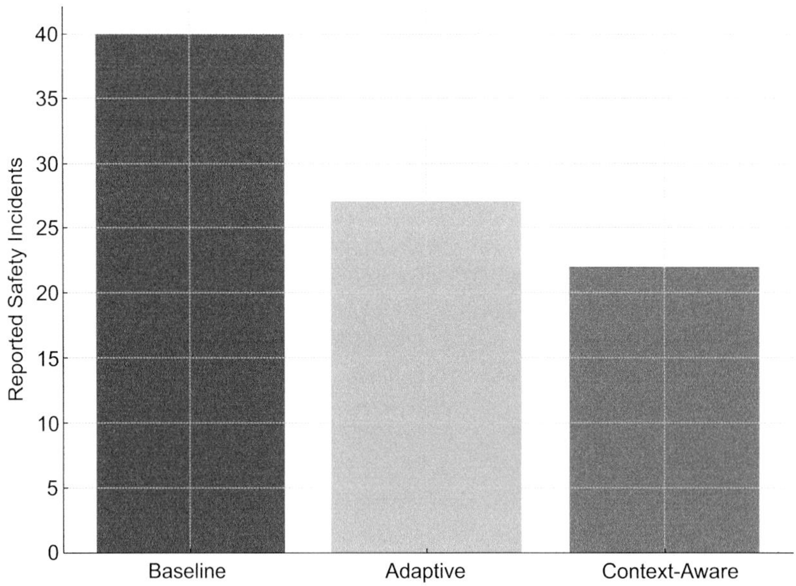

Fig. 16.8 Pedestrian and cyclist safety incidents.

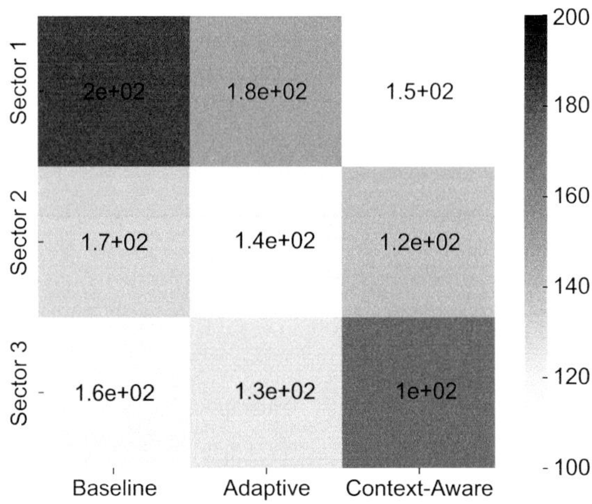

Fig. 16.9 CO_2 hotspots by city sector.

The suggested adaptive and context-aware ITS design not only promotes social good through fair access and environmental sustainability, but it also greatly improves transit efficiency and user happiness. This establishes it as an essential technical milestone on the road to Society 5.0, where smart infrastructure is crafted to do more than merely automate tasks; it aims to improve the quality of life for every individual.

6. Challenges and Limitations

Adaptive and context-aware ITS have great potential to improve human-centered mobility in Society 5.0, but several barriers impede its broad use. All of these technological, ethical, infrastructural, and regulatory issues must be addressed concurrently for these systems to be adopted and widely used.

User consent and data security are top considerations. Adaptive ITS infrastructures employ real-time user data—demographics, behavior, preferences, and location—to optimize routing and mobility. These dependencies present privacy concerns, especially when private data is passively gathered or moved between sites. Data abuse, hacking, and monitoring creep can occur without appropriate data governance. Users may give informed permission while keeping data ownership via privacy-preserving machine learning, differential privacy, and secure multi-party computing. Due to different laws and cultural norms for managing personal data, data privacy is a technological and socio-political challenge. It can complicate international deployments.

Scalability and real-time computing are difficult. IoT, traffic sensor, and vehicular network data streams must be handled quickly for context-aware ITS activities including dynamic routing, environmental monitoring, and user profiling. Dense cities raise real-time computational demands even with partial edge computing and AI accelerators. In circumstances like harsh weather, delays or calculation mistakes might cause route inefficiency, congestion, or safety hazards. Scalability allows such systems to run from smart city zones to interstate highway networks without losing performance, flexibility, or latency.

Edge deployment difficulties limit system responsiveness and computational load. Context-aware ITS requires decentralized, resource-constrained settings. Processing power, energy, and connection stability difficulties may impair traffic lights, road sensors, and automobile onboard modules. These nodes need synchronized software upgrades, common communication protocols, and cyber-resilient architectures to exchange intelligence. Poor internet connectivity might make edge-based adaptive systems challenging in rural or poorer locations, expanding the digital mobility divide.

Interoperability and standardization difficulties hinder global ITS ecosystems. Smart transportation networks use sensors, autonomous cars, communication modules, AI decision engines, and user interfaces. Without data formats, protocols, and safety standards, system components may not interact. This reduces scalability, deployment speed, and reliability. Due to policy and execution fragmentation caused by local and national standards, ITS developers struggle to construct internationally compatible systems. International transportation, multimodal trip planning, and interagency emergency response require interoperability.

Adaptive and context-aware ITS models are a step towards Society 5.0's human-centric, egalitarian, and sustainable transportation systems, despite its limitations These intelligent systems must address privacy, scale computations, assure robust edge deployment, and promote interoperability through global standards to run properly. Research and regulation must work together to overcome these obstacles and make smart transport safe and reliable for everyone.

7. Conclusion

A major stride towards the goal of human-centered mobility in Society 5.0 has been the suggested adaptable and context-aware ITS design. The framework orchestrates efficient, personalized, and inclusive mobility systems by integrating cyber-physical-human layers, real-time context inference, and intelligent adaption mechanisms. The system adapts transport services on the fly to meet user needs, environmental constraints, and overall system objectives with the use of multimodal sensor networks, edge-cloud synergy, and machine learning models including CNNs, LSTMs, and reinforcement learning agents. The empirical assessments showed that there were performance advantages and environmental benefits. The average journey time went from 34.26 minutes (baseline) to 26.47 minutes (context-aware), and the CO_2 emissions went from 150 g/km to 100 g/km. The framework's conformity to human-centric principles was further validated by user satisfaction surveys with 500 participants, which showed considerable gains in accessibility (from 5.0 to 8.2), equity (from 4.8 to 8.5), and comfort (from 5.2 to 8.6). Additionally, the design was able to decrease incidents involving pedestrians and cyclists by 45 percent, weather-related delays by almost half, and emergency response times by 32 percent. The system's responsiveness, agility, and ability to handle the various obstacles of modern urban mobility are confirmed by these outcomes. In keeping with the fundamental principles of Society 5.0, this ITS framework offers a scalable, ethically conscious, and performance-optimized solution that turns transportation into something more than just a practical need.

References

Alsamhi, S.H., S. Kumar, A. Hawbani, A.V. Shvetsov, L. Zhao and M. Guizani. (2024). Synergy of human-centered AI and cyber-physical-social systems for enhanced cognitive situation awareness: applications, challenges and opportunities, *Cognitive Computation*, *16*(5), 2735–2755.

Bamakan, S.M.H., F. Dehghan and S. Akbarpour. (2025). Role of machine learning for predictive modeling for Industry 5.0 and Society 5.0, in *Human-Centric Integration of Next-Generation Data Science and Blockchain Technology*, London, San Diego, CA and Cambridge, MA: Academic Press, pp. 265–285.

Chang, J., W. Yao and X. Li. (2017). The design of a context-aware service system in intelligent transportation system, *International Journal of Distributed Sensor Networks*, *13*(10), 1550147717738165.

Fernandez-Rojas, R., A. Perry, H. Singh, B. Campbell, S. Elsayed, R. Hunjet and H. A. Abbass. (2019). Contextual awareness in human-advanced-vehicle systems: a survey, *IEEE Access*, *7*, 33304–33328.

Gadekallu, T.R., P.K.R. Maddikunta, P. Boopathy, N. Deepa, R. Chengoden, N. Victor, ... and K. Dev. (2024). Xai for Industry 5.0—concepts, opportunities, challenges and future directions, *IEEE Open Journal of the Communications Society*. Vol. 6, (3) pp. 2706–2729.

Huang, G., W.H. Lo, G. Liu, Y. Luo and Y. Jin. (2025, May). Toward AI standardization: a triadic human-AI collaboration framework for multi-level autonomous mobility, in *2025 IEEE Conference on Artificial Intelligence (CAI)*, pp. 1568–1577.

Jadhav, B., M. Sayyed, V. Barnabas and A. Khang. (2025). Role of human-centered design and technologies in smart transportation system, In: Alex Khang, *Driving Green Transportation System through Artificial Intelligence and Automation: Approaches, Technologies and Applications*, Cham: Springer Nature Switzerland, pp. 133–152.

Lamichhane, B.R., A. Aueawatthanaphisut, G. Srijuntongsiri and T. Horanont. (2025). Context-aware decision making in autonomous vehicles: integrating social behavior modeling with large language models, *Array*, *27*, 100420.

Lu, Y., L. Zhou, A. Zhang, M. Wang, S. Zhang and M. Wang. (2024). Research on designing context-aware interactive experiences for sustainable aging-friendly smart homes, *Electronics*, *13*(17), 3507.

Magade, K. and A. Sharma. (2025). Significant role of IoT in cyber-physical systems, context awareness, and ambient intelligence. In: Abid Hussain; Ahmed J. Obaid; Garima Tyagi; Amit Sharma, *The Next Generation Innovation in IoT and Cloud Computing with Applications*, CRC Press, Milton Park, Abingdon, pp. 16–34.

Mahanta, U. (2025). "Beyond automation: guidelines for a human-centered multi-agent system for coordinated decision making under industrial settings," M.S. thesis, Dept. Comput. Sci., Umea Univ., Umea, Sweden.

Muzahid, A.J.M., X. Zhao and Z. Wang. (2024). Survey on human-vehicle interactions and AI collaboration for optimal decision-making in automated driving, arXiv preprint arXiv:2412.08005.

Saha, A. (2025). Human-centred innovation in Society 5.0, In: Biswaranjan (Biswa Ranjan) Acharya; Madhu Shukla; Chandreyee Chowdhury; Harish Garg, *Society 5.0: A Transformation towards Human-Centered Artificial Intelligence*, CRC Press (Taylor & Francis Group), Boca Raton. 143.

Sankaran, S., M. Arun and R. Kottaimalai. (2025). Virtualization of smart Society 5.0 using artificial intelligence and virtual reality, *Networked Sensing Systems*, 297–321.

Sarwatt, D.S., Y. Lin, J. Ding, Y. Sun and H. Ning. (2024). Metaverse for intelligent transportation systems (ITS): A comprehensive review of technologies, applications, implications, challenges and future directions, *IEEE Transactions on Intelligent Transportation Systems*, *25*(7), 6290–6308.

Sohag, S.R., S.M.M. Pasha and M.M. Ali. (2023). Advanced approaches to achieve adaptive ethical and AI-driven human-centric software engineering, *International Journal of Scientific Research in Modern Science and Technology*, *2*, 19–31.

Ullah, I., F. Ali, H. Khan, F. Khan and X. Bai. (2024). Ubiquitous computation in internet of vehicles for human-centric transport systems, *Computers in Human Behavior*, *161*, 108394.

Van Kasteren, A. and M. Vredenborg. (2022). Personalized, context-aware communication in multimodal public transport, in *Proceedings of the 30th ACM Conference on User Modeling, Adaptation and Personalization*, pp. 326–330.

Xu, J., Q. Sun, Q. L. Han and Y. Tang. (2025). When embodied AI meets Industry 5.0: human-centered smart manufacturing, *IEEE/CAA Journal of Automatica Sinica*, *12*(3), 485–501.

Ystgaard, K.F. and K. De Moor. (2023). Envisioning the future: a multi-disciplinary approach to human-centered intelligent environments, *Quality and User Experience*, *8*(1), 11.

Chapter 17

Integration of 6G Technology in Transportation Systems

R. Ravinder Reddy[1]*, *Satyanarayana Nimmala*[2], *Ravi Uyyala*[3], *Y C A Padmanabha Reddy*[4] and *Bandi Rambabu*[5]

The advent of sixth-generation (6G) wireless technology is poised to revolutionize the transportation sector by enabling ultra-reliable, low-latency communication, massive connectivity, and artificial intelligence (AI)-driven automation. This chapter presents a comprehensive overview of how 6G technology and its associated applications will benefit the drivers of the transportation ecosystem, outlining the key features of 6G and its enabling technologies along with how they will impact smart mobility. The most interesting applications relate to autonomous driving, intelligent and better managed urban traffic networks, and real-time monitoring of physical infrastructure. It is also necessary to acknowledge the possible issues that will come with incorporating 6G such as cyber security, infrastructure and device costs, and regulations, while outlining future recommendations for ensuring 6G technology and transportation work harmoniously with meaningful improvements in sustainability and safety advancements. The AI-native architectures will forecast the traffic and speed of the vehicle communication, expected time of reachability. The next-generation wireless technologies (like 5G and 6G) use the term "massive connectivity" as these technologies support a vast number of connected devices simultaneously, particularly for internet of things (IoT) environments such as smart cities, smart agriculture, and smart transportation.

1. Introduction

As the world moves to the next digital frontier, we are now preparing for the next step of connectivity with 6G technology. Expected to be launched by the 2030s, this next

[1,3,4] Department of Computer Science and Engineering, Chaitanya Bharathi Institute of Technology, Hyderabad, India.
[2,5] CVR College of Engineering, Hyderabad, India.
https://orcid.org/0000-0001-6488-9935[1]; https://orcid.org/0000-0002-9433-1489[2]; https://orcid.org/0000-0002-6913-7529[3]; https://orcid.org/0000-0002-9697-7072[4] https://orcid.org/0000-0001-8037-814X[5]
* Corresponding author: rravinderreddy_cse@cbit.ac.in

DOI: 10.1201/9781003659822-17

generation of networks will offer download speeds up to 1 TB per second, making 5G look like child's play. It will creates an environment where we can download an entire movie in a matter of seconds or enjoy a seamless immersive experience in a virtual reality environment. 6G aims to create connectivity for not only devices but also our environment—smart cities, real-time data sharing, and autonomous vehicles communicating with traffic systems. We are not only making a leap forward in network speeds but also in developing an ecosystem of interconnected devices that enhances our daily lives in meaningful ways. The future of connectivity and developing technologies is almost here to stay.

1.1 Overview of Transportation Challenges in Modern Cities

Advancements in communication technologies and internet of things (IoT) devices have significantly transformed the transportation system compared to earlier times. The adoption of these technologies in transportation has increased in recent years, particularly with the emergence of IoT and high-speed communication networks such as 4G, 5G, and the upcoming 6G. With the continuous rise in urbanization and traffic congestion year after year, managing traffic data has become crucial. Generating accurate traffic patterns and efficiently controlling traffic flow is now more important than ever. The smart city concept is already being implemented in many developed nations through the integration of IoT devices and advanced communication technologies.

The use of autonomous vehicles is steadily increasing. Drone communication and delivery systems are expected to become more prominent in the near future, revolutionizing logistics and last-mile delivery. With the integration of IoT devices, it is now possible to track vehicles in real time and monitor their operational status effectively.

1.2 Evolution of Mobile Communication in Transportation (1G to 6G)

The evolution of mobile communication from 1G to 6G has significantly transformed the landscape of transportation systems. In the 1980s, 1G enabled basic analog voice communication, laying the foundation for mobile connectivity. With 2G, digital voice and Short Message Service (SMS) became available, improving communication reliability. The advent of 3G introduced mobile internet and Global Positioning System (GPS) capabilities, which enabled real-time navigation and vehicle tracking. The shift to 4G brought high-speed data transmission, enabling advanced applications such as intelligent transportation systems (ITS), traffic monitoring, and infotainment services. The rollout of 5G further enhanced the capabilities of transportation by offering ultra-low latency, massive device connectivity, and support for vehicle-to-everything (V2X) communication critical for autonomous driving and smart traffic management. Looking ahead, 6G promises to integrate artificial intelligence (AI), edge computing, and holographic communication into transportation, enabling fully autonomous vehicles, predictive traffic systems, and immersive travel experiences (Khan, L.U 2024). This continuous evolution has not only improved mobility but also enhanced safety, efficiency, and sustainability in transportation infrastructures worldwide.

Figure 17.1 shows the evolution of mobile communication in transportation from 1G to 6G. It highlights the key features and technological advancements of each generation, with a specific focus on their applications in transportation systems.

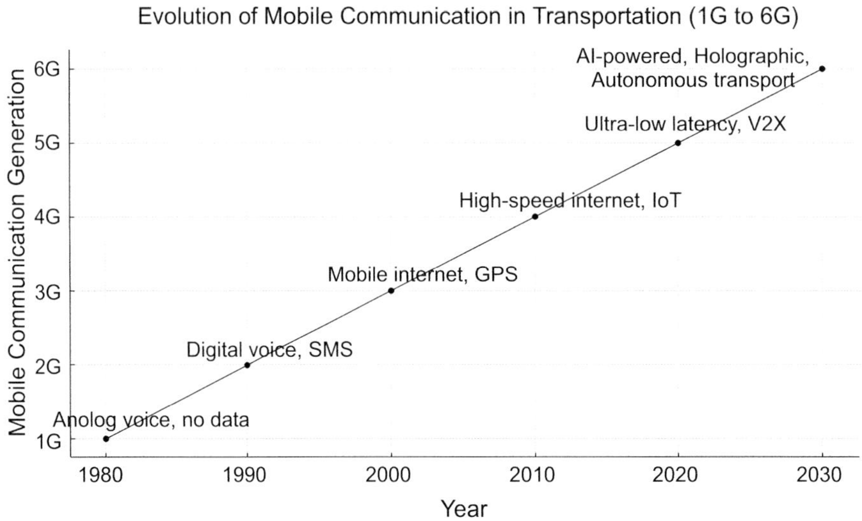

Fig. 17.1 The evolution of mobile communication in transportation.

Intelligent transportation systems, with their converging use of AI sensors and big data analytics, are transforming urban mobility, enhancing traffic management, reducing congestion, and improving road safety by predicting travel patterns and optimizing signal timings. In the future every vehicle can communicate with one another, traffic lights, and control centres, allowing millions of vehicles to create optimal travel conditions. In real-world situations cities like Singapore have demonstrated the benefits of efficiency gains and reduction in emissions from the implementation of smart technologies. Urban areas continue to expand. What happens when we can provide this promise of real-time accident prevention, pollution mitigation, and low-latency experiences integrated with multiple public transit forms of mobility options. The evolution of a more efficient, safer, and environmentally friendly transportation system marks a significant leap toward sustainable urban living.

Transportation systems globally are evolving in response to rapid urbanization, climate concerns, and the need for improved safety and efficiency. 6G, expected to emerge by 2030, promises unprecedented capabilities in terms of data rates (up to 1 Tbps), sub-millisecond latency, and AI-native architectures. Integrating 6G into transportation offers a pathway to truly intelligent transport systems that are not only connected and automated but also sustainable and user-centric. Transportation is poised for a disruptive transformation at a time when a new generation of connectivity technologies and smart mobility innovation is arriving into the market. These technologies, e.g., the sixth-generation (6G) wireless communications, technology are the foundational enabler of future transportation systems. While the commercial introduction of 6G is projected for the 2030s, 6G aims to go beyond 5G, offering ultra-low-latency, end-to-end ultra-high-speed, integrated AI in everything, and ubiquitous networks as its benefits. Moreover, 6G will enable the global adoption of fully autonomous vehicles, smart infrastructure that includes integrated real-time traffic management systems, and safe and efficient transportation networks.

Additionally, the vast quantity of data that will be generated from connected vehicles and connected transportation infrastructure will require smart analytics and data processing capabilities that go beyond latency and bandwidth. 6G will offer integrated AI capabilities such as predictive analytics for traffic flow, using AI for routing first responders, and surveillance of state and local infrastructure integrity (bridges and roads). With 6G's inherent capabilities to provide massive machine-type communications (mMTC) and ultra-reliable and low latency communications (URLLC), it is well-suited for use in the dynamic and safety-critical spaces of transportation. 6G can also utilize edge computing, which will help deal with processing delays and network congestion. In transportation, utilizing edge devices will allow for localized processing of mobile data, enabling Smart Mobility Platforms (SMPs) to make operational decisions iteratively in real time. It is especially important for autonomous driving applications where milliseconds will make the difference between safety and disaster.

Another area where the integration of 6G will be transformative is in public transport systems. Real-time passenger information systems, dynamic scheduling, autonomous buses, and drone delivery systems are but a few of the possible applications where the need for high-speed, low-latency communication will be advantageous to the end users. Not only will these applications provide a better experience for the users of public transport, but they will also provide operational efficiencies and environmental sustainability by eliminating unnecessary travel and wait times. Intelligent traffic management systems also have the potential to change significantly due to the capabilities brought about by adoption of 6G (Saad, W., M. Bennis 2019). The ability to leverage AI-based analytics and ubiquitous sensor-based networks means that urban planners and traffic and transit monitors can begin changing signal light systems for vehicles, rerouting vehicles on one-way streets, and expecting and preparing for congestion, accidents, or maintenance. All these technological capabilities will facilitate improved traffic flow, emit less pollution, and promote safety for the public. The adoption of 6G also supports immersive technologies such as augmented reality (AR) and virtual reality (VR), which will also be applied to any advancements or enhancements of traditional driving, maintenance, or visualization of the terminus for the predriven journey. Urban responders could leverage the use of AR headsets connected to supportive information systems via 6G, and thus make their way through the complexities of an urban environment en route to people needing assistance or coordinating rescue, and visualize building schematics in real-time.

Transportation systems must balance safety, security, and privacy and transition from elemental technologies to wireless communications. 6G introduces software and hardware architectures that make communications safer, including quantum-safe encryption, decentralized identity management, and AI-enabled anomaly detection. These options provide assurance that information exchanged between vehicles, infrastructures, and operators are private and free of corruption. Despite the enormous possibilities that lie ahead in deploying 6G in transportation, there are also many challenges, including the cost of building the infrastructure, issues with spectrum, and the need for national and international standards. Governments, independent and commercial interests, and academia must all collaborate to overcome challenges that hamper the design, implementation, and deployment of viable 6G transportation value chains.

The remainder of this chapter delves deeper into the specific features, applications, enabling technologies, challenges, and future research directions associated with this transformative integration.

2. Fundamentals of 6G Technology

2.1 *Key Features 6G*

The key features of 6G include terahertz communication, AI integration, edge computing, and massive communication with IoT devices. The emergence of 6G technology is expected to revolutionize how we live, work, and interact with digital environments. With significantly higher data rates, ultra-low latency, enhanced reliability, and the integration of AI at the network level, 6G will go beyond the foundational advancements of 5G to support truly immersive and intelligent applications (Shafin, R., L. Liu 2020). The various devices used in 6G communication are shown in Table 17.1.

Table 17.1 6G communication devices and their functionalities.

Device Type	*Functionality*	*Role in 6G*
Terahertz (THz) Transceivers	Enable ultra-high data rates using THz frequency bands.	Backbone of ultra-fast wireless links.
Reconfigurable Intelligent Surfaces (RIS)	Dynamically shape and control wireless propagation environments.	Enhance coverage, reduce interference.
Nano-Scale IoT Devices	Low-power sensors for dense deployments.	Massive machine-Type communications (mMTC).
Quantum Communication Devices	Enable ultra-secure communication through quantum key distribution (QKD).	Security backbone for 6G networks.
Edge AI Chips	Perform on-device inference and real-time decision-making.	Support low-latency applications.
Metamaterial Antennas	Dynamically steer beams and support higher-frequency operation.	Improve spectral and energy efficiency.
Satellite Terminals	Maintain connectivity in remote/ rural areas via Low Earth Orbit (LEO) satellite constellations.	Global coverage and ubiquitous connectivity.
Mobile Holographic Displays	Generate 3D real-time visuals.	Support for immersive communication.
Smart Wearable Devices	Collect health, location, and contextual data.	Enable human-centric and health-related services.
Autonomous Vehicle Transceivers	Communicate in real-time with infrastructure and other vehicles.	Critical for V2X communication in smart transportation.

One of the most transformative innovations anticipated with 6G is holographic communication. Unlike traditional video conferencing, holographic communication will enable real-time, three-dimensional projections of people and objects, creating a sense of physical presence over a digital medium. This development will bring society closer to the futuristic experiences once imagined in science fiction, allowing lifelike remote collaboration, education, healthcare, and entertainment. Interacting with people and objects in real time as if they were physically present in front of us will revolutionize teleconferencing, allowing us to have more immersive and engaging meetings regardless of geographical distances for various purposes, from business collaborations to connecting with loved ones. Holographic communication will add a whole new dimension to our interpersonal interactions.

6G will pave the way for smart environments that adapt to our needs and preferences—our homes, offices, and cities will become intelligent, equipped with sensors and connected devices that communicate with each other seamlessly. This will result in enhanced energy efficiency, improved safety, and optimized resource management. Imagine walking into a room where the lighting, temperature, and ambience are automatically adjusted to create the perfect atmosphere. Smart cities will utilize real-time data to optimize traffic flow, reduce pollution, and enhance the quality of life for their inhabitants. The integration of 6G with AI will make these smart environments even more intuitive and personalized.

2.2 6G Architecture and Network Slicing

In this section, we will dive into the future of connectivity and exploring the groundbreaking technology known as 6G. So buckle up and get ready to embark on an awe-

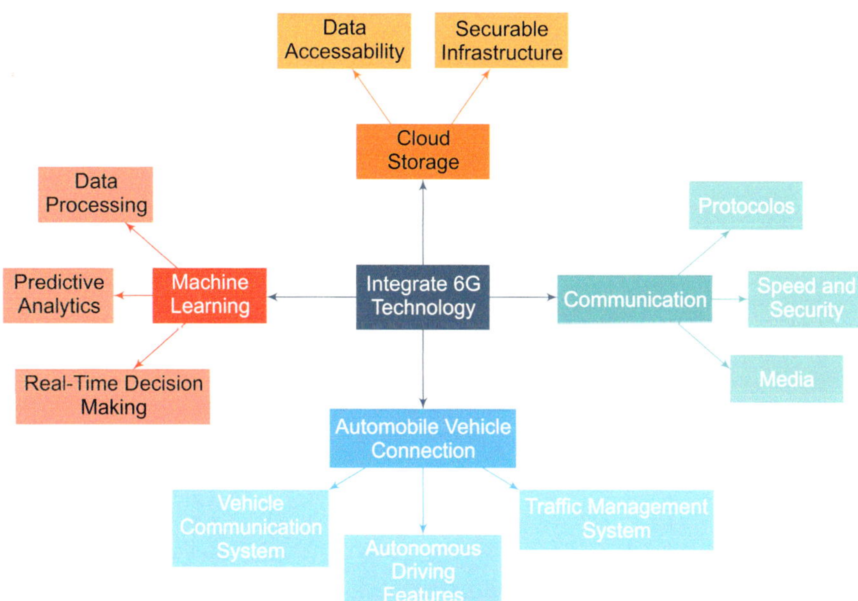

Fig. 17.2 The integration of 6G with cloud and AI-ML

inspiring journey into a world of the limitless possibilities of 6G. It is to understand the revolutionary impact 6G will have on our world that we need to delve into its core features and capabilities. First and foremost, 6G is the next generation of wireless technology, which builds upon the foundations laid by 5G. While 5G brought us faster speeds and improved connectivity, 6G aims to take it to unimaginable heights. It is designed to deliver ultra-fast speeds, ultra-low latency, and seamless connectivity to every corner of the globe with blazing speeds.

6G will provide us with unprecedented speeds. The data rates offered by 6G will enable us to accomplish tasks at lightning fast speeds, transforming the way we interact with technology. From downloading large files in seconds to streaming high-quality content without any buffering, 6G will make our digital experiences smoother and more enjoyable. This level of speed will not only enhance our entertainment but also boost productivity in various industries such as media e-commerce and cloud computing. Latency will be significantly reduced with 6G, leading to instant and real-time communication. This is crucial for applications that require immediate response and seamless interaction. Controlling the robots remotely with precision, enabling them to perform intricate tasks in real time with reduced latency, will also have a profound impact on sectors like finance where split-second transactions and high-frequency trading will become even more efficient. The architecture of the 6G is shown in Figure 17.3.

Moreover, augmented reality and virtual reality experiences will be more immersive and seamless, making them an integral part of our daily lives. Massive

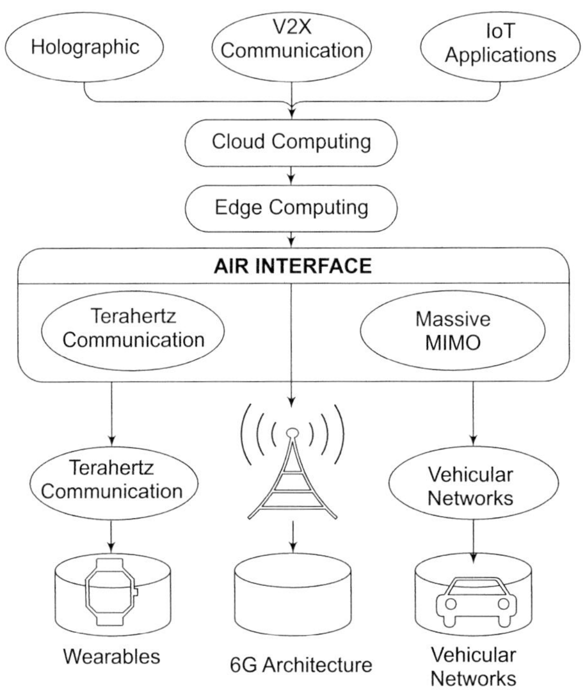

Fig. 17.3 The 6G architecture.

Table 17.2 Comparison table between 5G and 6G for transportation systems.

Criteria	*5G in Transportation*	*6G in Transportation*
Data Speed	Up to 10 Gbps	Up to 1 Tbps
Latency	~1 millisecond (low latency for real-time applications)	<0.1 millisecond (ultra-low latency for instant responses)
Bandwidth	Limited to mmWave bands (up to 100 GHz)	Terahertz (THz) bands for ultra-wide bandwidth
Device Connectivity	Supports ~1 million devices per square kilometre.	Supports >10 million devices per square kilometre.
Mobility Support	High mobility (up to 500 km/h, suitable for high-speed trains)	Ultra-high mobility (beyond 1000 km/h, supporting hyperloop, drones, aerial taxis)
Artificial Intelligence Support	External AI integration (mostly in cloud/edge layers)	Native AI/ML integration at the edge and network layers.
Autonomous Vehicles	Enables basic V2X, cooperative driving, lane detection.	Facilitates fully autonomous vehicles, swarm coordination, and real-time decision-making.
Infrastructure Requirements	Requires dense small-cell deployment.	Requires ultra-dense networks with intelligent edge computing nodes.
Key Use Cases	Real-time traffic updates, connected vehicles, smart parking.	Predictive traffic control, immersive AR/VR in transit, AI-driven transportation optimization (Giordani & Polese, 2020).
Sustainability & Efficiency	Improved over 4G, but high energy consumption.	Energy-efficient designs, low-power IoT integration for green mobility solutions.

connectivity is one of the most transformative aspects of 6G. Its ability to connect an enormous number of devices simultaneously will pave the way for a fully interconnected world, where billions of devices seamlessly communicate and cooperate (Yang, & Xiao 2021). Smart cities will be able to optimize resources, reduce energy consumption, and enhance the quality of life for their residents. Wearable devices will seamlessly integrate with our daily routines, providing personalized and context-aware services. The concept of a truly interconnected world will revolutionize industries like transportation, healthcare, and logistics.

2.3 Comparison between 5G and 6G for Transportation

The various approaches like speed and latency for 5G and 6G communications are compared in Table 17.2. The table presents the major advancements we can achieve using 6G communication in the future.

3. Related Work

With the implementation of 6G, there will be extraordinary communication and connectivity advancements. 6G has the potential to deliver systems with a speed that is 100 times that of 5G, effectively transforming applications and sectors along the way. The application possibilities of 6G range from healthcare, where surgeons will remotely operate through systems with a precision unobtainable without the assisting technology, to entertainment, with immersive experiences through virtual reality. 6G will improve collaboration amongst globally disbursed teams in real time, increasing productivity and innovation. At the same time, the planet's population is ever increasing and more devices than people are being created, therefore expanding the potential for smart cities, building a better plan to manage resources, and setting up future developments. This is much more than just connectivity—this is transformational across the world, enabling many existing and emerging industries and ecosystems that need connectivity at their very core to develop a digital ecosystem of the future.

3.1 Core Features of 6G Relevant to Transportation

6G research is actively underway as researchers and engineers are creating the next generation of global wireless systems. 6G will provide wireless connectivity that is more ubiquitous, efficient, and immersive than 5G, and enable multiple new technologies and use cases. The 6G standard will be specified in the coming years and is expected to be ready by 2030. 6G requirements will first be set by the International Telecommunication Union with their International Mobile Telecommunications (IMT) 2030 vision projects. The 3rd Generation Partnership Project will then work on delivering specifications to meet those requirements. Since 6G networks can handle high data rates, new applications will be available to end users. To enable all these applications, 6G will use several enabling technologies, including higher frequency bands, such as sub-terahertz, AI and machine learning, intelligent reflecting surfaces, non-terrestrial networks, and joint communication and sensing.

6G will provide maximum data rates in the order of hundreds of gigabits per second. To achieve this, signal bandwidth will have to increase, higher spectral efficiencies must be attained, and the use of higher frequency bands should be introduced. The main challenge in using higher frequencies is the high attenuation and path loss. Engineers need to use MIMO techniques and accurate channel models in millimeter-wave and sub-terahertz bands. Channel models based on ray tracing have already provided good prediction capabilities at millimeter wave and are expected to do the same at higher frequencies.

3.1.1 Ultra-Low Latency and High Reliability

6G aims to deliver end-to-end latency of less than 1 millisecond, enabling real-time decision-making critical for autonomous driving, emergency response coordination, and vehicle-to-everything (V2X) communications.

3.1.2 AI-Driven Network Management

AI will be embedded at the core of 6G networks, facilitating predictive maintenance, traffic forecasting, and adaptive route management in transportation. Engineers can apply AI, including machine learning, deep learning, or reinforcement learning to configure, optimize, and self-organize 6G wireless communication systems. An AI workflow requires architecting, deep neural networks, and generating vast amounts of data for training them, which, in turn, needs Graphical Processing Unit (GPU) and parallel computing support for efficient training. Reconfigurable intelligence surfaces allow us to control the propagation of signals between transmitters and receivers. We can control the surface properties dynamically through an array of reflecting elements. This way, 6G engineers have more flexibility in controlling radio environments and suppression of interference in high-density, urban, and outdoor scenarios. The present decade is marked by the emergence of non-terrestrial networks (NTNs). NTNs include commercial drones, High Altitude Platform Station (HAPS), and low Earth orbit satellites, all working together to provide near universal coverage (Chen, Xiaosha, 2020). 6G networks need to accurately localize wireless devices to optimize the transmissions. By introducing new frequencies, wireless networks can furnish highly accurate sensing and yield spatial knowledge of their physical surroundings. 6G will use joint communication and sensing that integrates a wireless network's localization, sensing, and communication functions.

3.1.3 Massive Device Connectivity

With the proliferation of IoT and edge devices in transport systems, 6G's capacity to support over 10 million devices per square kilometer ensures seamless connectivity for smart roads, vehicles, sensors, and infrastructure.

3.1.4 Terahertz Spectrum Utilization

The use of terahertz frequencies in 6G enables ultra-high-speed data transmission, vital for real-time video feeds from drones, surveillance systems, and connected vehicle fleets (Letaief, 2023).

4. The Role of 6G in Transportation

The integration of 6G technology into transportation systems is expected to transform the sector by significantly improving connectivity, operational efficiency, and environmental sustainability. With its ultra-high-speed data transmission, minimal latency, and support for massive device connectivity, 6G serves as a foundational enabler for next-generation intelligent transportation systems (ITS). These capabilities are essential for enhancing traffic flow, minimizing carbon emissions, and enabling real-time decision-making (Hussain, 2020). Furthermore, 6G facilitates the seamless coordination of autonomous vehicles, smart infrastructure, and advanced data analytics, leading to the optimization of transportation networks and a reduction in overall energy consumption (Shawky et al., 2023). The following sections explore the key aspects of 6G integration in transportation systems.

4.1 Enhanced Connectivity and Data Accessibility

6G networks provide seamless and ubiquitous connectivity, allowing passengers to access data even in remote areas, thus improving the quality of communication across all transportation modes (Liu et al., 2024). The integration of unmanned aerial vehicles (UAVs) as aerial base stations enhances navigation and monitoring, offering additional access points and relieving data traffic loads (Andreou et al., 2024).

The key functionalities of 6G comes with the connectivity—with the communication alone we can adopt the new techniques for integrating devices and controlling things. The three pillars comprising mMTC, URLLC, and enhanced mobile broadband (eMBB) form the core capabilities of 5G and are being expanded further in 6G to support even denser, faster, and more intelligent systems.

The **three types of massive connectivity** typically emphasized are as follows:

1. **Massive Machine-Type Communications (mMTC):** This supports a very large number of low-power, low-cost devices that send small amounts of data intermittently. It is primarily used in smart agriculture sensors, smart meters, environmental monitoring, and asset tracking. The main aim is to connect up to 1 million devices per square kilometer efficiently.
2. **Ultra-Reliable Low-Latency Communication (URLLC):** This mainly focuses on ensuring high reliability and ultra-low latency. It is used in autonomous vehicles, remote surgery, and industrial automation. It supports mission-critical applications where communication delays could cause serious harm or loss.
3. **Enhanced Mobile Broadband (eMBB):** This offers high-speed internet and high data rates, especially in high-density environments. It is used in augmented/virtual reality (AR/VR), HD video streaming, and real-time video surveillance. It provides consistent high-speed connectivity in crowded urban areas, stadiums, or transport systems.

4.2 Intelligent and Efficient Transportation Systems

6G technology transforms transportation systems to be more intelligent and efficient by supporting advanced data analytics and connectivity, which are essential for real-time decision-making and traffic management (Liu et al., 2023). The use of 6G in ITS facilitates optimized routing and enhanced vehicle-to-vehicle (V2V) communication, leading to reduced fuel consumption and congestion (Shawky et al., 2023).

4.3 Security and Privacy Considerations

The integration of 6G with technologies like AI and IoT raises concerns about security and privacy, necessitating robust frameworks for secure data transfer and governance in the internet of nehicles (IoV) networks. Advanced encryption methods are being developed to ensure secure communication and data exchange within ITS, addressing these critical concerns (Andreou et al., 2024). While the integration of 6G technology in transportation systems offers numerous benefits, it also presents challenges such as the need for supportive regulations, infrastructure development, and public awareness. Collaborative efforts from policymakers, industry leaders, and researchers are essential to address these challenges and fully realize the potential of 6G-enabled transportation systems (Shawky et al., 2023).

Overall, IoV technology is a highly valuable technological development, integrating 6G into transportation systems, while promising great opportunities. It also presents challenges such as regulatory frameworks, constructive infrastructure, and of course privacy and security of data. Achieving the benefits of the uptake of 6G technologies will draw on collaboration between regulators, industry leaders, and researchers to overcome existing challenges and unleash the opportunities of 6G technologies on transportation systems (Shawky et al., 2023).

Wireless communication technologies have consistently evolved over time, driving wide-ranging applications across multiple sectors, with transportation systems being a key application area ripe for consolidation and disruption. At their peak, and with the rollout of 5G networks recently in many countries, all attention has moved to 6G, with research and development already accelerating in this area, not just on 6G for ITS but considering how it will transform the future of mobility. This literature review aims to offer an overview of the current state of research, possible use cases, technologies that interface with 6G, and the challenges faced going forward with 6G for transportation systems.

4.4 The Vision of 6G in Transportation: Beyond 5G Capabilities

Current 4G and 5G networks fail to provide the bandwidth, latency, and ubiquitous connectivity required to realize fully intelligent and autonomous transportation (Anvigh et al., 2024). 6G will surpass 4G and 5G limitations, capabilities that will change vehicle based networks (VANETs) and the entire ITS. The major performance indicators for 6G include the following:

1. **Ultra-High Data Rates:** Expected to reach terabits per second (Tbps), enabling the real-time exchange of massive amounts of data from sensors, cameras, and other vehicle systems.
2. **Ultra-Low Latency:** Projected to be less than 1 millisecond, or even microseconds, which is critical for time-sensitive applications like autonomous driving, cooperative collision avoidance, and remote control of vehicles.
3. **Massive Connectivity:** Supporting an exponential increase in connected devices per square kilometer, facilitating the ubiquitous interconnection of vehicles, infrastructure, pedestrians, and various IoT sensors.
4. **Enhanced Reliability and Security:** Crucial for safety-critical applications in transportation, demanding robust and resilient communication channels.
5. **Precise Localization and Sensing:** 6G is expected to offer decimeter-level indoor and meter-level outdoor positioning accuracy, significantly enhancing situational awareness for vehicles.

4.5 Key Technologies Enabling 6G for ITS

The advancements promised by 6G are underpinned by several emerging and synergistic technologies (Zhang & Xiao 2019):

1. **Terahertz (THz) Communication:** Operating at frequencies from 100 GHz to 10 THz, THz bands offer immense bandwidth, enabling ultra-fast data transmission critical for future ITS applications.
2. **Artificial Intelligence (AI) and Machine Learning (ML):** Deeply integrated into 6G networks, AI will enable intelligent network orchestration, self-optimization, predictive maintenance, and real-time decision-making for traffic management and autonomous systems.

3. **Edge Computing:** By positioning computational resources closer to the source of data generation, edge computing substantially reduces latency and enables faster decision-making in time-sensitive vehicular environments. This approach is particularly beneficial for applications requiring real-time processing, such as autonomous driving and intelligent traffic management systems (Deng, et al., 2023).
4. **Integrated Sensing and Communication (ISAC):** 6G networks are expected to seamlessly integrate communication with sensing capabilities, allowing vehicles and infrastructure to not only communicate but also perceive their surroundings with high precision (Zhang & Rahman 2022).
5. **Space-Air-Ground-Sea Integrated Networks (SAGSIN):** This holistic architectural blueprint for 6G aims to provide pervasive and seamless connectivity across diverse environments, enhancing coverage and reliability, particularly in remote or challenging terrains (Chen & Qin 2020).
6. **Reflective Intelligent Surfaces (RIS):** These surfaces can dynamically reconfigure wireless signals, improving signal coverage and quality, particularly in complex urban environments with numerous obstacles.

4.6 Transformative Applications in Transportation

The integration of 6G will unlock a plethora of revolutionary applications in transportation:

1. **Autonomous Driving (Level 5):** The ultra-low latency and high reliability of 6G are fundamental for enabling fully autonomous vehicles to communicate in real time with each other, roadside infrastructure, and central traffic management systems, ensuring safety and efficiency. This includes advanced vehicle-to-everything (V2X) communication (V2V, V2I, V2P, V2N) for enhanced safety, platooning, and cooperative driving.
2. **Smart Traffic Management:** AI-powered 6G networks will facilitate dynamic traffic light synchronization, real-time route optimization, and proactive congestion management by analyzing data from millions of connected devices across an entire city or region.
3. **Cooperative Collision Avoidance:** Near-zero latency communication will allow vehicles to detect potential collisions and react instantaneously, significantly reducing traffic accidents (Anvigh et al., 2024).
4. **Enhanced Passenger Experiences:** 6G will enable immersive extended reality (XR) applications (Augmented Reality/Virtual Reality/Mixed Reality) within vehicles, offering advanced infotainment, remote work capabilities, and holographic telepresence for a truly connected travel experience.
5. **Drone Delivery and Urban Air Mobility (UAM):** The precise localization, high bandwidth, and low latency of 6G are crucial for safe and efficient operation of drones for delivery services and future UAM systems (Routis & Katsouris 2024).
6. **Real-Time Incident Management and Emergency Response:** 6G-enabled systems can instantaneously detect incidents, alert authorities, and reroute traffic, ensuring faster emergency response and minimal disruption (Saad, 2021).

5. Challenges of 6G Communication

Despite the immense potential, the integration of 6G into transportation systems faces several significant challenges that require extensive research and development (Zorzi & Zanella. 2022):

- **Cybersecurity:** The integration of 6G raises new threats such as signal spoofing, data interception, and AI model manipulation, necessitating advanced security protocols.
- **Infrastructure Costs:** Deploying 6G infrastructure, especially in rural or underdeveloped areas, requires significant investment in base stations, fiber optics, and edge nodes.
- **Spectrum Allocation and Management:** With new higher frequency bands (e.g., THz), there are complexities in signal propagation, path loss, and interference control, which provide opportunities for new spectrum management.
- **Network Security and Privacy:** The extensive interconnectivity of 6G-enabled intelligent transportation systems and the pervasive use of AI within them raise key factors of data privacy, cyber-related risks, and network resilience against disruptive threats (in particular, from malicious actors). Advanced protections, such as quantum-resistant cryptographic protections, will be required to accommodate any of these challenges (Osorio et.al., 2022)
- **Interoperability and Standardization:** Global standardization efforts are essential to facilitate communication and data exchange between different vehicle manufacturers, infrastructure providers, and network operators. Smart sensors linked through 6G networks will connect to monitor the structural conditions of bridges, roads, and tunnels, identifying a structural failure before an incident occurs, ultimately saving lives.
- **Resource Management:** Efficient allocation of computational, communication, and caching resources across the complex 6G architecture (including edge and cloud layers) is crucial for optimal performance.
- **Energy Efficiency and Sustainability:** The increased density of 6G infrastructure and higher data rates could lead to increased energy consumption. Balancing technological advancement with environmental sustainability is a critical research area.
- **Deployment Cost and Infrastructure:** The extensive deployment of new infrastructure, including high-frequency transmitters and potentially satellites, will involve significant financial investment and logistical complexities.
- **AI and ML Integration Challenges:** While AI is a cornerstone of 6G, challenges remain in channel estimation, network traffic prediction, intelligent routing, and dynamic radio resource management for highly complex and dynamic vehicular environments.
- **Challenges and Considerations:** 6G promises an incredible future, but there are challenges that we must overcome.
    1. **Infrastructure Development:** Implementing 6G globally will require significant investments in infrastructure for upgrading existing networks, and deploying new infrastructure will be essential for the seamless adoption of this ground-breaking technology.

2. **Security and Privacy:** As we embrace a hyper-connected world, ensuring the security and privacy of our data will be paramount. Robust measures must be in place to protect against cyber threats and safeguard personal information. In exploring the fascinating world of 6G, a technology that promises to revolutionize our lives in ways we can only dream of—from mind-blowing speeds to immersive experiences—the possibilities are endless. The various security threats are listed in Table 17.3.

Table 17.3 Security concerns in 6G networks.

Security Concern	*Description*	*Potential Mitigation Strategies*
Privacy Leakage	Massive data collection from IoT and wearables may expose personal data.	Federated learning, differential privacy, and encryption techniques.
AI-Based Threats	Malicious use of AI to manipulate network behavior or create deep fakes.	AI-based threat detection and explainable AI for transparency.
Quantum Attacks	Quantum computing may break current cryptographic methods.	Post-quantum cryptography and quantum-resistant algorithms.
Device Authentication	Billions of connected devices pose challenges for secure identification.	Lightweight authentication protocols and blockchain-based identity.
Edge Security	Vulnerable edge devices can be exploited to gain access to core systems.	Secure edge computing frameworks, trusted execution environments.
Network Slicing Attacks	Improper isolation may allow cross-slice attacks or resource hijacking.	Strict access control and dynamic slice isolation mechanisms.
Denial-of-Service (DoS)/ Distributed Denial of Service (DDoS)	Overloading networks or services to cause disruption.	AI-based anomaly detection and traffic shaping.
Malicious RIS Configuration	Reconfigurable surfaces may be hijacked to redirect or block communications.	Secure control protocols and RIS firmware integrity checks.
Supply Chain Attacks	Threats introduced via third-party hardware/software components.	Secure supply chain auditing and zero-trust architecture.
Satellite Link Vulnerabilities	Satellite communications can be intercepted or spoofed.	End-to-end encryption and secure satellite control protocols.

6. Applications in Transportation Systems

6.1 Autonomous Vehicles

6G technology greatly increases vehicle autonomy through the ability to share data seamlessly in real time amongst vehicles, infrastructure, and the cloud. This provides advanced functionalities such as cooperative driving, collision avoidance, and dynamic route planning. ITS refers to the seamless integration of advanced technologies into transportation infrastructure and vehicles to foster safety, efficiency, and mobility. It involves aspects of information and communication technology (ICT) to address a number of transportation problems such as traffic management, vehicle safety, and public transit. ITS also includes cooperative and user-friendly technologies, e.g., coordinated signals, real-time traffic status, and advanced technologies such as electronic tolling and smart sensors. ITS allows for the collection of various data pertaining to transportation purposes, thereby allowing for the optimization of traffic flow, mitigation of congestion, and improved road safety. Ultimately, ITS plays a key role in integrating the widespread use of autonomous vehicles but similarly contributes to sustainable and intelligent transportation system development. In summary its leverages technology to make transportation systems smarter safer and more efficient.

6.2 Intelligent Traffic Management

AI-enabled 6G networks support dynamic traffic light control, congestion prediction, and efficient incident response. Real-time analytics ensure smooth traffic flow and reduced emissions. ITS is a rapidly growing field that combines technologies including communications, electronics, and information systems to improve the efficiency, safety, and mobility of transportation systems. ITS are referred to as a wide range of advanced technologies and applications for the purpose of improving the efficiency, safety, and mobility of transportation networks. Specifically, each of these applications aims at improving traffic flow, for instance, by reducing accidents or improving the use of transportation infrastructure. It can include everything from advanced traffic management systems and connected and autonomous vehicles to smart transit systems. A significant component of ITS is advanced traffic management systems (ATMS), which utilize real-time traffic data and information to optimize signal timing, mitigate congestion, and enhance traffic flow. Another major ITS application is the implementation and expansion of connected and autonomous vehicles (CAVs), which use sensors, communications, and AI technologies to improve roadway safety, ease the burden of traffic congestion, and increase mobility for those unable to otherwise drive.

Furthermore, opportunities to utilize smart transit systems will also improve the efficiency, reliability, and user experience of public transportation services, utilizing technologies such as real-time passenger information systems or GPS tracking. These systems have been rapidly evolving, and the collective contributions of these and other systems have the potential to transform the future of our mobility patterns. As ITS technologies continue to advance, and even with a horizon of 20–30 years, we expect that more people will be transported in a more efficient and safe manner and that goods will move through urban areas better as well. Ultimately, with connected vehicles, smart infrastructure, data-driven traffic operations, and sustainability, we can expect shorter trips, fewer accidents, more equitable mobility options, and a renewed focus on low-carbon modes. The ITS landscape will be an important part of the future for smarter, safer, and more livable communities.

Emergency Vehicle Prioritization: As a part of emergency vehicle prioritization, the AI-driven 6G can easily prioritize emergency vehicles and send the information to nearby vehicles for route clearance, automating the route for vehicles like ambulances and VIP vehicles.

6.3 Infrastructure Monitoring

6G networks will connect to monitor the structural conditions of bridges, roads, and tunnels, identifying a structural failure before an incident occurs, ultimately saving lives.

With the advent of 6G, all the drones can easily communicate to the network and access the data for clearance and transmit the drones easily.

6.4 Smart Public Transport

6G will encompass transport systems that are integrated into a multi-modular system where all facets will be known in real time like passenger flow, future passenger numbers, and service adjustments to optimize efficiency by examining demand. By using AI and data analytics, we can easily predict traffic patterns and traffic in various routes. Traffic can be easily classified during the peak times and diverted to other routes for easily traffic management.

Smart public transport refers to public transport systems using advanced technologies such as the IoT, AI, big data analytics, and next generation wireless technology (5G/6G), together with smart infrastructure. The aim is to improve urban mobility by making public transport more efficient, accessible, safe, and environment-friendly.

The key features of smart public transport systems incorporate the following:

1. **Real-time tracking** of buses, trains, and other transit vehicles.
2. **Dynamic route optimization** based on traffic and passenger flow.
3. **Smart ticketing systems**, such as contactless payments and mobile apps.
4. **Passenger information systems** for live updates and scheduling.
5. **Predictive maintenance** using sensor data from vehicles and infrastructure.

The key technologies involved in the smart environment are listed as follows.

1. **IoT Sensors:** Monitor vehicle performance, traffic patterns, and environmental conditions.
2. **AI and ML:** Analyze data to predict delays, optimize routes, and improve service planning.
3. **Big Data Analytics:** Support decision-making and policy development through analysis of ridership, demand, and operations.
4. **5G/6G Connectivity:** Enables high-speed, low-latency communication for real-time control and monitoring.
5. **Cloud Computing:** Stores and processes vast volumes of data for smart city integration.

The key benefits of smart public transport include the following:

1. **Improved Efficiency:** Reduces congestion, wait times, and operational costs.
2. **Enhanced User Experience:** Real-time information and seamless ticketing improve passenger convenience.

3. **Sustainability:** Promotes a shift from private to public transport, reducing emissions and energy consumption.
4. **Safety and Security:** Integrated surveillance, emergency response systems, and predictive analytics enhance commuter safety.

The key challenges of smart public transport include the following.

1. **Data Privacy and Cyber Security:** Protecting user and operational data from breaches.
2. **High Initial Investment:** Requires significant infrastructure and system upgrades.
3. **Interoperability:** Ensuring diverse technologies and systems work together smoothly.
4. **Public Acceptance:** Ensuring adoption among various demographics and educating users.

7. Enabling Technologies

The revolution of 6G technology is taking shape through the integration of advanced technologies such as AI, edge computing, blockchain, and quantum communication. The convergence of these innovations enables ultra-reliable, high-speed communication and facilitates the delivery of high-quality services to end users across diverse domains.

7.1 Edge Computing

Edge computing enables data to be processed locally on edge devices, significantly reducing latency and alleviating the computational burden on centralized servers. This is particularly critical for time-sensitive operations within transportation systems. The integration of edge computing with 6G technology is reshaping modern mobility by facilitating ultra-low latency, real-time decision-making, and improved data security. By processing information closer to the data source such as within vehicles or roadside traffic sensors, edge computing minimizes dependence on cloud infrastructure and accelerates system responsiveness, thereby enhancing the efficiency and reliability of intelligent transportation networks to microseconds (critical to autonomous driving, smart traffic management, and V2X communications). 6G networks advance these capabilities with AI-based optimization, quantum-resistant data encryption, and connectivity that can support massive connected devices (IoT), providing scalability and resiliency. This dynamic duo will provide energy efficiency, data privacy, and distributed AI, forming a framework for the automated, sustainable, and intelligent transportation system ecosystem. This ensemble will yield breakthroughs in smart cities, drone traffic management, and holographic telematics, preparing the way for next-generation mobility solutions.

7.2 Blockchain

Blockchain ensures secure and transparent communication between autonomous agents in transportation, particularly for logistics and supply chain applications.

Incorporating blockchain with 6G-based transportation systems will create more secure, transparent, and efficient mobility systems. Blockchain powers an immutable, distributed ledger capable of creating without interference and verifying data. This is important for secure, verifiable data transfer between vehicles, infrastructure, and service providers, required for applications such as coordinating autonomous vehicles, smart tolling, and managing logistics in the supply chain. With 6G's ultra-reduced latency, combined with high bandwidth, the channels will enable real-time transactions through blockchain that will allow for micro transactions including vehicle-to-grid (V2G) for energy trading or smart contracts for ride-sharing designed through dynamic contracts. Furthermore, blockchain's smart contracts facilitate the automation of trustless agreements such as traffic violations, freight documents, to alleviate the issues of fraud and the absence of time with carrier admin. 6G's low latency and high-bandwidth capacity as well as block chain's security will also allow for decentralized traffic management as edge nodes collaborate to verify congestion data in the absence of a central authority. For transportation to evolve into a connected and fully autonomous environment, the integration of blockchain and 6G will ensure the integrity of the data, enable good cyber resilience, and permit seamless interoperability while creating the foundation for safer, smarter mobility networks.

7.3 Digital Twins

Digital replicas of transportation networks allow simulation and predictive modeling, enhancing system resilience and planning. Digital twins for transportation networks are a ground-breaking technological advancement that allow for the real-time digital virtualization of physical networks, vehicles, and traffic systems. Digital twins are created through assimilating data from sensors, IoT devices, and AI algorithms attached to, or tracking, real-world physical networks. Digital twins provide a dynamic and interactive model that orchestrates and showcases transportation systems wherever they are deployed in the built environment. Digital twins provide the capability for continuous monitoring, predictive analysis, and direct involvement in the data that can be incorporated into asset management studies and knowledge streams. Digital twins, in the context of smart cities, provide an important function for transportation development, maintenance management, and incident management, enabling adaptable, resilient, and sustainable urban transport networks.

7.4 Augmented Reality

Augmented reality will become an integral part of our daily lives with the advent of 6G by overlaying digital information onto the physical world, across sectors such as education, healthcare, and entertainment. Students will be able to interact with lifelike 3D models in their classrooms, making learning more engaging and interactive. In the medical field, surgeons will have access to real-time patient data and overlay. Furthermore entertainment experiences will be taken to new heights as we immerse ourselves in AR games, interactive. movies, and virtual tours. The transformative applications of 6G are truly awe-inspiring, from holographic communication to smart environments and augmented reality. This next generation technology will reshape our world in ways we can only dream of—the possibilities are endless and we are just scratching the surface of what 6G has in store for us.

8. Environmental Impact and Sustainability

The development of 6G provides sustainable and environment friendly solutions. People everywhere are demanding seamless mobility and safety and sustainability for the environment in transportation (Alsharif & Kim 2020). An unmatched transition in transportation systems and the development of smart cities is expected with this, given the high density in urban areas, growing populations, and that the strain in present transport systems. In this context, 6G presents a hopeful, disruptive enabler with improvements of existing 5G capabilities. It marks a new development direction along with AI potential for intelligent connectivity (hailed as 6G), machine learning, and edge computing with ultra-dense networks. The transformational potential of 6G in transportation is more than just high-speed data; it is about connecting everything—vehicles, users, infrastructure, and devices using low-latency secure communication in as close to real-time as possible. Autonomous vehicles need constant connection to their surroundings to make accurate driving decisions 6G is able to meet the bandwidth and latency requirements of these constant connections with V2X communications, which can help provide safer and better-coordinated mobility experiences.

1. **Green and Energy-Efficient Transport:** The integration of 6G technology into the transportation sector holds significant promise for advancing green and energy-efficient mobility solutions. With its ultra-low latency, high data rates, and AI-native architecture, 6G enables real-time optimization of traffic flow, dynamic route planning, and intelligent energy management in vehicles and transport infrastructure. These capabilities support the deployment of electric and autonomous vehicles by enhancing battery efficiency, minimizing idle time, and reducing fuel consumption. Additionally, 6G facilitates the widespread use of IoT sensors and edge computing to monitor environmental conditions, emissions, and energy usage, thereby enabling data-driven decisions that contribute to sustainable urban mobility. The seamless interconnection of smart transportation systems under 6G not only improves operational efficiency but also aligns with global climate goals by significantly lowering carbon footprints and promoting eco-friendly transportation networks.
2. **Reducing Congestion and Carbon Footprint through Optimization:** AI-driven 6G can estimate the traffic and avoid the congestion it leads to through the optimization of resources.
3. **Eco-friendly Routing and Smart Charging:** The advancement of traffic data and analysis of traffic patterns can estimate optimized routing.

9. Conclusion and Future Directions

Intelligent public transportation is key to intelligent, livable cities that mitigate the mobility-related challenges brought on by patterns of urbanization and innovation in urban public transport. With cities expanding and increasingly complex mobility needs, the future of urban transport depends on the smart mobility solutions we will invest in today to help ensure that our cities are sustainable and inclusive. 6G will be a key foundational technology in next-generation transport systems. It will enable new levels of capability not possible with 5G and past wireless communications. It will delight in the ability to extend pervasive intelligence, real-time responsiveness, and huge connectivity, thus fuelling the shift of mobility infrastructures from effective and efficient flows of people, goods, and vehicles to adaptive, resilient, and user-focused ecosystems. To bring about the full potential of 6G from a transport perspective will take strategic investment, strong cyber security, and effective collaborative governance.

The integration of 6G within transportation systems represents a future that could be fundamentally different from the present, providing unprecedented levels of connectivity, autonomy, safety, and efficiency. While the vision is enticing, delivering the potential of 6G involves addressing major technical, security, and economic constraints. Progress in core areas of research, such as THz communication, AI-assisted network management, strong mechanisms for security, and comprehensive standards, will be paramount to a future that includes intelligent and connected transport systems. The incorporation of 6G into transportation systems represents a transformational opportunity for smart, safe, and sustainable mobility. A 6G-based ITS provides the inspiration for new approaches to the function and form of moving people and goods in the 21st century.

In the future toward the development of green 6G solutions to reduce energy consumption in transportation networks, integration with quantum communication will enhance security, and AI-optimized mobility-as-a-service (MaaS) platforms can be implemented.

References

Akyildiz, I.F., A. Kak and S. Nie. (2020). 6G and beyond: the future of wireless communications systems, *IEEE Access*, *8*, 133995–134030.

Alsharif, M.H., J. Kim and J.H. Kim. (2020). Green and sustainable cellular base stations: an overview and future research directions, *Energies*, *13*(4), 833. https://doi.org/10.3390/en13040833.

Andreou, Andreas, et al. (2024). Enhancing secure communication in 6G-enabled IoV through UAV and control center integration, in *2024 International Wireless Communications and Mobile Computing (IWCMC)*.

Anvigh, A.A., Y. Khavan and B. Pourghebleh. (2024). Transforming vehicular networks: how 6G can revolutionize intelligent transportation? *Science, Engineering and Technology*, *4*(1), 80–93.

Chen, S., F. Qin, B. Hu and B. Ai. (2020). 6G: opening new horizons for integration of space–air–ground–sea communication, *National Science Review*, *7*(6), 988–1014. https://doi.org/10.1093/nsr/nwaa079.

Chen, Xiaosha, et al. (2020). Deep-learning-based intelligent intervehicle distance control for 6G-enabled cooperative autonomous driving, *IEEE Internet of Things Journal*, *8*(20), 15180–15190.

Deng, X., et al. (2023). A review of 6G autonomous intelligent transportation systems: mechanisms, applications and challenges, *Journal of Systems Architecture*, *142*, 102929.

Giordani, M., M. Polese, A. Roy, D. Castor and M. Zorzi. (2020). Toward 6G networks: use cases and technologies, *IEEE Communications Magazine*, *58*(3), 55–61. https://doi.org/10.1109/MCOM.001.1900411.

Hussain, F. and S.A. Hassan. (2020). Future vehicular networks: a review on integration of 6G and intelligent transportation system, *IEEE Access*, *8*, 165123–165144. https://doi.org/10.1109/ACCESS.2020.3022192.

Khan, L.U., et al. (2024). *Edge AI for Autonomous Vehicles: A 6G Perspective.* Springer Autonomous Systems.

Letaief, K.B., et al. (2023). 6G wireless networks: vision and challenges, *IEEE Wireless Communications.*

Liu, R. et al. (2024). 6G enabled advanced transportation systems, *IEEE Transactions on Intelligent Transportation Systems*, *25*(9), 10564–10580.

Osorio, D.P.M. et al. (2022). Towards 6G-enabled internet of vehicles: security and privacy, *IEEE Open Journal of the Communications Society*, *3*, 82–105.

Routis, G., G. Katsouris and I. Roussaki. (2024). Cryptography-based location privacy protection in the internet of vehicles, *Journal of Ambient Intelligence and Humanized Computing*, *15*(8), 3119–3139.

Saad, W., et al. (2021). 6G wireless systems: vision, requirements, and challenges, *IEEE Vehicular Technology Magazine.*

Saad, W., M. Bennis and M. Chen. (2019). A vision of 6G wireless systems: applications, trends, technologies, and open research problems, *IEEE Network*, *34*(3), 134–142. https://doi.org/10.1109/MNET.001.1900287.

Shafin, R., L. Liu, S. Chandrasekharan, Y. Chen and X. Wang. (2020). Artificial intelligence-enabled cellular networks: a critical path to 6G, *IEEE Wireless Communications*, *27*(2), 212–217. https://doi.org/10.1109/MWC.001.1900323.

Shawky, M.A., et al. (2023). Cooperative intelligent transport systems for net-zero, 167–191.

Yang, P., Y. Xiao, M. Xiao and S. Li. (2021). 6G wireless communications: vision and potential techniques, *IEEE Network*, *35*(2), 104–111. https://doi.org/10.1109/MNET.011.2000277.

Zhang, Z., Y. Xiao, Z. Ma, M. Xiao, Z. Ding, X. Lei and H.V. Poor. (2019). 6G wireless networks: vision, requirements, architecture, and key technologies, *IEEE Vehicular Technology Magazine*, *14*(3), 28–41. https://doi.org/10.1109/MVT.2019.2921208.

Zorzi, M. and A. Zanella. (2022). Intelligent vehicular communications in 6G: challenges and perspectives, *IEEE Internet of Things Journal*, *9*(4), 2346–2357. https://doi.org/10.1109/JIOT.2021.3108654.

Zhang, J.A., Rahman, M.L., Wu, K., Huang, X., Guo, Y.J., Chen, S. and Yuan, J. (2022). Enabling joint communication and radar sensing in mobile networks—a survey. *IEEE Communications Surveys & Tutorials 24*(1), 306–345

Chapter 18

Integration of Advanced Wireless Communication Technologies 5G, 6G and Beyond in Transportation Systems: Opportunities and Challenges

A. Vasuki [1] and *Vijayakumar Ponnusamy* [2*]

There are many potential improvements to intelligent transportation systems (ITS) with the arrival of sixth-generation (6G) wireless communication technologies. By examining the standards, technology, and implementation problems related to 6G, this chapter explores its potential to transform transportation systems. Building on the development of 5G, 6G delivers extraordinary Potentiality, including very low latency, high throughput, wide-ranging connections, and intelligent network harmony. By utilizing these features, 6G may enable smooth communication between automobiles, infrastructure, and people, opening the door to innovative uses such as cooperative collision prevention, smart traffic management, and autonomous driving.

1. Introduction to Intelligent Transport Systems

As a key component of modernizing transportation networks, intelligent transportation systems (ITS) are revolutionizing the flow of people and goods inside and between cities (Anvigh, et al., 2024). Vehicular ad hoc networks (VANETs) were created by integrating communication technologies into vehicular networks, allowing cars to communicate with the environment, infrastructure, and one another. Transportation operations now operate with more safety, efficiency, and sustainability thanks to this technical breakthrough. A major turning point in the development of VANETs was the

[1] Department of Electronics and Communication Engineering, SRM Institute of Science and Technology, Vadapalani, Chennai, India. https://orcid.org/0000-0002-4397-6244

[2] Department of Electronics and Communication Engineering, SRM Institute of Science and Technology, Kattankulathur, Chengalpattu, India. https://orcid.org/0000-0002-3929-8495

* Corresponding author: vijayakp@srmist.edu.in

DOI: 10.1201/9781003659822-18

introduction of 5G technology, which offered more connections, lower latency, and quicker data rates (Guo et al., 2022; Kaul & Altaf, 2022). But as the need for intelligent transportation keeps increasing, 5G's shortcomings become more apparent, which encourages companies and academics to investigate the revolutionary possibilities of 6G technology. The future technology of 6G has been developed for the unrelenting pursuit of technical innovation.

1.1 Essential Elements of ITS

By combining cutting-edge technology, ITS has completely transformed transportation.

ITS has developed steadily since the 1970s to provide transportation users with better services. ITS includes innovative technology into the transportation framework, together with electronic sensors, data transmission, and intelligent control (Guerrero-Ibáñez et al., 2018).

ITS is a collection of components that cooperate to strengthen the sustainability, protection, and efficiency of ITS networks. The ITS components are illustrated in Figure 18.1 (Jha et al., 2024).

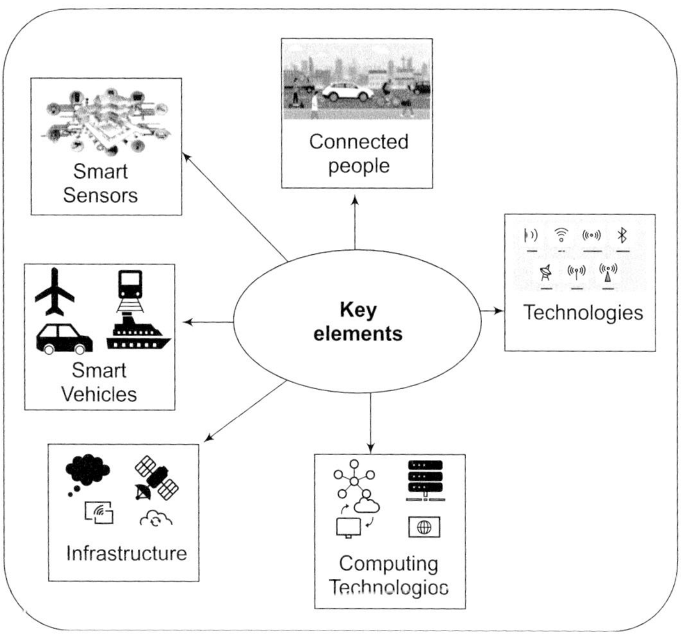

Fig. 18.1 Key elements of ITS.

- **Intelligent Framework and Sensors:** ITS is equipped with innovative tools to detect and interpret different traffic conditions on roads, bridges, traffic signals, and lighting. Additionally, the transportation network is equipped with cameras, lidar, radar, and environmental sensors. The different data on weather, traffic flow, and road conditions are gathered to give practical data for managing traffic and decision-making.

- **Networking Pedestrians:** By giving pedestrians access to information and connection, this component aims to ensure their safety and convenience. This may include technologies or applications that assist pedestrians in safely navigating metropolitan areas, as well as pedestrian-friendly infrastructure like crosswalks and pedestrian signals. Vehicle to pedestrian (V2P) communication allows pedestrians and cars to exchange information.
- **Smart Vehicles:** Advanced technologies installed in smart cars allow them to connect with the transportation infrastructure as well as with one another. These advancements boost safety and traffic management.
- **Communication Framework:** This component includes the basic communication networks that provide data sharing between different ITS components. Drones and satellite communication may both serve as communication infrastructures. For cars, infrastructure and central control systems to share data in real time, high-speed, low-latency networks are necessary.
- **Communication Methods:** Communication methods encompass the standards and protocols used for transferring data within the ITS. Examples include Wi-Fi, cellular networks, and wide-area networks. This technique facilitates information transfer between vehicles and infrastructure.
- **Computing Techniques:** Artificial intelligence (AI) and machine learning (ML) handle the fundamental functions, including traffic forecast, anomaly detection, and traffic flow optimization. To handle large amounts of data transfer while ensuring reliability and scalability, cloud computing is utilized. Edge computing is used to lower the delay and make a faster decision-making process. Big data analytics collects and examines vast amounts of data from a variety of sources, such as sensors, Global Positioning System (GPS) units, and traffic cameras.

2. Requirements of ITS

ITS needs might change based on their unique aims and objectives. Nonetheless, the following are some standard needs that are usually taken into account while developing ITS (Patel et al., 2019).

- **Data gathering and evaluation:** Data must be collected from different sensors, cameras, and cars for the ITS to compile information on traffic, road circumstances, and related parameters. This gathered information is then analyzed to enhance decision-making and provide insights in real time.
- **Communication Structure:** For ITS to facilitate data and information sharing across various system components, a strong communication infrastructure is necessary. Internet connection, wireless networks, and specialized communication channels are a few examples of this infrastructure.
- **Integration with Existing Infrastructures:** To increase efficiency and reduce disturbance, ITS should be built to interface with the current transportation systems and infrastructures.
- **Advanced Technologies:** AI, ML, and data analytics are used by ITS to evaluate and comprehend the collected data using emerging technology. These

technologies help predict congestion, streamline traffic, and improve system performance in general.

- **Incorporation and Compatibility:** To ensure smooth coordination and communication, it is crucial that ITS components are interconnected and compatible. This enables efficient data sharing and collaboration across many systems and stakeholders.
- **Scalability:** As the system grows, ITS should be able to manage growing numbers of users and data.
- **Practical Capabilities:** For many ITS applications to make judgments and react fast, real-time data processing and transmission are essential.
- **Accessible Links:** ITS should provide easily accessible applications and links for passengers, transit providers, and other users. These links must be handy, unambiguous, and quick to deliver pertinent data and services. Real-time data processing and transmission are necessary for many ITS applications to support quick decisions and reactions.
- **Safety and Security:** It is essential to guarantee the security and safety of ITS. To safeguard data and stop harmful activity or unauthorized access, measures like access control, authentication, and encryption should be put in place.

There is a good chance that 6G technology will meet many of the ITS needs. For instance, it may provide reduced latency, increased data throughput, and quicker and more dependable communication networks. This would enable practical data interpretation and transfer, which is crucial for self-governed vehicles and traffic control applications. Furthermore, by offering a more reliable and adaptable communication infrastructure, 6G technology may enhance the scalability and interoperability of ITS systems. Additionally, blockchain could be used to enhance data integrity in ITS. In conclusion, the development of dependable, effective, and user-focused ITS may greatly benefit from 6G.

3. Development of 6G

From 1G to 6G, cellular technologies have significantly advanced in terms of speed, latency, applications, and other factors.

1G: Although they had limited coverage, the initial analogue cellular networks (1G) offered simple voice calling.

2G: With the introduction of digital transmission brought about by 2G in the 1990s, voice quality and capacity were enhanced.
In contrast to 1G networks, which transmitted data and speech using analogue signals, 2G networks utilized digital signals., making it possible to send data in addition to speech.

3G: In 2000, 3G marked the shift from 2G narrowband data services to mobile broadband, opening up services like internet access and multimedia messaging that went beyond voice. 3G networks enabled worldwide data roaming, allowing customers to use data services when traveling overseas, much like 2G's global roaming features for voice calls.

4G: With a connection capacity of 600 users per cell, 4G offered higher data speeds (up to 3 GB/s for downloads and 1.5 GB/s for uploads), reduced delay (around 30–50 ms), and compatibility for multimedia apps.

5G: With 5G, it was feasible to connect a huge number of devices (10^6 users per square kilometre), achieve ultra-low delay (10 ms), and achieve very high data speeds (up to 20 Gbps for downlink and 10 Gbps for uplink). These networks have recently improved existing data services and supported several vertical businesses. Enhanced mobile broadband (eMBB), which offers higher data speeds together with modest latency improvements, is the main emphasis of the first phase of 5G implementation. It seems sensible that transportation systems will depend more and more on communication strategies as they get bigger and more complex.

6G: Beyond the current 5G standard, 6G, or sixth-generation wireless communication technology, is the next significant advancement in mobile networks. Although 5G has already started to change a number of sectors, 6G promises to push the envelope even farther with its unmatched capabilities and game-changing potential.

Although the exact definition and requirements of 6G are still being developed since it is a developing industry, it is anticipated to provide significant advancements in data throughput, latency, connection, and intelligent network management. The three layers that make up the 6G architecture are shown in Figure 18.2.

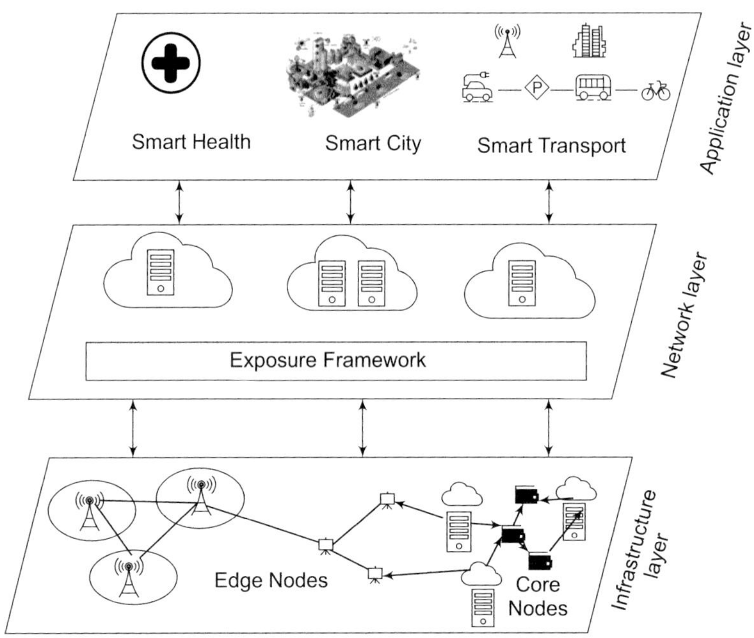

Fig. 18.2 Architecture of 6G

- **Infrastructure Layer:** This physical layer houses the equipment needed to construct the network, including data centres, cloud infrastructure, switches, routers, radio equipment, and communication cables. This layer provides the required resources for the application and network layer.
- **Network Layer:** It offers logical resources for application layer components. The application layer receives network services from this layer. 6G uses microservices to perform network operations, processes, and applications, resulting in an intelligent and software-defined architecture. It is made up of the virtualization layer, the logical network layer, and management orchestration features including security, risk management, and inter-domain administration.
- **Application Layer:** It comprises the end-user services and apps that are enabled by the network infrastructure and auxiliary services. This layer provides the services that end users interact with, such as smart homes, smart transportation, and healthcare.

4. Integration of 6G and Intelligent Transport System

Network Edge Solutions towards better V2X Communications: In real-world applications of ITS, edge-located solutions will be crucial (Kilanioti et al., 2022). As traffic volume increases, more data will need to be electronically sent across several road users. It is necessary to use such shared V2X delay-constrained material for passenger entertainment, road safety, traffic management, and environmental monitoring. As seen in Figure 18.3, the vehicular edge computing (VEC) paradigm focuses on content management via effective offloading (task processing, content distribution) and storage to meet the need for efficient manipulation of massive volumes of data.

Linked intelligence is the foundation of most 6G's evolving solutions for delayed content delivery. Big data techniques and ML, which efficiently and without explicitly stated programmed rules extract knowledge from noisy data with non-stationary features, enable emerging technologies in dynamic environments of multi-modal and time-varying data being produced for high-definition maps, infotainment materials, dynamic planning information, etc. ML-based approaches, such the optimization or inference of latent variables for data-driven models, are supported by the proliferation of low-cost storage and processing power available in the cloud, the expansion of data sources, and advancements in network softwarization.

On the edge level, popular material might be proactively gathered, and the car could broadcast to other cars. The choice of what should be cached may be made collaboratively. Qin et al. (2020) offer a hierarchical edge-to-edge system combining data transmission, computation offloading, and content caching, finding a suboptimal solution using a deep deterministic policy gradient-based resource allocation system. Wu et al. (2021) distribute work offloading using multi-RATs and a fronthaul-aware upper confidence bound methodology.

Kuo et al. (2022) collaborate to improve communication, processing, and caching between vehicles and the vehicular network's roadside units (RSUs) to guarantee the quality of video. It is discovered that the quality may be significantly improved by using a cache replacement strategy.

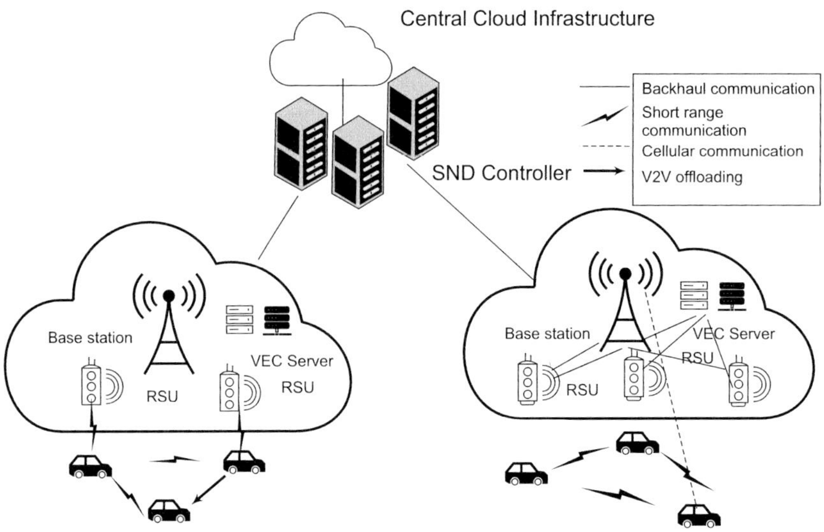

Fig. 18.3 Vehicular edge 6G-enabled ITS structure.

5. 6G Technologies for ITS

Intelligent transportation systems' open architecture, which uses wireless connectivity, causes problems with computing, communication, security, and privacy. Despite the fact that 5G communication technology was created to solve these issues, high mobility and transportation-related issues are not addressed by contemporary 5G designs and advancements. For instance, 5G allows for 10 Gb/s download and 20 Gb/s upload data rates, with a latency of 10–50 ms. However, since cars move quickly, AI-enabled car transportation systems need more dependable communication. To reduce latency to 1 ms and optimize the data rate to 1 Tb/s, 6G must be included into the transportation system. Imagine an ITS powered by 6G, where smart automobiles can interact with existing infrastructure, other vehicles, and people's devices. Our transportation system would become safe and interoperable as a result. Numerous advantages, new business opportunities, and a fresh perspective on the 6G-enabled transportation system are all present.

Case 1: Modern cars can see in all directions within 300 meters and confidentially store personal data. Drivers may get prompt congestion notifications by integrating 6G into the ITS. They may sometimes get it by vibration, tone, or visual display, which aids drivers in preventing possible collisions. Conversely, cars may connect to 6G-enabled roadside infrastructure to promptly notify other drivers when another vehicle is crossing. This also helps notify drivers in inclement weather, on snowy roads, or when the roadways are slick.

Furthermore, faster information transmission regarding natural hazards from several cars may be possible with the incorporation of 6G into the vehicular network, warning drivers before they come into contact with them.

Case 2: When 6G-enabled roadside equipment is deployed, vehicle weather data may be instantly sent to the traffic management system, giving real-time information to help manage or track the efficiency of the transportation system. Traffic and speed light adjustments are two ways that centres might respond. Before leaving their homes, drivers may check the weather and road conditions in real time.

It is possible to raise awareness of medical emergencies and police activities. Using 6G connectivity, an accident zone warns cars of the risk area and recommends braking and route changes in a matter of seconds. Additionally, the digital transportation system with 6G connectivity links passengers and public transit more effectively.

The primary computing and communication technologies that are expected to influence the 6G-ITS communication environment are discussed in this section.

The foundation of contemporary connectedness and information sharing is made up of communication technologies, both wired and wireless. Through physical connections, wired communication techniques like fibre optics, while conventional copper wires provide dependable and secure data delivery. The foundation for reliable phone lines, strong network infrastructures, and fast Internet is provided. The wireless techniques like Wi-Fi and satellite offer independence and the convenience of an untethered connection. Mobility, wide-ranging device support, and ubiquitous information access are all made possible by these wireless technologies.

Their complementary functions aid in the creation of a smooth and all-encompassing communication ecology, which serves as the foundation for the 6G-ITS environment's functionality. Terrestrial and non-terrestrial communications are types of communication technology.

Non-terrestrial Communication: One of the main communication technologies in 6G-ITS is non-terrestrial communication (Yu et al., 2024a). Signal transmission using channels other than land-based infrastructure is known as non-terrestrial communication. It usually refers to communication systems, such as satellites and space-based networks, that function outside of the Earth's surface. Satellite communication (Giordani & Zorzi, 2020) is one of the non-terrestrial techniques used in 6G systems.

In 6G-ITS networks, satellite communication will be crucial, particularly for bridging the digital divide and offering worldwide coverage (Araniti et al., 2021). When terrestrial infrastructure is scarce, satellites may be utilized to enhance terrestrial networks, provide connectivity to far-flung locations, and facilitate smooth communication. Now, the research works are focusing on providing high-capacity and low-latent communication via 6G by achieving high-throughput satellites and advanced communication protocols.

- **Terrestrial communication:** Terrestrial communication systems are those that transmit signals using infrastructure located on land. These systems send data signals over comparatively small distances using fibre optics, wires, and cables [Jha et al., 2021]. Similar to the 4G network, 6G will be used for national, regional, and local communication (Yastrebova et al., 2019).
- **Multiple-Input Multiple-Output (MIMO):** It is expected that 6G would significantly enhance and optimize MIMO technology, which is now extensively used in 4G and 5G networks. MIMO enhances system performance, capacity, and

spectral efficiency by using several antennas at the sender and receiver (Affan et al., 2021). Higher data speeds, greater coverage, and better interference control in 6G networks may be made possible by advanced MIMO methods as intelligent beamforming and massive MIMO (Dilli et al., 2021).

- **Quantum Key Distribution:** Unmatched levels of encryption are provided by quantum key distribution protocols, which allow for safe key exchange between communicating parties. For crucial applications in 6G-ITS networks, quantum communication may provide strong security by guarding against new dangers like quantum hacking and guaranteeing the integrity and secrecy of data transmission (Mehic et al., 2020).
- **Modulation Techniques:** Wireless communication systems rely on modulation methods, and improvements in modulation schemes keep boosting spectral efficiency and data throughput (Chen et al., 2020). Higher-order Quadrature Amplitude Modulation (QAM) and Orthogonal Frequency Division Multiplexing (OFDM) variations are two novel modulation techniques being investigated for Sixth Generation (6G) networks in 6G-ITS. Higher data throughput and enhanced 6G-ITS performance in difficult propagation situations is made possible by these sophisticated modulation methods, which enable more effective use of the available frequency spectrum.
- **Terahertz (THz) Communication:** The frequency range in which terahertz communication works is between microwave and infrared waves. Because of the wider range of spectrum, THz is a key candidate for providing ultra-high data rates in 6G (Huq et al., 2022). However, air absorption and propagation losses present difficulties for THz communication. So, there is a demand for developing effective transceivers, antenna structures, and signal processing in THz.

5.1 Key Approaches for Vehicular Networks

- **AI-Powered Wireless Network Architecture:** AI is the key element for any intelligent system. Numerous domains and applications are included in this system (Adhikari et al., 2022). Similar to this, AI in the transportation system reduces travel expenses by determining the ideal distance, assisting drivers in making choices more quickly, and offering an accident and road condition warning system. Additionally, a driver with this level of intelligence may operate a vehicle without the need for human assistance. DL and ML models are being utilized extensively in 2019 and 2030 for predictive maintenance, intelligent decision-making, and device health monitoring, which lowers additional driver overhead.
- **Ultra-Reliable Low-Latency Communication:** Lower delay and reliability of wireless sys1tems are difficult for delay-limited services, such as ITS, health services, and industry applications, where data must be transmitted to the appropriate device within the given delay. This calls for updating current standards and using cutting-edge technologies like Visible Light Communication (VLC) and THz communication. 6G technology opens up new possibilities by offering fast data rates for real-time apps and continuous communication. The reduction of information loss and transfer latency between the devices is achieved with the help of 6G technologies.

- **Greater Dynamicity and Quicker Mobility:** One of the main objectives of the transportation system is mobility management, or the effective tracking and monitoring of automobiles. Since smart cars are so dynamic and active, it might be challenging to communicate with other cars and roadside base stations. Additionally, the industrial sector's high level of dynamicity limits vehicles' ability to communicate with one another, which hinders the exchange of knowledge and expertise. Another approach to the mobility problem may be 6G communication technology, which would allow cars to use mobility management tools and strategies directly from the system operator.
- **Internet of Vehicle (IoV) Security and Privacy:** Smart sensors and internet of things (IoT) devices installed in IoV cars assist in gathering environmental data and sending it to distant computing devices for further analysis using cutting-edge communication technologies like 6G. This data may aid with traffic management, route optimization, obstacle detection, and safe navigation. Data vulnerability is a major issue when using vehicle data. Therefore, it is necessary to create a number of effective techniques and protocols that protect vehicle data from harmful organizations and provide security, credence, and confidentiality to both communicating entities. It is difficult to guarantee the security and privacy of sent data since so many linked gadgets in cars have direct access to private information.

  The sector has recently had to deal with ransomware assaults, cargo system manipulation, and GPS jamming (Aceto et al., 2019). Furthermore, security is one of the most important issues in vehicular networks because of their open nature, limitations on data privacy, high vehicle mobility, and lack of infrastructure. Therefore, establishing a reliable transportation system requires both protecting the data saved at remote servers for analysing the data or decision-making, as well as guaranteeing the security of vehicle network traffic utilizing state-of-the-art communication technology.
- **Dynamic Networks for Transportation:** The latest arrival of new methodologies in vehicular networks, such as shared mobility services, intelligent incentives and routing systems, and autonomous and connected automobiles, has a considerable influence on traffic flow in road networks. The autonomous transportation system and sophisticated vehicle sensors with new data collecting and transmission capabilities provide new difficulties and a huge potential for successfully monitoring and regulating vehicular networks as they rapidly evolved. It is crucial to use dynamic traffic assignment models and dynamic transportation network analysis to integrate new computing and communication technologies, such as edge computing and 6G, into large-scale computation. Additionally, multi-modal dynamic transportation networks with cutting-edge technology like 6G connectivity as crucial components are the subject of new research problems brought about by dynamic transportation. Another crucial research issue for the transportation system is expanding the application of the dynamic traffic assignment model to new trends in vehicle network mobility.

5.2 6G Communications-Enabled Intelligent Connected Transportation System Powered by Unmanned Aerial Vehicle (UAV)

The many IoT devices on 6G networks and the linked cars in IoV systems form an ITS. An IoV system is made up of connected cars that communicate with one another to share or exchange data. Interactions between adjacent IoT devices and linked cars inside ITS create computationally or data-intensive jobs. This network simultaneously deploys edge servers and UAVs to provide job offloading for IoT devices. Assume that a wireless sensor network is made up of k IoT devices spread over a rectangular area with side length l. An edge computing server inside this sensor network offers edge computing services to IoT devices. A UAV must fly over IoT devices to gather data and transfer it to the edge computing server for analysis. IoT devices are used in the sensor network for a variety of tasks, including processing and handling large amounts of data. The goal of this work (Liu et al., 2021) is to develop an offload technique for wireless sensor networks, which are complex networks, in order to decrease system energy consumption and increase their lifespan. This study proposes the UAV-CCIO (computing-communications intelligent offloading) algorithm to accomplish the purpose.

The typical offload strategy groups IoT devices per deployment scenario, with cluster-head nodes in the center. Data is sent to cluster-head nodes by other IoT device nodes via single or multiple hops. Next, to gather data for each cluster, the UAV alternately goes by each cluster-head node. After data gathering, the UAV transfers it to the edge computing server for processing. In this work, the process of offloading created tasks to the edge server is analyzed. This work analyses the procedure of offloading tasks to the edge server, noting that the results obtained are inadequate compared to the volume of data the task demands.

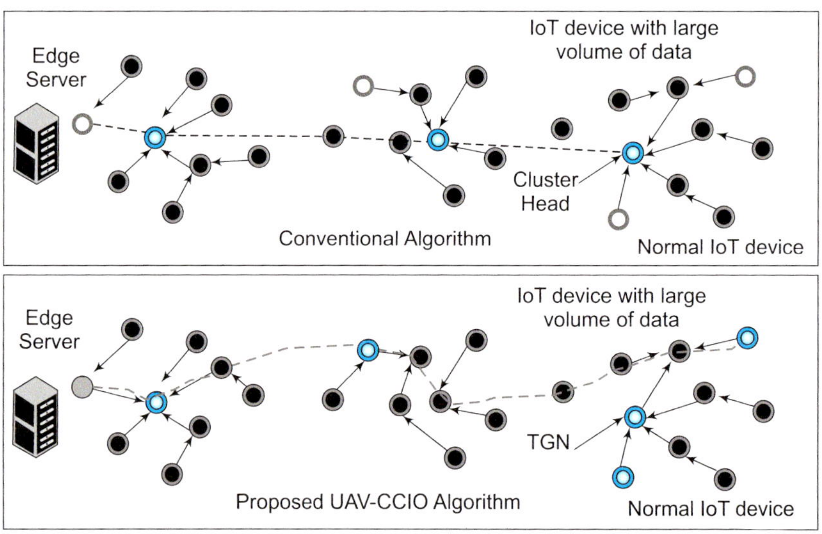

Fig. 18.4 UAV-CCIO algorithm.

Each tiny network has IoT devices with substantial data quantities, as shown in Figure 18.4. In the figure, these nodes are shown as blue nodes. Ordinary nodes with little data and a lot of processing or other circumstances are represented by the black nodes in the illustration.

In the UAV-CCIO method in Figure 18.4, Temporal Graph Networks (TGNs) are calculated from the data volume of tasks performed by IoT devices, and the direction of data transmission is determined by the relative locations of TGNs and regular nodes. It chooses the route for data transfer. After sending data from regular nodes to the big data node via the arrow, the UAV goes across the big data node to gather it. In this scenario, transmitting massive volumes of data to other nodes is not necessary for IoT devices with high data volume duties. As a result, network life is extended, energy use is efficient, and data transmission energy consumption is decreased.

5.3 6G for Autonomous Intelligent Transport Systems (AITS)

A three-dimensional 6G-AITS (Deng et al., 2023) is shown in Figure 18.5. It is dedicated to achieving three-dimensional connectivity and deeply integrates the transportation industry with 6G+ABCDE. Perception and communication operate independently of one another between 1G and 5G. This integration is made feasible in the 6G by the expansion of the spectrum to include visible light, terahertz, and millimetre waves, as well as overlap with the conventional perception spectrum. Three methods for uplink broadband communication, real-time broadband communication, and integrated communication and sensing have been formulated using 5G technologies, including eMBB, massive machine type communication (mMTC), and ultra-reliable and low latency communications (URLLC), to meet the requirements of the internet of everything (IoE) (Patzold, 2019) in 6G-AITS.

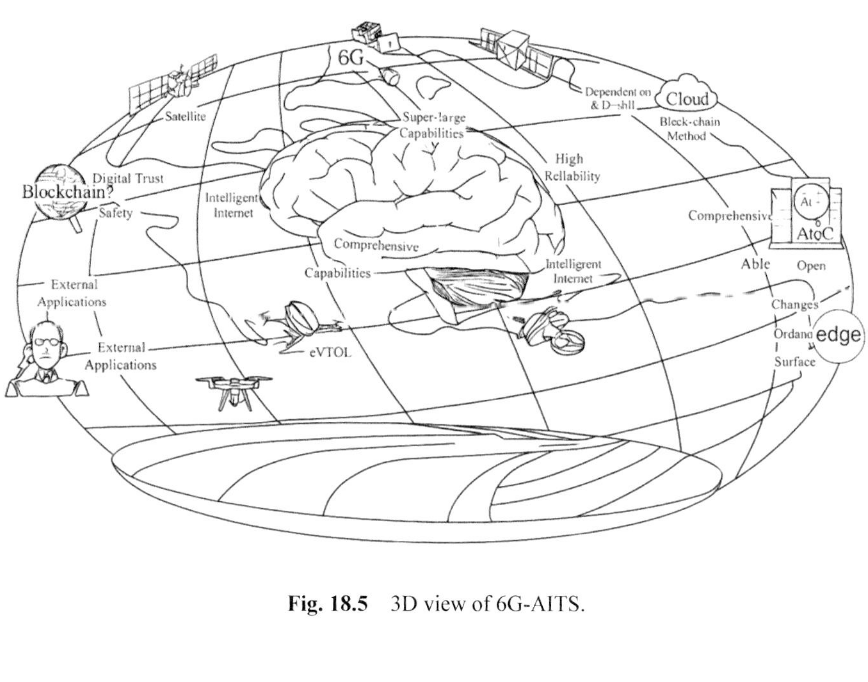

Fig. 18.5 3D view of 6G-AITS.

5.3.1 System Architecture

The following modifications will be made to the 6G-AITS system architecture (Figure 18.6) in addition to 6G+ABCDE:

- 6G-AITS is a collaborative system that spans space, air, and ground (SAG) domains and operates in a distributed manner. 6G-AITS, which relies on IoE, combines multi-domain traffic data to provide resource combining and exchanging of information.
- Sensor data may be swiftly processed and fused by 6G-AITS using 6G+ABCDE from multiple domains. Highly automated abnormal condition identification, categorization, tracing, and selection of the right target are features of 6G-AITS. Safety services are enhanced by transforming data into information that supports independent decision-making through the use of automated technologies.
- 6G-AITS can establish a strong, secure, reliable, and resilient federated dispersed network through the utilization of multiple levels and channels of resilient communication technologies. To ensure the continuous transportation forum, other channels could be used to transfer the information, and 6G-AITS can auto-recover from the fragmented nodes.
- Simple integration of new technologies or capabilities, platforms, and sensors.

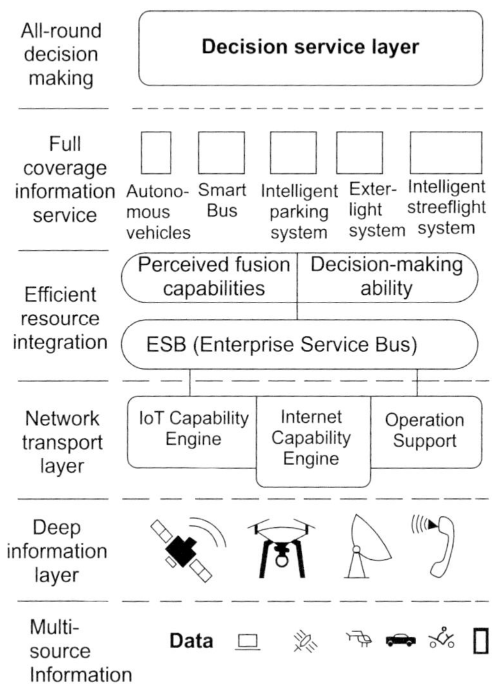

Fig. 18.6 6G-AITS system architecture.

5.4 SAGIN (Space–Air–Ground Integrated Network) for 6G-AITS

The SAGIN (Zhang et al., 2015) plan and deployment provide more extensive assistance for the successful advancements of 6G-AITS. High coverage, transmission resources, and increased reliability are achieved by SAGIN compared to ground information networks.

The several satellites that make up the SAGIN constellation have different types, orbits, and capacities. SAGIN establishes intimate connections between users, aeroplanes, and other communication equipment. To establish a unified high-speed broadband and extensive information network for accurate information gathering, rapid processing, and efficient transmission of optical, infrared, and various spectral information resources, it employs IP as the medium for information exchange along with advanced high-speed on-board processing, switching, and routing technologies.

5.4.1 Architecture of SAGIN

The three elements of SAGIN consist of a terrestrial network, an aerial network, and a space network, as illustrated in Figure 18.7.

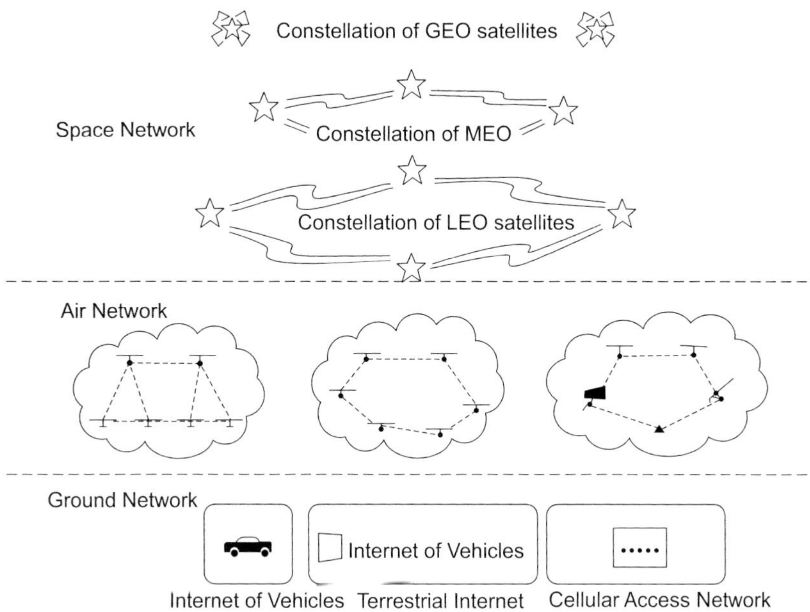

Fig. 18.7 SAGIN structure.

- **Space Network:** Satellites in space networks orbit at different heights. They are categorized into low Earth orbit (LEO), medium Earth orbit (MEO), and geostationary Earth orbit (GEO) satellites based on the orbital height. GEO satellites have an orbital height of 35,860 kilometres. The coverage area of GEO satellites is relatively fixed around the Earth, and it is bigger than that of the other two types of satellites. Global coverage is possible with three or four satellites, irrespective of the polar regions, whereas 42 percent of the planet's surface is covered by a single GEO satellite. The coverage of a single MEO satellite is substantially less than that of GEO satellites. MEO spacecraft have orbital heights of 2,000 to 20,000 km. More MEO satellites are required for global coverage. LEO satellites have an orbital height of less than 2,000 kilometres and a smaller coverage area than the others. The composition of several satellites is needed to give coverage around the planet. Furthermore, LEO satellites have a shorter transmission delay—typically approximately 30 ms—than GEO and MEO satellites, in addition to being smaller and less expensive.

- **Air Network:** UAVs, flying vehicles, and high-altitude platform stations (HAPS) (Yan et al., 2018) are all part of an air network. To provide communication services, HAPS install wireless base stations aboard aeroplanes that have a lengthy flight duration. It utilizes current communication technology to establish direct connections with mobile users on the ground. UAVs have the potential to significantly enhance wireless network coverage and capacity as airborne base stations. The benefits of HAPs and UAVs, which may effectively compensate for the limitations of ground networks, include flexible deployment, extensive service coverage, and reduced sensitivity to ground conditions.
- **Ground Network:** The ground communication technologies include cellular networks, mobile satellite terminals, ground stations, and information processing centres. Wide coverage and fast data rates are features of cellular wireless networks, which may provide a range of communication services to consumers.

Cellular wireless networks now handle enormous connections and provide consumers greater data transfer speeds thanks to their evolution from 1G to 5G mobile communication technology. However, ground networks are sensitive to natural disasters due to limited coverage and fail to provide services in remote areas.

5.5 6G in Smart Public Transport System

People are forced to look for more intelligent and effective answers to today's urban transportation problems. 6G technology is seen as a major turning point for the next generation of communication networks in the quest for a quicker and more dependable communication age. It is also essential to the ITS.

The primary components of this system include user services and experience, intelligent scheduling and optimization, real-time monitoring and management, and data collecting and processing. The data collecting and processing component is in charge of obtaining and processing a variety of data pertaining to bus operations, such as vehicle positions, passenger needs, and traffic circumstances, in real time. In order to enable real-time monitoring and management of bus operations, the real-time monitoring and management component keeps track of the location, speed, and operational status of buses in real time as well as the traffic conditions on routes. In order to increase operational effectiveness and service quality, the intelligent scheduling and optimization component strategically schedules and optimizes bus operations using cutting-edge algorithms and models. The user services and experience component, which includes real-time bus information searches, route planning, vehicle arrival time prediction, and other features, is dedicated to providing passengers with easy travel

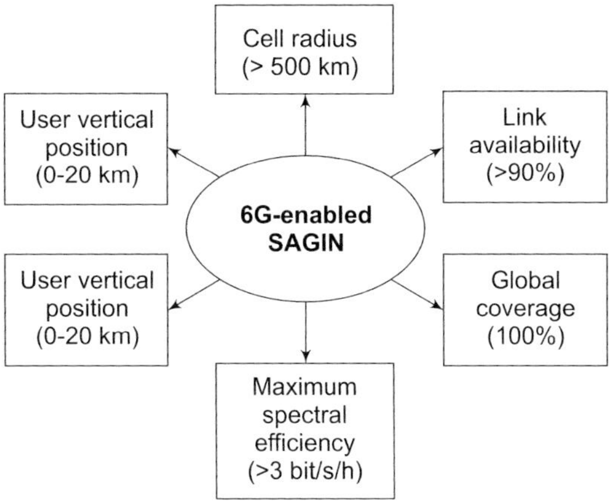

Fig. 18.8 Key techniques of 6G.

services and excellent user experiences. These fundamental elements work together to form the overall structure of the smart public transportation system, enabling its intelligent and contemporary evolution via mutually beneficial interactions. Figure. 18.9 shows the general layout of the ITS [Yu et al., 2024b].

By obtaining data gathered by GPS devices and bus sensors, the information service system in Figure 18.9 updates real-time bus information and offers services to users. Through the data gathered by sensors and cameras, the monitoring management system keeps an eye on the condition of the vehicle and relays information about unusual circumstances to the dispatch management system for processing. Using the real-time and historical data supplied by the monitoring system, the dispatching management system plans, optimizes, and communicates the dispatching outcomes to the bus. After receiving the user's trip requests, the travel management system uses an advanced algorithm to determine the optimal route and sends the route details to the customer. By using the aforementioned data flow and interaction method, the suggested system may successfully implement intelligent public transportation services.

By obtaining data gathered by GPS devices and bus sensors, the information service system in Figure 18.9 updates real-time bus information and offers services to users. Through the data gathered by sensors and cameras, the monitoring management system keeps an eye on the condition of the vehicle and relays information about unusual circumstances to the dispatch management system for processing. Using the real-time and historical data supplied by the monitoring system, the dispatching management system plans, optimizes, and communicates the dispatching outcomes to the bus. After receiving the user's trip requests, the travel management system uses an advanced algorithm to determine the optimal route and sends the route details to the customer. By using the aforementioned data flow and interaction method, the suggested system may successfully implement intelligent public transportation services.

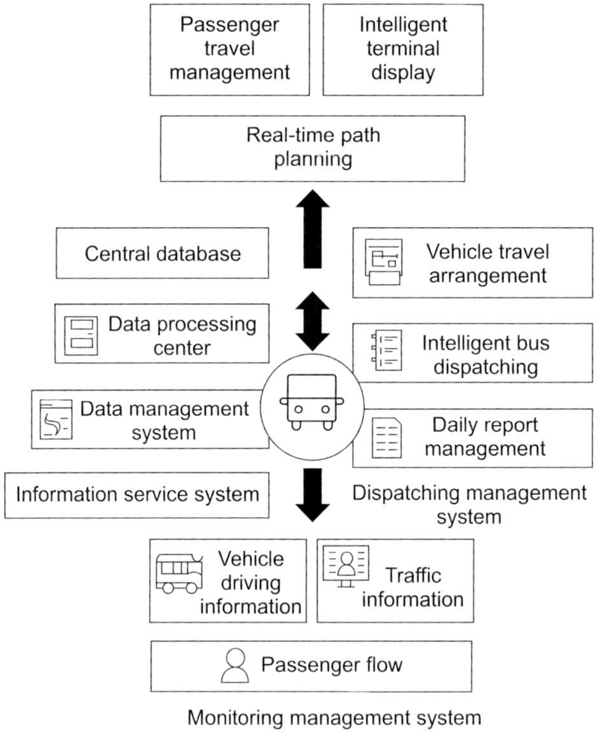

Fig. 18.9 Smart transport system

The intelligence level of public transportation service may be significantly raised by the smart system via the natural integration of the four main modules mentioned earlier. The scheduling of the bus vehicles and the travel experience of the passengers are directly impacted by the accuracy of the vehicle travel time prediction in these modules. Thus, the schedule management module of the intelligent public transportation system is the main topic of this study. Additionally, it is dedicated to researching and refining the bus segment travel time forecast technology in order to increase the system's operational effectiveness and service quality.

5.6 DL-Based Model for Security Problems in 6G-ITS

In big cities, traffic congestion has been a real issue that has led to both environmental pollution and human annoyance. Therefore, ITS applications are developed to resolve this problem by tracking the traffic conditions in practical situations. To develop a truly intelligent transportation environment, quick, dependable, and secure solutions are thus preferred. To better understand traffic behavior from a security standpoint, DL techniques have been developed. As a result, we employ a DL approach to optimize the safety concerns and offer a new gradient approach for vehicle misuse detection. To maximize the network convergence speed, the suggested model employs a technique called triple network replay. To maximize the policy network variables, three networks are chosen. To produce a quicker model, the replay algorithm is finally divided.

Figure 18.10 shows the suggested structure, which incorporates the 5G and 6G features for vehicle-to-everything (V2X) communication (Ribeiro et al., 2023). Services like expanded sensors and improved driving assistance are supported by the used V2X technology. Sensors are used to gather and identify the ambient conditions, allowing for the modification of certain automotive attributes to ensure the safety of the vehicle for drivers. The 5G-V2X and 6G-V2X communication systems are shown in Figure 18.10. A vehicle sends a navigation query to the closest node and answers to the target node using a method called hop-by-hop relaying. Vehicles and nodes may therefore acquire real-time traffic data via reports. By simulating traffic duties and cutting down on network training time, the triple network based on Deep Deterministic Policy Gradient (DDPG) operates continuously, leading to a quicker ITS situation and safety in traffic networks. A basic network, a double network, and various network topologies were used in earlier studies. However, the triple-network model produced the greatest outputs to reduce delay during the training time of the network.

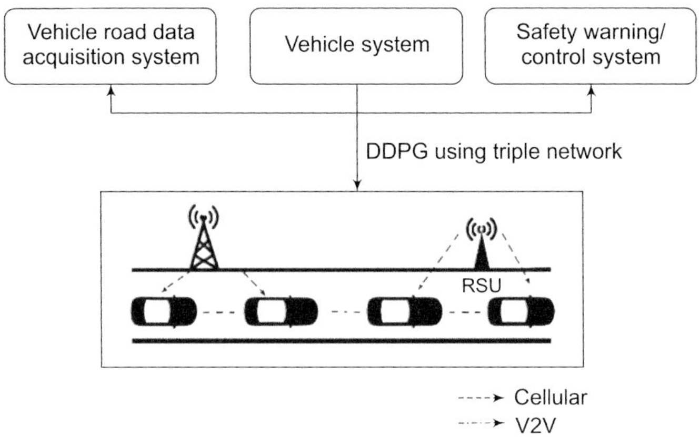

Fig. 18.10 DDPG algorithm for 5G-V2X and 6G-V2X systems.

6. Future Scope Challenges in the Integration of 6G in ITS

- **Interoperability:** Interoperability across various devices, protocols, and systems is essential for smooth communication and teamwork in 6G-ITS. Standardized interfaces and protocols are necessary due to the variety of 6G-ITS components, including automobiles, infrastructure of roads, and traffic control systems. For effective and efficient 6G-ITS integration, research should focus on interoperability frameworks and standardized protocols. In addition, further research is required to ensure backwards compatibility and enable a smooth transition between current ITS standards and emerging 6G standards.
- **Heterogeneous Hardware Constraints:** 6G uses many communication system types, including spectrum, communication architectures, and service transfer.

Mobile terminal and access point hardware configurations will also vary greatly. Additionally, it will make the algorithm and communication protocol design more difficult. Nonetheless, the communication will make use of AI and ML.

- **Autonomous Wireless Systems:** Automation systems like driverless automobiles, AI-powered Industry 4.0, and UAVs will all be fully enabled by the 6G network. The sub-systems of diversified networks, such as autonomous computing, interoperable processes, networking systems, and autonomous cloud, must converge for autonomous wireless systems to function. This complicates the process of developing the whole system. For example, autonomous cars developed in 6G must outperform the human-controlled cars, but the development of driverless cars is very difficult.
- **Security Issues in the Physical Layer:** Many of the applications of 6G are particularly important for human-centric communications. Consequently, 6G systems ensure security, confidentiality, and privacy. Numerous security concerns, such as those about network privacy, data interoperability, openness, and decentralization, still plague 5G networks. The present privacy and security protocols in 6G do not guarantee the physical layer security. The conventional Rivest-Shamir-Adleman (RSA) algorithm provides security in 5G. However, the RSA cryptosystems are strained by big data and AI technologies, rendering them dangerous.
- **Challenge in V2X:** Edge caching in 6G V2X content delivery struggles to maintain a strong connection in a rapidly changing environment. Large dimensionality of the obtained raw data, changing parameters of several operators and vendor proprietary data, incomplete data collection, difficulties in data harvesting and user privacy protection, and unreasonable bandwidth needs for training are some of the challenges.
- **Handling Interference and Spectrum Allocation:** The 6G spectra must be effectively controlled because of the limited spectrum resources and interference problems, especially via resource allocation and management systems. To optimize Quality of Service (QoS) while using the most resources possible, effective spectrum management is required. In future, issues in spectrum management and sharing in diversified networks that synchronize transmissions on the same frequency at 6G must be dealt. There is a need to develop approaches to eliminate the interference in a parallel and sequential manner.

7. Conclusion

Every iteration of communication systems introduces intriguing new features. With its amazing qualities, 5G communication technology is set to make its formal global debut. In future, 5G may not be able to meet the increasing requirements of wireless systems. Thus, the subsequent technology, 6G, has to be developed. The development of 6G is in progress. This chapter covered the opportunities and approaches of integrating 6G in ITS. The various information architecture, communication technologies, and computing approaches of ITS and 6G were discussed in detail in this chapter.

References

Aceto, G., V. Persico and A. Pescapé. (2019). A survey on information and communication technologies for Industry 4.0: state-of-the-art, taxonomies, perspectives, and challenges, *IEEE Communications Surveys & Tutorials*, *21*(4), 3467–3501.

Adhikari, M., A. Hazra, V.G. Menon, B.K. Chaurasia and S. Mumtaz. (2022). A roadmap of next-generation wireless technology for 6G-enabled vehicular networks', *IEEE Internet of Things Magazine*, *4*(4), 79–85.

Affan, A., S. Mumtaz, H.M. Asif and L. Musavian. (2021). Performance analysis of orbital angular momentum (OAM): a 6G wave form design, *IEEE Communications Letters*, *25*(12), 3985–3989.

Anvigh, A. Aghaei, Y. Khavan and B. Pourghebleh. (2024). Transforming vehicular networks: how 6G can revolutionize intelligent transportation? *Science, Engineering and Technology*, *4*(1), 80–93.

Araniti, G., A. Iera, S. Pizzi and F. Rinaldi. (2021). Toward 6G non-terrestrial networks, *IEEE Network*, *36*(1), 113–120.

Chen, S., Y. -C. Liang, S. Sun, S. Kang, W. Cheng and M. Peng. (2020). Vision, requirements, and technology trend of 6G: how to tackle the challenges of system coverage, capacity, user data rate and movement speed, *IEEE Wireless Communications*, *27*(2), 218–228.

Deng, X., L. Wang, J. Gui, P. Jiang, X. Chen, F. Zeng and S. Wan. (2023). A review of 6G autonomous intelligent transportation systems: mechanisms, applications and challenges, *Journal of Systems Architecture*, *142*, 102929.

Dilli, R., R. Chandra and D. Jordhana. (2021). Ultra-massive multiple input multiple output technologies for 6G wireless networks, *Engineered Science, 16*(19), 308–318.

Giordani, M. and M. Zorzi. (2020). Non-terrestrial networks in the 6G era: challenges and opportunities, *IEEE Network*, *35*(2), 244–251.

Guerrero-Ibáñez, J., S. Zeadally and J. Contreras-Castillo. (2018). Sensor technologies for intelligent transportation systems, *Sensors*, *18*(4), 1212.

Guo, C.H., X. Zhou, J. Liu and Y. Zhang. (2022). Vehicular intelligence in 6G: networking, communications, and computing, *Vehicular Communications*, *33*, 100399.

Huq, K.M.S., Otung, I.E. and Rodriguez, J. (2022). A study of coverage probability-based energy-efficiency analysis for UAV-aided THz-enabled 6G networks. *IEEE Transactions on Intelligent Transportation Systems, 24*(7), pp. 7404–7411.

Jha, A.V., B. Appasani, M.S. Khan, S. Zeadally and I. Katib. (2024). 6G for intelligent transportation systems: standards, technologies, and challenges, *Telecommunication Systems*, *86*(2), 241–268.

Jha, A.V., S.K. Mishra, B. Appasani and A.N. Ghazali. (2021). Communication networks for metropolitan E-health applications, *IEEE Potentials*, *40*(2), 34–42.

Kaul, A. and I. Altaf. (2022). Vanet-TSMA: a traffic safety management approach for smart road transportation in vehicular ad hoc networks, *International Journal of Communication Systems*, *35*(9), e5132.

Kilanioti, I., G. Rizzo, B.M. Masini, A. Bazzi, D.P.M. Osorio, F. Linsalata, M. Magarini, D. Löschenbrand, T. Zemen and A. Kliks. (2022). Intelligent transportation systems in the context of 5G-beyond and 6G networks, in *2022 IEEE Conference on Standards for Communications and Networking (CSCN)*, pp. 82–88.

Kuo, T. -Y., M. -C. Lee and T. -S. Lee. (2022). Quality-aware caching, computing and communication design for video delivery in vehicular networks, in ICC IEEE 2022, *IEEE International Conference on Communications*, pp. 261–266.

Liu, R., A. Liu, Z. Qu and N.N. Xiong. (2021). A UAV-enabled intelligent connected transportation system with 6G communications for internet of vehicles, *IEEE Transactions on Intelligent Transportation Systems*, *24*(2), 2045–2059.

Mehic, M., M. Niemiec, S. Rass, J. Ma, M. Peev, A. Aguado, V. Martin, S. Schauer, A. Poppe, C. Pacher and M. Voznak. (2020). Quantum key distribution: a networking perspective, *ACM Computing Surveys (CSUR)*, *53*(5), 1–41.

Patel, P., Narmawala, Z. and Thakkar, A. (2019). A survey on intelligent transportation system using internet of things. *Emerging Research in Computing, Information, Communication and Applications: ERCICA 2018*, Volume 1, pp. 231–240.

Patzold, M. (2019). 5G is coming around the corner [mobile radio], *IEEE Vehicular Technology Magazine*, *14*(1), 4–10.

Qin, Z., S. Leng, J. Zhou and S. Mao. (2020). Collaborative edge computing and caching in vehicular networks, in *2020 IEEE Wireless Communications and Networking Conference (WCNC)*, pp. 1–6.

Ribeiro, D.A., D.C. Melgarejo, M. Saadi, R.L. Rosa and D.Z. Rodríguez. (2023). A novel deep deterministic policy gradient model applied to intelligent transportation system security problems in 5G and 6G network scenarios, *Physical Communication*, *56*, 101938.

Wu, B., T. Chen, K. Yang and X. Wang. (2021). Edge-centric bandit learning for task-offloading allocations in multi-RAT heterogeneous networks, *IEEE Transactions on Vehicular Technology*, *70*(4), 3702–3714.

Yan, X., Xiao, H., Wang, C.X. and An, K. (2018). Outage performance of NOMA-based hybrid satellite-terrestrial relay networks. *IEEE Wireless Communications Letters, 7*(4), pp. 538–541.

Yastrebova, A., M. Höyhtyä, R. Kirichek and A. Serebryakova. (2019). Airborne-terrestrial integrated architecture for self-driving vehicles realization, in *2019 11th International Congress on Ultra Modern Telecommunications and Control Systems and Workshops (ICUMT)*, pp. 1–6.

Yu, H., T. Taleb, K. Samdanis and J. Song. (2024a). Toward supporting holographic services over deterministic 6G integrated terrestrial and non-terrestrial networks, *IEEE Network*, *38*(1), 262–271. doi: 10.1109/MNET.133.2200509.

Yu, M., Zhang, H., Li, J., Kang, S., Gao, M., Zhu, J. and Fu, J. (2024). The Application of 6G Technology and Back Propagation Neural Network in the Smart Public Transport System. *IEEE Transactions on Intelligent Transportation Systems*.

Zhang, N.T., K.L. Zhao and G.L. Lin. (2015). Thought on constructing the integrated space-terrestrial information network, *J. China Acalerng Hetronics Inf. Technol.*, *10*(3), 223–230.

Index

About the Editor

Amit Kumar Tyagi has a PhD from Pondicherry Central University, Puducherry, India. He is presently working as an Assistant Professor at National Forensic Sciences University, Gandhinagar, India. Prior to this he was as an assistant professor at several institutes like Lord Krishna College of Engineering, Ghaziabad (LKCE) (for the periods of 2009-2010, and 2012-2013, Lingaya's Vidyapeeth (formerly known as Lingaya's University), Faridabad (2018-2019), VIT Chennai (2019-2022) and NIFT New Delhi (2022-2025). His supervision experience includes more than 10 Masters' dissertations and one PhD thesis. He has contributed to several projects such as "AARIN" and "P3- Block" to address some of the open issues related to the privacy breaches in Vehicular Applications like parking and Medical Cyber Physical Systems. His research focuses on Next Generation Machine Based Communications, Blockchain Technology, Smart and Secure Computing and Privacy. He is a regular member of the ACM, MIRLabs, Ramanujan Mathematical Society, Cryptology Research Society, and Universal Scientific Education and Research Network, and ISTE. He is a senior member of IEEE.